Petrophysical Characterization and Fluids Transport in Unconventional Reservoirs

Petrophysical Characterization and Fluids Transport in Unconventional Reservoirs

Edited by

Jianchao Cai
China University of Geosciences, Wuhan

Xiangyun Hu
China University of Geosciences, Wuhan

ELSEVIER

Elsevier
Radarweg 29, PO Box 211, 1000 AE Amsterdam, Netherlands
The Boulevard, Langford Lane, Kidlington, Oxford OX5 1GB, United Kingdom
50 Hampshire Street, 5th Floor, Cambridge, MA 02139, United States

Library of Congress Cataloging-in-Publication Data
A catalog record for this book is available from the Library of Congress

British Library Cataloguing-in-Publication Data
A catalogue record for this book is available from the British Library

ISBN: 978-0-12-816698-7

Publisher: Candice Janco
Acquisition Editor: Amy Shapiro
Editorial Project Manager: Ali Afzal-Khan
Production Project Manager: Nilesh Kumar Shah
Cover Designer: Christian Bilbow

Typeset by SPi Global, India

Contents

Contributors

Reza Barati Chemical & Petroleum Department, University of Kansas, Lawrence, KS, United States

Jianchao Cai Hubei Subsurface Multi-Scale Imaging Key Laboratory, Institute of Geophysics and Geomatics, China University of Geosciences, Wuhan, P.R. China

Guanglei Cui Key Laboratory of Ministry of Education on Safe Mining of Deep Metal Mines, Northeastern University, Shenyang, P.R. China

Hassan Dehghanpour Department of Civil and Environmental Engineering, School of Mining and Petroleum Engineering, University of Alberta, Edmonton, Alberta, Canada

Derek Elsworth Department of Energy and Mineral Engineering, EMS Energy Institute and G3 Center, Pennsylvania State University, University Park, PA, United States

Yiren Fan School of Geosciences, China University of Petroleum; CNPC Key Well Logging Laboratory in China University of Petroleum, Qingdao, Shandong, P.R. China

Ruimin Feng Chemical & Petroleum Department, University of Calgary, Calgary, AB, Canada

Quan Gan Department of Petroleum Geology & Geology, School of Geosciences, University of Aberdeen, United Kingdom

Xinmin Ge School of Geosciences, China University of Petroleum (East China); CNPC Key Well Logging Laboratory in China University of Petroleum, Qingdao, Shandong, P.R. China

Ali Habibi Department of Civil and Environmental Engineering, School of Mining and Petroleum Engineering, University of Alberta, Edmonton, Alberta, Canada

Sheng He Key Laboratory of Tectonics and Petroleum Resources, Ministry of Education, China University of Geosciences, Wuhan, P.R. China

Xiangyun Hu Hubei Subsurface Multi-scale Imaging Key Laboratory, Institute of Geophysics and Geomatics, China University of Geosciences, Wuhan, P.R. China

Qinhong Hu Department of Earth and Environmental Sciences, The University of Texas at Arlington, Arlington, TX, United States

Ali Javaheri Department of Civil and Environmental Engineering, School of Mining and Petroleum Engineering, University of Alberta, Edmonton, Alberta, Canada

Bao Jia Chemical & Petroleum Department, University of Kansas, Lawrence, KS, United States

Bikai Jin Harold Vance Department of Petroleum Engineering, College of Engineering, Texas A&M University, College Station, TX, United States

Junqian Li School of Geosciences, China University of Petroleum (East China), Qingdao, P.R. China

Yu Liang Hildebrand Department of Petroleum and Geosystems Engineering, Cockrell School of Engineering, The University of Texas at Austin, Austin, TX, United States

Kouqi Liu Department of Petroleum Engineering, University of North Dakota, Grand Forks, ND, United States

Jun Liu School of Energy Resources, China University of Geosciences, Beijing, P.R. China

Yongsheng Liu State Key Laboratory of Geological Processes and Mineral Resources, Faculty of Earth Sciences, China University of Geosciences, Wuhan, P.R. China

Dameng Liu School of Energy Resources, China University of Geosciences, Beijing, P.R. China

Cheng Lu Oil & Gas Survey, China Geological Survey, Beijing, P.R. China

Shuangfang Lu School of Geosciences, China University of Petroleum (East China), Qingdao, P.R. China

Wanjing Luo School of Energy Resources, China University of Geosciences (Beijing), Beijing, P.R. China

Bo Ning Research Institute of Petroleum Exploration and Development, PetroChina, Beijing, P.R. China

Yang Ning Department of Petroleum Engineering, Cullen College of Engineering, The University of Houston, Houston, TX, United States

Mehdi Ostadhassan Department of Petroleum Engineering, University of North Dakota, Grand Forks, ND, United States

Zhejun Pan CSIRO Energy Business Unit, Clayton South, VIC, Australia

Guan Qin Department of Petroleum Engineering, Cullen College of Engineering, The University of Houston, Houston, TX, United States

Harpreet Singh National Energy Technology Laboratory, Morgantown, WV, United States

Yuling Tan State Key Laboratory of Geomechanics and Geotechnical Engineering, Institute of Rock and Soil Mechanics, Chinese Academy of Sciences, Wuhan; University of Chinese Academy of Sciences, Beijing, P.R. China; CSIRO Energy Business Unit, Clayton South, VIC, Australia

Changfu Tang Exploration Research Institute, Anhui Provincial Bureau of Coal Geology, Hefei, P.R. China

Jyun-Syung Tsau Tertiary Oil Recovery Program (TORP), University of Kansas, Lawrence, KS, United States

Fuyong Wang Research Institute of Enhanced Oil Recovery, China University of Petroleum, Beijing, P.R. China

Zhenlin Wang Research Institute of Exploration and Development, Xinjiang, Oilfield Company, PetroChina, Karamay, P.R. China

Hongyan Wang PetroChina Research Institute of Petroleum Exploration & Development, Beijing, P.R. China

Rui Yang Key Laboratory of Tectonics and Petroleum Resources, Ministry of Education, China University of Geosciences, Wuhan, P.R. China

Yanbin Yao School of Energy Resources, China University of Geosciences, Beijing, P.R. China

Li Zhang Petroleum Exploration and Development, Jianghan Oilfield Branch Company, Sinopec, Wuhan, P.R. China

Peiqiang Zhao Hubei Subsurface Multi-scale Imaging Key Laboratory, Institute of Geophysics and Geomatics, China University of Geosciences, Wuhan, P.R. China

Jier Zhao School of Geosciences, China University of Petroleum, Qingdao, Shandong, P.R. China

Shangwen Zhou PetroChina Research Institute of Petroleum Exploration & Development, Beijing, P.R. China

Preface

Unconventional reservoirs, which primarily include shale, coal, and tight sandstone reservoirs, have recently found an important role in the production of oil and gas as other reservoirs are either depleted or too expensive to produce during times of low oil prices. Industry's lack of experience with these reservoirs, relative to conventional resources, requires attention from the perspective of technology and theoretical understanding to fill the gap between classical theory of fluid flow in porous media and the unconventional nature of these reservoir rocks. One of the major issues with unconventional reservoirs is their low recovery factor; therefore, enhancing oil and gas recovery from these reservoirs is crucial for their economic development. Some of the important challenges in enhancing oil and gas recovery from unconventional reservoirs are accurate characterization of the microstructure and the ability of fluid transport, especially in understanding which transport mechanisms play important roles in the flow of oil and gas at different scales and rock types.

This book presents a comprehensive overview of new methods and technologies for petrophysical characterization of unconventional reservoirs, recent theoretical advances and modeling of fluid flow and transport, multiscale modeling, rock reconstruction, and upscaling approaches in unconventional reservoirs.

The book is intended for graduate students, geoscientists, and engineers engaging in research and development of unconventional oil and gas reservoirs in academia and industry. This book is divided into two parts: Part 1 (Petrophysical Characterization) and Part 2 (Porous Flow Dynamics). In Part I (Petrophysical Characterization), Chapter 1 focuses on characterizing pore size distributions of shale, in which three current widely used techniques and methods, i.e. field emission scanning electron microscope, nitrogen adsorption and mercury intrusion capillary pressure, are introduced and summarized to analyze pore microstructure properties of shale rocks. Chapter 2 presents petrophysical characterization of the pore structure of coal by means of low-field nuclear magnetic resonance technology. In Chapter 3, the characteristics of tight sandstone petrophysical properties are reviewed and pore structure characterization techniques as well as different methods of calculating fractal dimensions are introduced and discussed. In addition, mathematical models for predicting permeability of tight sandstones are also briefly introduced. Chapter 4 further studies the pore structure and heterogeneity of unconventional tight oil reservoirs based on multifractal analysis and low-field nuclear magnetic resonance technology. In Chapter 5, an integrated method to quantify organic pores using low-field nuclear magnetic resonance is introduced, combined with backscattered electron images and energy dispersive X-ray spectrometer. Relationships between the organic-related porosity and geochemical parameters are analyzed using petrophysical data. Chapter 6 presents a critical overview of methods available to predict shale permeability from experimental and modeling perspectives at both laboratory and field-scales, and also presents a nonempirical model for two-phase relative permeability in shale fractures. Chapter 7 introduces the characteristics of pore structure, connectivity, and wettability of several typical core samples from organic-rich shale formation as well as their coupled effects on fluid migration. Chapter 8 aims at understanding the wettability behavior of tight rocks by performing spontaneous imbibition experiments and contact angle measurements.

In Part 2 (Porous Flow Dynamics), Chapter 9 presents a general fracture wing model to analyze the transient pressure response of complex fracture systems in low permeability reservoirs. Chapter 10 explores the impact of fluid circulation rates on the heterogeneity of thermal drawdown and its potential impact on the timing and magnitude of induced seismicity, and develops a dimensionless semianalytical model to determine thresholds for the evolution of uniform or shock-front distributions of thermal drawdown within rock comprising the reservoirs. Chapter 11 establishes a basic overview of the applicability of some commonly used pore-scale modeling methods in shale formations and discusses the hydrocarbon transport and storage mechanisms in shale nanopores. Chapter 12 presents high-pressure methane adsorption experiments on shale samples and proposes a novel adsorption model by combining the micropore filling and monolayer coverage theories to describe methane adsorption process in shale. Chapter 13 proposes a coal permeability model based on the nonconstant vertical stress condition by considering roof deformation acting to hold part of the overburden load. Chapter 14 develops a model to predict adsorbing-gas (e.g., CO_2) permeability change by coupling pore size evolution with permeability change during gas pressure depletion, and also proposes two parameters (change ratio of permeability and total loss of flow rate) to quantitatively evaluate the dynamic process of CO_2 flow in different coal ranks. Chapter 15 introduces a workflow to integrate all the physics affecting the dynamic permeability evolution. The complex relationships between apparent porosity, adsorption, and apparent permeability based on the pulse-decay experiment are also described.

We would like to thank the individual chapter authors of the book for their inspiring contributions and diligent work. We also specially acknowledge the Elsevier press for providing this opportunity to bring this book to readers. Special thanks is also reserved for Amy M. Shapiro, Ashwathi P. Aravindakshan, Ali Afzal-Khan, and Nilesh K. Shah of Elsevier's book production group for editing this book. We also acknowledge the National Natural

Science Foundation of China (41722403, 41630317, 41572116), and the Hubei Provincial Natural Science Foundation of China (2018CFA051) for supporting the series of studies on petrophysical characterization and flow and transport in porous media.

Jianchao Cai, Xiangyun Hu
Wuhan, China

PART 1

Petrophysical Characterization

CHAPTER 1

Characterizing Pore Size Distributions of Shale

Kouqi Liu*, Mehdi Ostadhassan*, Jianchao Cai†

*Department of Petroleum Engineering, University of North Dakota, Grand Forks, ND, United States
†Hubei Subsurface Multi-scale Imaging Key Laboratory, Institute of Geophysics and Geomatics, China University of Geosciences, Wuhan, P.R. China

Chapter Outline

1 Introduction

Since commercial production from shale-gas reservoirs began, research on shale has increased significantly. Understanding the pore structure of unconventional reservoirs, such as shale, can assist in estimating their elastic transport and storage properties, thus enhancing the hydrocarbon recovery from such massive resources.

Petrophysical Characterization and Fluids Transport in Unconventional Reservoirs. https://doi.org/10.1016/B978-0-12-816698-7.00001-2

Porosity is defined as the pore fraction. The pores can either be closed, isolated or open and connected. The matrix-related pores (with a size range from a nanometer to micrometer) are commonly found in shale rocks. Based on the pore classification suggested by Loucks et al. [1], pores in shale can be separated into three main pore types: interparticle pores, intraparticle pores, and organic pores. According to the International Union of Pure and Applied Chemistry (IUPAC) [2], pores can be divided into three categories: micropores (<2 nm), mesopores (2–50 nm), and macropores (>50 nm). In the past decades, many shale-gas reservoirs in the world have been studied extensively by researchers, such as the Barnett shale (USA) [1, 3], Marcellus shale (USA) [4, 5], Longxi shale (China) [6, 7], Toarcian shale (Germany) [8], Boom Clay (Belgium) [9], Posidonia shale (Germany) [10], Longmaxi shale [11], and Silurian shale [12]. Multiple methods have been employed to study the pore structure of these reservoirs, including mercury intrusion porosimetry (MIP) [13], nuclear magnetic resonance [14], gas adsorption [6], small angle neutron scattering [15], and scanning electron microscopy (SEM) image analysis [1, 10, 16, 17].

In this chapter, we summarize the current widely used techniques that can assist in future experiment research work on the pore structures of shale formations. Three commonly used methods will be explained in this chapter and some real sample analysis will also be introduced. It should be kept in mind that these three methods are based on different working mechanisms and each method has its own strengths and limitations. Different methods can yield different pore structure analysis results and the combination of these methods may elucidate different features of the pore structure and provide more information about the pores.

2 Scanning Electron Microscopy

SEM is one of the most popular imaging techniques. SEM, unlike conventional light microscopy, produces images by recording various signals resulting from interactions of an electron beam with the sample as it is scanned in a raster pattern across the sample surface [18]. Field emission SEM (FESEM), which has a much higher resolution than conventional SEM, has a greater ability in characterizing the nanopores in shale formations.

2.1 Sample Preparation

Mechanical polishing is widely applied by researchers for sample preparation. Usually, the shale samples are resin coated. When the resin is solid, sand paper (with different sizes of grit from 600 to 1200) is used to polish the sample surface. Finally, we used diamond polishers, with various grain sizes (5, 3, and 0.5 μm), to polish the sample [19]. This procedure can produce surface topographic irregularities because of the differential hardness of the components (Fig. 1A). These irregularities sometimes greatly exceed the size of the nanopores in the shale and therefore, this method of sample preparation is inadequate for pore

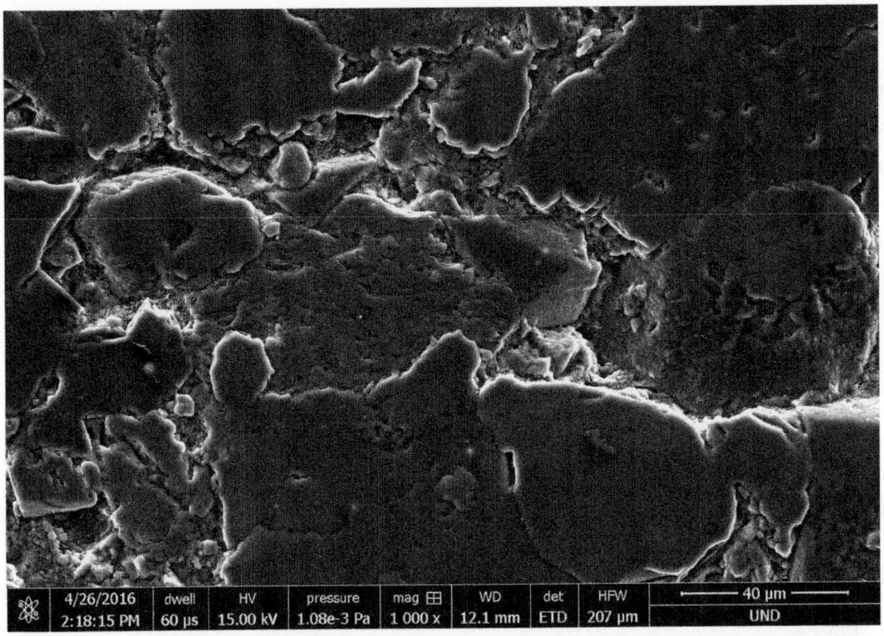

(A)

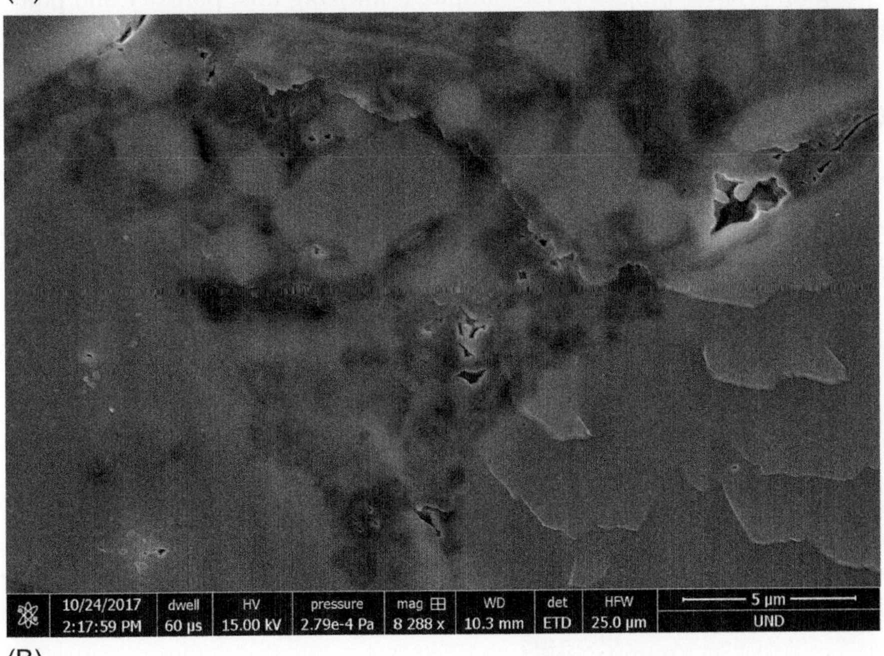

(B)

Fig. 1

SEM images of the topography surfaces of samples by (A) mechanical polishing and (B) argon ion milling.

quantification using the SEM technique. To eliminate the limitations of mechanical polishing, argon ion milling was used to produce a much flatter surface (Fig. 1B). Samples, which were taken parallel to the bedding, were trimmed down to a 0.5 cm square cube. Then all faces of the cube were smoothed out by hand with a Buehler polishing wheel using 600-grit silicon carbide grinding paper. The samples were mounted to the ion mill sample holder and placed in the Leica EM TIC 3 × argon ion mill. Thereafter, all samples were milled at an accelerating voltage of 8 kV, with a gun current at 3 mA for 8 h. Samples were then removed from the ion mill's sample holder and mounted on a clean SEM stub using carbon paint [20].

2.2 Image Analysis

In order to investigate the pore structure from the grayscale SEM images, ImageJ, a powerful commercial computer program that is widely used in image analysis in other fields, was used [21, 22]. For the backscattering electron images or the secondary images of shale, the pores have low grayscale values while the solid components have a high grayscale value. After selecting an appropriate threshold value, the images were segmented into a binary image where the black pixels represent the pores and the white pixels show the solid matrix. Fig. 2 shows the imaging process for one shale sample. Following this, porosity and pore size were quantified.

2.3 Bakken Example Analysis

Samples from the Bakken Formation were collected for the real case study. To quantitatively analyze the pore structure of the samples, the representative elementary area (REA) method was used. The REA for a porous medium has a minimum averaging area that will yield

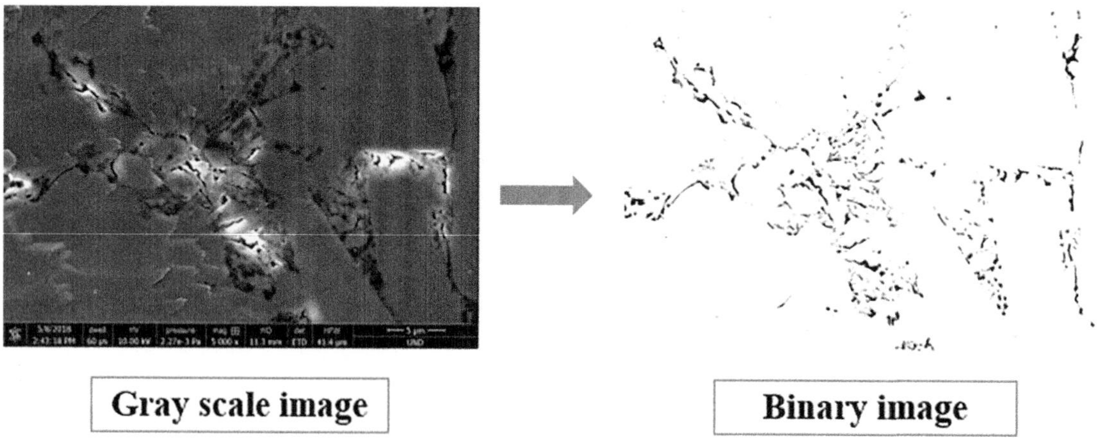

Fig. 2
Image process for one shale sample.

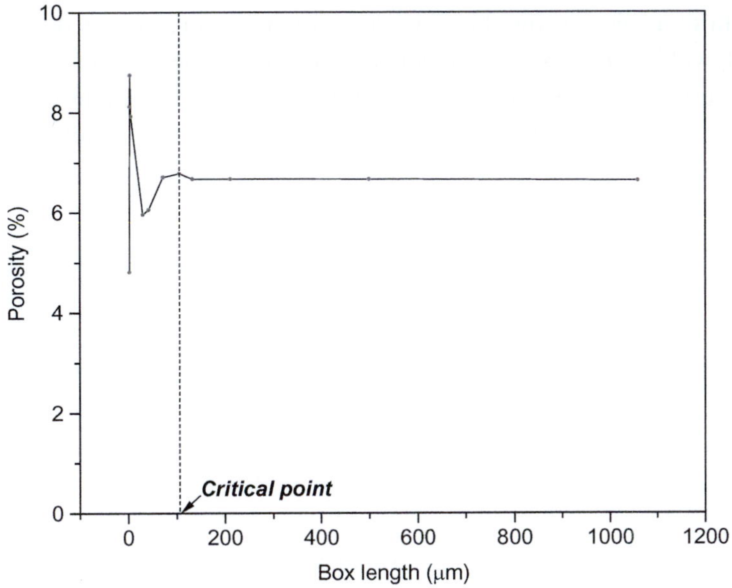

Fig. 3
Determination of the REA of Bakken sample (#9) using a porosity indicator.

a value representative of the whole. The porosity indicator was used to determine the REA by analyzing the influence of the magnification on porosity [23]. For the samples studied, it was found that as the magnification was reduced (scan area increased), the porosity value oscillated until it remained steady under a critical magnification. The scan area under this critical magnification point was used as the REA. Fig. 3 shows the process of determining the REA of one Bakken sample.

Based on the REA, the surface porosity of the sample was calculated. Table 1 shows that the samples had abundant pore counts and most of the samples had very low porosity values

Table 1 Porosities of the samples calculated from SEM images of their REA

Sample	Porosity (%)	Pore Counts
#1	9.71	9823
#2	8.58	14,895
#3	3.89	8968
#4	9.59	21,196
#5	7.26	18,443
#6	9.97	5752
#7	4.70	9818
#8	9.63	12,435
#9	6.70	16,837
#10	7.67	10,334
#11	11.56	19,495

(<10%). The data in Table 1 show that samples with high pore counts do not necessarily have high porosity values. This is because pore counts refer to the number of pores in the sample while porosity refers to the ratio of the pore volume to the total sample volume. The study by Saraji and Piri [23] shows that samples with large pore size and small pore counts can still have large porosity values. Thus, porosity should be determined by the combination of pore size and pore counts, not by pore counts alone.

The correlations between mineralogical composition and TOC (total organic carbon) content on one hand and the pore counts and porosity on the other were also investigated. The results are shown in Table 2. The data illustrate that the samples used in this study contain the same minerals but in different weight ratios. In order to not eliminate the influence of one parameter on the other based on the bivariate plots, when in fact the parameter may exert a significant influence when it is joined by another independent parameter, partial least squares (PLS) regression was applied. The composition of the samples (quartz, pyrite, clay, feldspar, and TOC) was treated as the independent parameter, and porosity and pore counts were used as the dependent variables. The fit parameters of the PLS model are shown in Table 3. The results show that for the porosity/pore counts, clays and feldspar have a positive influence on the porosity/pore counts while quartz and pyrite have negative effects on the porosity/pore counts. The increase of TOC can decrease the

Table 2 Mineralogical composition of the samples

Samples	TOC (wt%)	Quartz (wt%)	Clays (wt%)	Feldspar (wt%)	Pyrite (wt%)
#1	12.72	39.01	17.11	26.18	2.45
#2	17.45	47.88	18.82	8.42	2.15
#3	14.16	55.97	23.18	2.83	2.43
#4	12.12	41.3	29.88	9.67	5.01
#5	14.92	27.31	2.04	51.90	1.36
#6	14.71	39.92	39.23	3.33	2.30
#7	13.87	50.82	22.39	4.91	4.65
#8	10.10	52.14	26.97	2.25	5.30
#9	11.62	43.66	32.70	2.74	2.65
#10	17.73	36.20	27.15	5.43	4.28
#11	10.57	16.10	62.60	3.13	0.05

Table 3 PLS model fitting results

Variable	Porosity	Pore counts
Intercept	13.11	21894.92
Quartz (%)	−0.07	−113.5
Pyrite (%)	−0.17	−295.43
Clays (%)	0.03	44.93
Feldspar (%)	0.01	19.23
TOC (%)	−0.19	−316.77

porosity/pore counts due to the extensive occurrence of nonporous organic matter. Of all of the minerals, TOC affected the porosity and pore count the most [20].

We used the SEM quantitative analysis method to determine the pore parameters that could cause uncertainty due to resolution limitation. The minimum pore size that we could analyze was 10 nm, which means that pores smaller than 10 nm may exist in the shale samples but cannot be detected by the SEM.

3 Gas Adsorption

Gas adsorption is of major importance in measuring the pore structure over a wide range of porous materials. Since Dewar [24] reported the adsorption of nitrogen and other gases at liquid air temperature when studying the composition of atmospheric gases, nitrogen has become a potentially available adsorption material. The monumental work on the monolayer adsorption by Langmuir [25] attracted great interest from researchers for the interpretation of adsorption data. In the 1930s, Benton and White [26] published on the existence of the multilayer adsorption of nitrogen at the temperature of 77 K. Brunauer and Emmett applied gas adsorption to analyze the surface area of samples [27]. In 1938, the publication of the Brunauer-Emmett-Teller (BET) theory [28], which is the extension of the Langmuir monolayer adsorption model to a multilayer adsorption model, provided researchers the theoretical method to determine the surface area of porous media. In the late 1940s, by using the Kelvin equation, Barrett, Joyner, and Halenda (BJH) proposed a method [29] to derive the pore size distributions (PSDs) from an appropriate nitrogen isotherm. The BJH method is still one of the most popular methods used to date. In the early 2000s, based on the established principles of statistical mechanics and assuming a model solid structure and pore topology, the density functional theory (DFT) method was proposed and has been an important tool in characterizing the PSD of porous samples [30].

3.1 Sample Preparation

The samples were crushed to <0.2 mm and then out-gassed at 423 K (301.73°F) for 12 h to remove existing bound water that is absorbed by clay minerals. Then, nitrogen adsorption isotherms were obtained for relative pressure (P/Po) (P and P_0 are the equilibrium and the saturation pressure of adsorbates at the temperature of adsorption), varying from 0.01 to 0.995 at 77 K (−321 °F) [31].

3.2 Longmaxi Sample Analysis

Fig. 4 depicts nitrogen adsorption isotherms in four samples from the Longmaxi Formation [32]. All nitrogen adsorption isotherms exhibited the same characteristics. For the adsorption branch, as the relative pressure increased from 0, the adsorption quantity increased

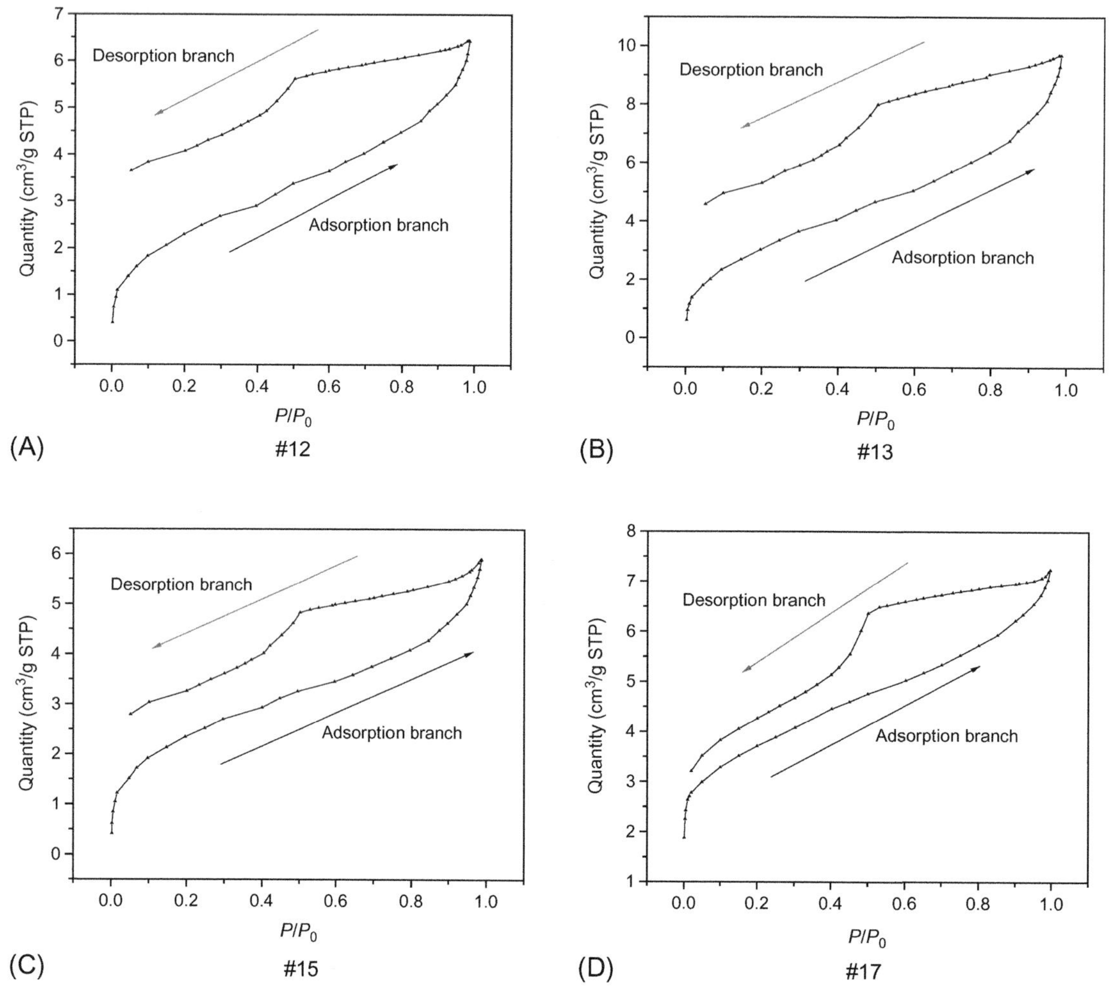

Fig. 4
Examples of nitrogen adsorption isotherms in Longmaxi shale samples.

rapidly and then followed a slow increase. The deflection point was used to separate the monolayer and multilayer adsorption phase [31]. When the relative pressure was larger than 0.8, the rate of the gas adsorption increased until the relative pressure approached 1. Hysteresis was observed for the samples, indicating the existence of mesopores in the Longmaxi samples. Based on IUPAC [2], the adsorption isotherms that were detected in the samples would belong to type IV and the shape of the pores in the Longmaxi Formation are mainly ink-bottle.

The BJH model (BJHAD (Barrett, Joyner, and Halenda adsorption branch) and BJHDE (Barrett, Joyner, and Halenda desorption branch)) and DFT model were applied to the data to analyze pore structure from the isotherms. The results can be found in Table 4, which show different outcomes from each method for surface area and pore volume calculation. Surface

Table 4 Pore information calculated from different methods

	# 12	# 13	# 14	# 15	# 16	# 17	# 18
Surface area (m^2/g)							
BJHAD	3.498	5.051	4.187	2.582	3.307	2.613	1.817
BJHDE	4.251	7.827	13.120	4.388	8.337	12.988	3.004
DFT	6.204	8.483	12.860	6.206	16.227	16.441	8.599
Pore volume (10^{-2} cc/g)							
BJHAD	0.722	1.127	1.190	0.589	0.800	0.600	0.436
BJHDE	0.477	0.894	1.506	0.526	0.900	0.900	0.378
DFT	0.849	1.265	1.483	0.772	1.300	1.000	0.645

area of the samples that was calculated from the BJHAD model varied from 1.817 to 5.051 m^2/g, which is smaller than the surface area calculated from the BJHDE and DFT models, which varied from 3.004 to 13.120 m^2/g and 6.204 to 16.441 m^2/g, respectively. Considering Sample #17 for example, the surface area calculated using the DFT method was around five times of the surface area calculated from the BJHAD method, while the surface area calculated from BJHDE model was around four times the surface area calculated from the BJHAD model. Regarding pore volume, the DFT model yielded, on average, a larger pore volume than both BJHAD and BJHDE models. The pore volume estimated for Sample #17 is 1×10^{-2} cc/g while the pore volume from the same sample calculated from the BJHAD and BJHDE models was 0.6×10^{-2} and 0.9×10^{-2} cc/g, respectively. There was no clear correlation between pore volume derived from the BJHAD and BJHDE methods. This discrepancy in the results will be explained later in the discussion.

Considering the BJHAD and BJHDE models, the modified Kelvin equation can be employed to analyze PSDs [33]. If the adsorption branch is used, the pore size can be calculated by

$$\log\left(\frac{P}{P_0}\right) = \frac{2\gamma V_m}{(r - t_A)RT} \tag{1}$$

While if the desorption branch is used, the pore radius can be estimated from

$$\log(P/P_0) = \frac{2\gamma V_m}{(r - t_D)RT} \tag{2}$$

where γ is the surface tension, V_m is the molar volume of the liquid, R is the universal gas content, r is the radius of the capillary, T is the absolute temperature, t_A and t_D are the thickness of the adsorbed multilayer film for the adsorption and desorption, respectively.

Additionally, in the DFT method, calculation of PSD is based on the solution of the general adsorption isotherm equation (2):

$$N(P/P^0) = \int_{W_{\min}}^{W_{\max}} N(P/P^0, W) f(W) dW \tag{3}$$

where $N(P/P^0)$ is the experimental adsorption isotherm, $f(W)$ is the PSD, $N(P/P^0, W)$ is the kernel of theoretical adsorption or desorption isotherms. In this study, the PSD of all the samples was calculated using all three different methods. These results were then compared and are presented in Fig. 5. The PSD results derived from the BJHAD, BJHDE, and DFT methods all exhibited a multimodal characteristic. However, the calculated PSD curves of the same samples that were obtained from the BJHAD, BJHDE, and DFT methods do not overlap, which indicates different results and the impact of utilizing different models.

The PSD curves of all the samples derived from the BJHDE method showed noticeable peaks for pore sizes around 5 nm. This is due to the existence of ink-bottle shaped pores in these shale samples, which was confirmed by gas adsorption isotherms (Fig. 4). In these types of pores, during the desorption process, as the pressure of the capillary evaporation is achieved for the smallest pore, the gas will abruptly leave the pore, which can be observed in the desorption branch of the isotherms [34]. Therefore, a narrower PSD with sharp peaks will be obtained from the desorption calculations. Using the DFT model, pores with a smaller radius can be detected, which is beyond the ability of the BJHAD and BJHDE methods. A few potential reasons can explain this phenomenon: (1) the BJH models separate the adsorbed film and the capillary condensate, which does not happen in reality within the pores; (2) the BJH models neglect the fluid-wall interactions; and (3) the BJH models are based on the Kelvin equation but the Kelvin equation is not valid in significantly narrow pores [2, 35].

3.3 Model Choice

The BJH and DFT models are the two major methods that are being used extensively to quantify mesopore structures. However, it was shown that these different models can lead to different result for the same samples. Therefore, what is the best method to obtain the most accurate results in term of pore structure information?

The BJH model is based on the Kelvin equation, which means both adsorption and desorption branches of isotherms can be used for pore structure analysis. Regarding the adsorption branch, the condensation can be delayed, and thus thermodynamic equilibrium might not happen. The cylindrical meniscus for the adsorption process is not stable. Therefore, adsorption is not recommended for the analysis of cylindrical pores. However, for the desorption branch, the evaporation process is not delayed and takes place under the thermodynamic equilibrium state [36]. Based on these two phenomena, scientists prefer to use the desorption branch for the analysis of PSD curves over the adsorption branch. Nevertheless, if the pores in the samples are ink-bottle shaped, such as in our samples in this study, the desorption branch of isotherms will not be in thermodynamic equilibrium. This can be due to pore blockage, which results in the desorption being controlled by the percolation process. Moreover, PSD curves that are obtained from the analysis of desorption branch can become artificially narrow around specific pore sizes (short dash dot curve in Fig. 5) [34]. Under the influence of such conditions, the adsorption must be utilized instead of the more commonly used desorption. Consequently, if

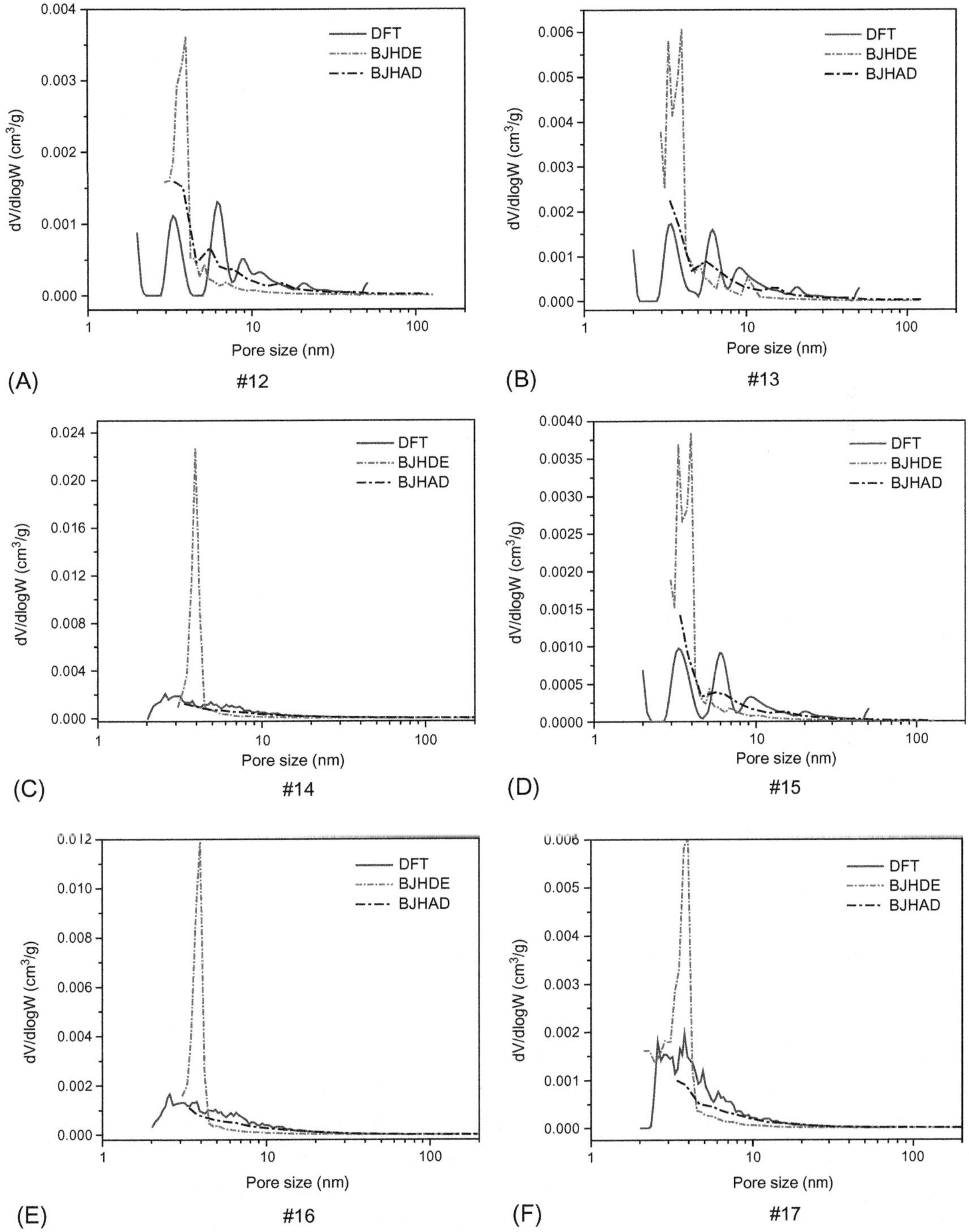

dV/dlogW (cm^3/g)
DFT
BJHDE
BJHAD
Pore size (nm)
(A)
#12
(B)
#13
(C)
#14
(D)
#15
(E)
#16
(F)
#17

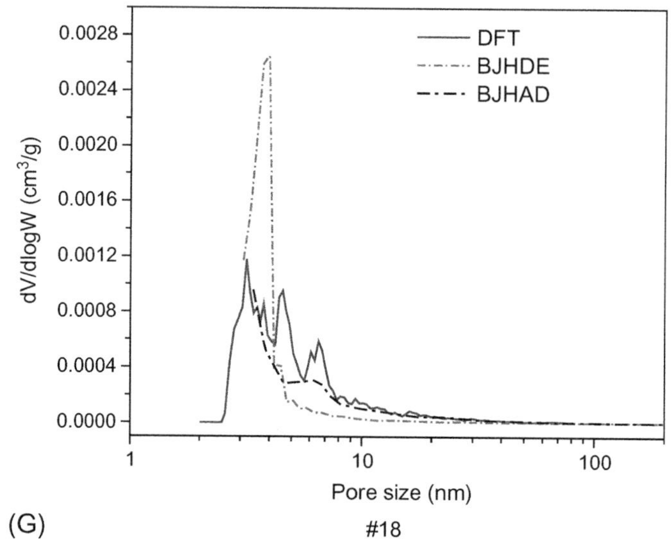

Fig. 5
Comparison of the PSDs using different methods.

either BJHAD or BJHDE models are used, the following steps should be taken to obtain accurate results: (1) adsorption data should be acquired with their isotherms; and (2) pore type should be determined based on the IUPAC suggestions. If pores are found to be ink-bottle shaped, the BJHAD method should be applied, otherwise the BJHDE method is preferred. Considering our samples from the Longmaxi Formation, the BJHAD method was more accurate than the BJHDE method.

The BJH model assumes a bulk condensation behavior, i.e., formation of a liquid in the pore with a hemispherical meniscus, which is not true for micropores. Hence, the BJH model can significantly underestimate the pore size for narrow mesopores, especially when the pore size is below 10nm [37]. Regarding the Longmaxi samples, the average pore volume that was derived using the BJH model (BJHAD and BJHDE) was smaller than the pore volume calculated by the DFT model. It was found that pores <10nm were widely distributed in the Longmaxi samples, which was also reported by other researchers [7, 38, 39]. However, the limitations that exist in the Kelvin equation can be avoided by applying microscopic methods based on molecular simulation or the DFT model. Based on the fundamental principles of statistical mechanics in explaining the molecular behavior of confined fluids in pore spaces, the DFT model can be used to describe the adsorption and phase behavior of fluids confined in pore structures. Thus, the DFT model can better define the thermodynamic behavior and density profiles of such fluids at a molecular level compared to other methods, such as BJH. As a result, DFT, in comparison with other common techniques, can capture the essential features of a wider range of pore sizes (micropores and mesopores) in a complete adsorption and desorption process to generate more reliable assessment [2, 40, 41].

4 Mercury Intrusion Capillary Pressure Test

Mercury is considered a typical nonwetting phase. It will not enter pores by capillary action, and it can only access interconnected pores. For a typical MIP test, a dry sample is placed into a container and then the container is evacuated to remove contaminant gases and vapors (usually water) and, while still evacuated, mercury is allowed to fill the container. This step creates an environment consisting of a solid (the sample), a nonwetting liquid (mercury), and mercury vapor. Next, pressure is increased while the volume of mercury entering larger openings in the sample bulk is monitored. The volume of mercury that can enter pore space is limited by the maximum pressure attained during analysis, which for many instruments is 60,000 psi (414 MPa). This high pressure will force mercury into pores, down to approximately 0.003 μm in diameter. The volume of mercury that intrudes into the sample due to an increase in pressure can be calculated and the PSD can be derived from the Washburn's equation [42]:

$$\Delta P = \frac{2\gamma \cos\theta}{r_{pore}} \tag{4}$$

where ΔP is the incremental pressure; γ is the surface tension of mercury and θ is the contact angle between the solid and mercury. r_{pore} is the pore radius.

The surface area can be calculated using the following equation [43]:

$$A = -\frac{1}{\gamma \cos\theta} \int_0^V P dV \tag{5}$$

4.1 Case Study

Mercury intrusion curves (cumulative pore volume occupied by mercury with an increase in pressure) for two Bakken samples are presented in Fig. 6. The intrusion branch of the curve shows that the volume of mercury increased as soon as the injection pressure increased and then steadily increased after the pressure reached the critical pressure point. At a very low pressure, the mercury easily accessed the larger pores, which leads to an instant increase in the injection volume. With an increase in pressure, mercury enters into the smaller pores and when all accessible pores are filled by mercury molecules, any increase in injection pressure will not add any volume to the detected volume and the curves become steady.

PSDs of two samples were calculated using the mercury intrusion capillary pressure (MICP) data and the results are shown in Fig. 7. It can be seen that similar to gas adsorption, the PSD curve of both samples exhibited a multimodal characteristic. Table 5 summarizes the pore structures and derived information for both samples. First, it can be deduced from the results that both samples have an overall very low porosity. The results derived from MICP data confirms that Sample #19 has a larger pore surface area and smaller average pore diameter compared to Sample #20.

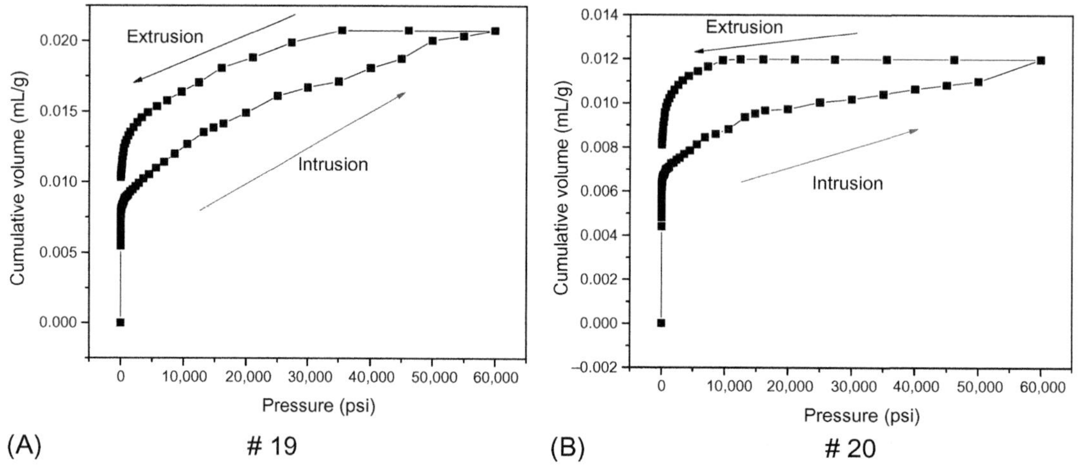

Fig. 6
Mercury injection and extrusion curves of two samples.

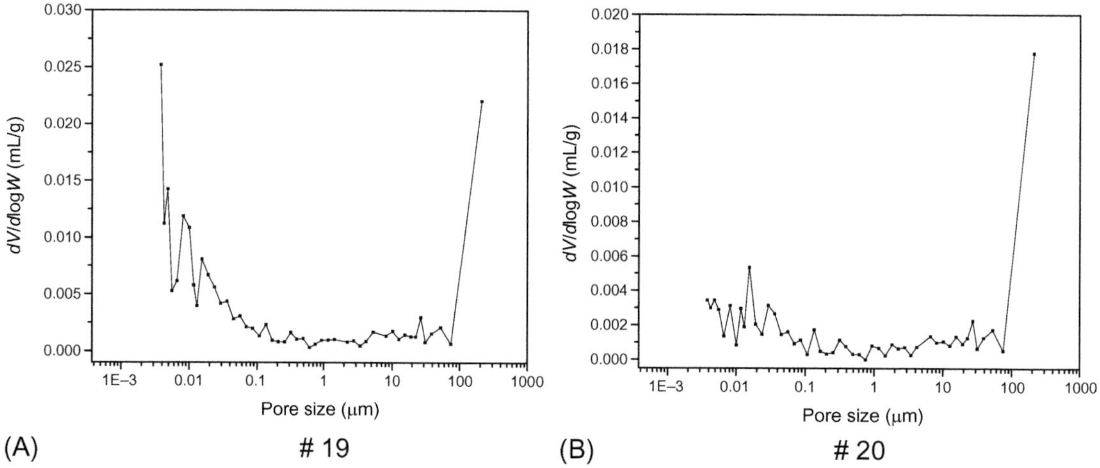

Fig. 7
PSDs of two samples from the MICP test.

Table 5 Pore information summary of two samples from the MICP test

Sample	Total Pore Area (m^2/g)	Average Pore Diameter (nm)	Porosity (%)
# 19	6.05	13.70	3.27
# 20	2.45	19.60	2.69

4.2 Comparison of the PSD from MICP and Gas Adsorption

Both MICP and nitrogen gas adsorption methods are fluid invasion techniques. These methods are considered to be the two major experimental techniques for pore structure analysis in unconventional reservoirs. However, the comparison between these two methods provides us with more in-depth information about the pore structure of low porosity samples. The comparison can also tell us if different measuring techniques can influence the results. The difference in PSDs can be seen in Fig. 8 and is mainly based on the difference in the working mechanisms of gas adsorption and MICP.

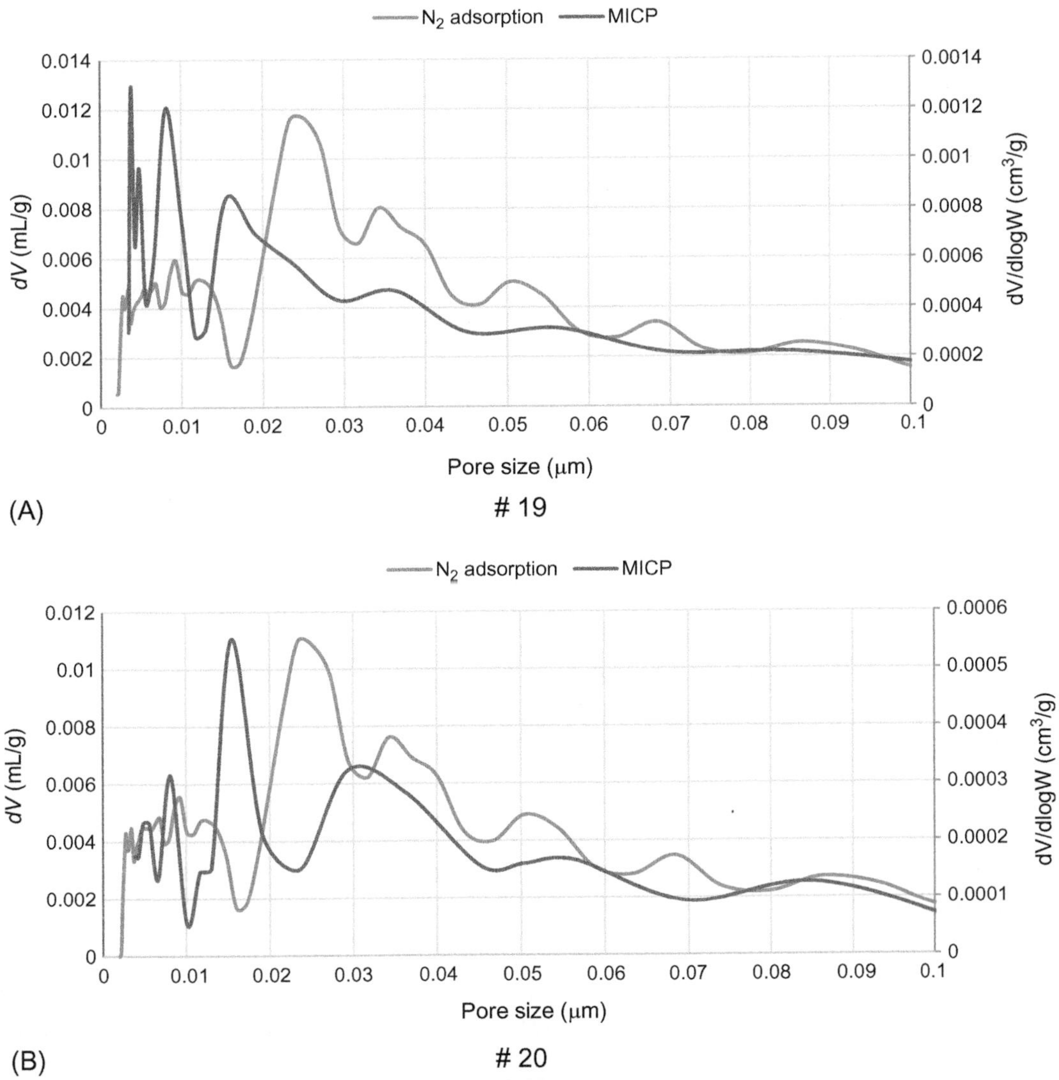

Fig. 8

PSD from MICP and nitrogen gas adsorption.

Considering gas adsorption, when nitrogen comes into contact with a solid, some gas molecules do not fully permeate into the pores but stay attached to its surface due to the weak van der Waals forces with the surface (the adsorbent). As the pressure increases, which is equivalent to an increase in the volume of injected nitrogen, gas molecules occupy the surface and start to fill in the pores. Therefore, pore information that is obtained from gas adsorption is basically a combination of both connected pore body and pore throats. However, in MICP testing, mercury is forced into the pores by applying an external pressure. Therefore, a slight increase in pressure is required to force mercury into larger macropores, whereas to force mercury into smaller pores, a significant amount of pressure is required as can be seen in Fig. 8. The injection pressure necessary for mercury to intrude into the sample's pores is inversely proportional to the size of the pores. Consequently, it can be said that in MICP, instead of measuring the internal pore sizes directly, the connection of the pores throughout the sample is being analyzed or, to be more specific, pore throat distribution is measured [18].

5 Conclusions

In this chapter, we used three methods to analyze the pore structures of shale and reached the following conclusions:

1. SEM is a direct method to identify the existence of the pore structures in shale. If the sample is argon ion milled, image analysis can be applied to quantify pore information, such as pore counts and porosity. The pores from the SEM are shown in two dimension and small pores cannot be detected due to its limiting resolution.
2. Gas adsorption can describe the pore information in one dimension. The limitation pore size for nitrogen adsorption is 200 nm. Applying different models to analyze the sorption isotherms can result in different PSDs. Strong attention is needed when choosing a suitable model for data analysis.
3. MICP can provide more pore size information than the gas adsorption method. The difference between the pore size information from MICP and gas adsorption is due to the different working mechanisms. MICP reveals information about the pore throat.

Acknowledgments

The authors would like to thank Ms. Hallie Fowler, Core Laboratories (Houston) for performing the milling process and for taking the FESEM photographs, Dr. Wang Liang from Southwest Petroleum University for sharing the gas adsorption data from Longmaxi and Dr. Zoya Heidari from University of Texas at Austin for running the MICP tests. Support from the China Scholarship Council (CSC) and the National Natural Science Foundation of China (41722403; 41572116) are also gratefully acknowledged.

References

[1] R.G. Loucks, R.M. Reed, S.C. Ruppel, et al., Morphology, genesis, and distribution of nanometer-scale pores in siliceous mudstones of the Mississippian Barnett shale, J. Sediment. Res. 79 (2009) 848–861.

[2] J. Rouquerolb, D. Avnir, C.W. Fairbridge, et al., Recommendations for the characterization of porous solids (technical report), Pure Appl. Chem. 66 (1994) 1739–1758.

[3] R.G. Loucks, R.M. Reed, S.C. Ruppel, U. Hammes, Spectrum of pore types and networks in mudrocks and a descriptive classification for matrix-related mudrock pores, AAPG Bull. 96 (2012) 1071–1098.

[4] V. Marcon, C. Joseph, K.E. Carter, et al., Experimental insights into geochemical changes in hydraulically fractured Marcellus shale, Appl. Geochem. 76 (2017) 36–50.

[5] X. Gu, D.R. Cole, G. Rother, et al., Standarization of reflectance measurements in dispersed organic matter: results of an exercise to improve interlaboratory agreement, Mar. Pet. Geol. 59 (2015) 22–34.

[6] X. Shao, X. Pang, Q. Li, et al., Pore structure and fractal characteristics of organic-rich shales: a case study of the lower Silurian Longmaxi shales in the Sichuan Basin, SW China, Mar. Pet. Geol. 80 (2017) 192–202.

[7] K. Jiao, S. Yao, C. Liu, et al., The characterization and quantitative analysis of nanopores in unconventional gas reservoirs utilizing FESEM–FIB and image processing: an example from the lower Silurian Longmaxi shale, upper Yangtze region China, Int. J. Coal Geol. 128 (2014) 1–11.

[8] M.E. Houben, A. Barnhoorn, L. Wasch, Microstructures of early Jurassic (Toarcian) shales of northern Europe, Int. J. Coal Geol. 165 (2016) 76–89.

[9] S. Hemes, G. Desbois, J.L. Urai, et al., Multi-scale characterization of porosity in boom clay (HADES-lever, Mol, Belgium) using a combination of X-ray μ-CT, 2D BIB-SEM and FIB-SEM tomography, Microporous Mesoporous Mater. 208 (2015) 1–20.

[10] J. Klaver, G. Desbois, R. Littke, et al., BIB-SEM pore characterization of mature and post mature Posidonia shale samples from the Hils area, Germany, Int. J. Coal Geol. 158 (2016) 78–89.

[11] S. Zhou, Y. Ning, H. Wang, et al., Investigation of methane adsorption mechanism on Longmaxi shale by combining the micropore filling and monolayer coverage theories, Adv. Geo-Energy Res. 2 (2018) 269–281.

[12] L. Chen, Z. Jiang, K. Liu, et al., Quantitative characterization of micropore structure for organic-rich lower Silurian shale in the upper Yangtze platform, South China: implications for shale gas adsorption capacity, Adv. Geo-Energy Res. 1 (2017) 112–123.

[13] M, Mastalerz, A. Schimmelmann, A. Drobniak, et al., Porosity of Devonian and Mississippian New Albany shale across a maturation gradient: insights from organic petrology, gas adsorption, and mercury intrusion, AAPG Bull. 97 (2013) 1621–1643.

[14] S. Karimi, M. Saidian, M. Prasad, Reservoir rock characterization using centrifuge and nuclear magnetic resonance: a laboratory study of Middle Bakken Cores, SPE 175069, in: Presented at SPE Annual Technical Conference and Exhibition in Houston, Texas, 28–30 September, 2015.

[15] C.R. Clarkson, N. Solano, R.M. Bustin, et al., Pore structure characterization of north American shale gas reservoirs using USANS/SANS, gas adsorption, and mercury intrusion, Fuel 103 (2013) 606–616.

[16] M.E. Curtis, C.H. Sondergeld, R.J. Ambrose, et al., Microstructural investigation of gas shales in two and three dimensions using nanometer-scale resolution imaging, AAPG Bull. 96 (2012) 665–677.

[17] H. Deng, X. Hu, H. Li, et al., Improved pore-structure characterization in shale formations with FESEM technique, J. Nat. Gas Sci. Eng. 35 (2016) 309–319.

[18] L.M. Anovitz, D.R. Cole, Characterization and analysis of porosity and pore structures, Rev. Mineral. Geochem. 80 (2015) 61–164.

[19] K. Liu, M. Ostadhassan, B. Bubach, Applications of nano-indentation methods to estimate nanoscale mechanical properties of shale reservoir rocks, J. Nat. Gas Sci. Eng. 35 (2016) 1310–1319.

[20] K. Liu, M. Ostadhassan, T. Gentzis, et al., Characterization of geochemical properties and microstructures of the Bakken shale in North Dakota, Int. J. Coal Geol. 190 (2018) 84–98.

[21] K. Liu, M. Ostadhassan, Multi-scale fractal analysis of pores in shale rocks, J. Appl. Geophys. 140 (2017) 1–10.

[22] Y. Chen, M. Mastalerz, A. Schimmelmann, Heterogeneity of shale documented by micro-FTIR and image analysis, J. Microsc. 256 (2014) 177–189.

[23] S. Saraji, M. Piri, The representative sample size in shale oil rocks and nano-scale characterization of transport properties, Int. J. Coal Geol. 146 (2015) 42–54.
[24] J. Dewar, The adsorption and thermal evolution of gases occluded in charcoal at low temperatures, Proc. R. Soc. Lond. 74 (1904) 122–127.
[25] I. Langmuir, The constitution and fundamental properties of solids and liquids. II. Liquids, J. Am. Chem. Soc. 39 (1917) 1848–1906.
[26] F. Benton, A. White, The sorption of gas by iron, J. Am. Chem. Soc. 54 (1932) 1820–1830.
[27] P.H. Emmett, S. Brunauer, The use of low temperature van der Waals adsorption isotherms in determining the surface area of iron synthetic ammonia catalysts, J. Am. Chem. Soc. 59 (1937) 1553–1564.
[28] S. Brunauer, P.H. Emmett, E. Teller, Adsorption of gases in multimolecular layers, J. Am. Chem. Soc. 60 (1938) 309–319.
[29] E.P. Barrett, L.G. Joyner, P.P. Halenda, The determination of pore volume and area distributions in porous substances. I. Computations from nitrogen isotherms, J. Am. Chem. Soc. 73 (1951) 373–380.
[30] P.I. Ravikovitch, A. Vishnyakov, R. Russo, et al., Unified approach to pore size characterization of microporous carbonaceous materials from N_2, Ar, and CO_2 adsorption isotherms, Langmuir 16 (2000) 2311–2320.
[31] K. Liu, M. Ostadhassan, J. Zhou, et al., Nanoscale pore structure characterization of the Bakken shale in the USA, Fuel 209 (2017) 567–578.
[32] K. Liu, M. Ostadhassan, L. Kong, Multifractal characteristics of Longmaxi Shale pore structures by N_2 adsorption: a model comparison, J. Petrol. Sci. Eng. 168 (2018) 330–341.
[33] R. Roque-Malherbe, Adsorption and Diffusion in Nanoporous Materials, CRC Press, Boca Raton, FL, 2007.
[34] Y. Zeng, C. Fan, D.D. Do, et al., Evaporation from an ink-bottle pore: mechanisms of adsorption and desorption, Ind. Eng. Chem. Res. 53 (2014) 15467–15474.
[35] K. Sing, The use of nitrogen adsorption for the characterisation of porous materials, Colloids Surf. A Physicochem. Eng. Asp. 187 (2001) 3–9.
[36] V. Bolis, Fundamentals in adsorption at the solid-gas interface. Concepts and thermodynamics, in: Calorimetry and Thermal Methods in Catalysis, Springer, Berlin, Heidelberg, 2013, pp. 3–50.
[37] S. Lowell, J. Shields, M.A. Thomas, et al., Characterization of Porous Solids and Powders: Surface Area. Porosity and Density, Springer, 2004.
[38] L. Wang, Y. Fu, J. Li, et al., Mineral and pore structure characteristics of gas shale in Longmaxi formation: a case study of Jiaoshiba gas field in the southern Sichuan Basin, China, Arab. J. Geosci. 9 (2016) 733.
[39] R. Yang, S. He, Q. Hu, et al., Applying SANS technique to characterize nano-scale pore structure of Longmaxi shale, Sichuan Basin (China), Fuel 197 (2017) 91–99.
[40] M. Thommes, K.A. Cychosz, Physical adsorption characterization of nanoporous materials: progress and challenges, Adsorption 20 (2014) 233–250.
[41] J. Landers, G.Y. Gor, A.V. Neimark, Density functional theory methods characterization of porous materials, Colloids Surf. A Physicochem. Eng. Asp. 437 (2013) 3–32.
[42] H. Giesche, Mercury porosimetry: a general (practical) overview, Part. Part. Syst. Charact. 23 (2006) 9–19.
[43] H.M. Rootare, C.F. Prenzlow, Surface areas from mercury porosimeter measurements, J. Phys. Chem. 71 (1967) 2733–2736.

CHAPTER 2

Petrophysical Characterization of the Pore Structure of Coal

Yanbin Yao, Jun Liu

School of Energy Resources, China University of Geosciences, Beijing, P.R. China

Chapter Outline

1 Introduction

Coalbed methane (CBM), an eco-friendly component in the energy portfolio, can help meet the energy demand all over the world, as well as enhance coal mining safety and reduce greenhouse gas [1, 2]. As an unconventional gas reservoir, coal is usually defined as a special and complicated porous medium when compared to conventional gas reservoirs (e.g., sandstone) [3]. Therefore, accurate investigation of the pore structure of coal is a challenge for commercial CBM exploration and development, especially for the evaluation of CBM resources in a coal reservoir and for designing the best CBM extraction method.

Petrophysical Characterization and Fluids Transport in Unconventional Reservoirs. https://doi.org/10.1016/B978-0-12-816698-7.00002-4

Many methods were employed to characterize pore structure in coal, such as mercury intrusion porosimetry (MIP), low-temperature N_2 adsorption (LTNA), scanning electron microscopy (SEM), small-angle X-ray scattering and small-angle neutron scattering [4, 5]. Although the traditional methods for pore structure characterization of coal are widely adopted, all these methods have shortcomings, such as slow data acquisition, inefficient testing, limited aperture range, and damage to samples. During the MIP measurement, the number of large pores decreased with the high-pressure intrusion by mercury because of the shielding effects of smaller pores [6, 7]. The LTNA investigation can only determine limited information on micropores and transitional pores in coal [8]. The SEM observation can only provide a local picture of pores and thus is usually used as a supplementary mean to evaluate the pore structure in coal [9]. The techniques of small-angle X-ray scattering and small-angle neutron scattering cannot accurately determine the pore size distribution (PSD) in coal, even though these methods have the ability to identify both connected pores and close pores in coal [5, 10]. In view of the limitations with traditional techniques, it was necessary to explore a new method that can characterize the coal pore structure more effectively.

In recent years, the low-field nuclear magnetic resonance (NMR) technique was introduced into petrophysical analysis and well logging, particularly for the investigation on reservoir properties such as porosity, permeability, viscosity and saturation of hydrogen fluids in oil or gas reservoirs [11–13]. Compared with other traditional methods, low-field NMR measurement is a nondestructive, easy and less time-consuming method for characterizing complex porous media, and it also covers the widest aperture range [3].

In this chapter, 12 coal samples were taken from 11 raw coal blocks. The low-field NMR experiments were performed on each coal sample, both at 100% water saturation and at irreducible water condition. The results indicate that the pore structure, including porosity, pore size distribution, pore geometry, pore connectivity, and permeability of coals, can be better and more effectively characterized by low-field NMR when compared to conventional methods.

2 Samples, Experiments and Methods

2.1 Samples

Eleven raw coal blocks were collected from underground mines, amongst which five blocks were from Qinshui basin, three from the Ordos basin and the rest were from the Hongyang coalfield, China. Two samples (Lvb1 and Lvb2) were subsamples cut from the same coal block to test the heterogeneity of the coal and to prove the reproducibility of NMR measurements. For each sample, one or more cylindrical core plugs with a diameter of 2.5 cm were cut parallel to the bedding planes. The core plugs were analyzed for helium porosity and air permeability using routine core analysis methods, while the remaining core cuttings were analyzed for vitrinite reflectance and coal maceral composition using routine methods.

The coal types that were sampled had a wide coal ranks (from subbituminous to anthracite) with mean maximum vitrinite reflectance in oil (R_o) of 0.39%–3.57%. The helium porosity (φ_g) ranged from 1.56%–9.00%, with a mean of 4.56%. The porosity, using the water-saturated method, (φ_w) ranged from 1.03%–5.88%, with a mean of 3.02% (Fig. 1). For the air permeability (K_a) of selected coals, the highest value was 82.1 mD and the lowest value was 0.00032 mD.

2.2 NMR Theory and Experiments

For the theory of low-field NMR relaxation, the NMR relaxation can be demonstrated when a substance containing magnetic nuclei (e.g., hydrogen proton) is immersed in a magnetic field [14]. Since quantification of multiphase methane is directly related to the number of hydrogen atoms from the methane molecules in coal, in a typical T_2 measurement of H^1 nuclear from methane in coal, the T_2 is affected by the relaxation times of bulk, surface and diffuse relaxations [15]:

$$\frac{1}{T_2} = \frac{1}{T_{2B}} + \frac{1}{T_{2S}} + \frac{1}{T_{2D}} \tag{1}$$

where subscripts B, S, and D refer to bulk, surface and diffuse relaxation, respectively. The bulk relaxation (T_{2B}) is an intrinsic property of the hydrogen-containing fluid and is affected by the physical properties (e.g., viscosity), temperature, and pressure of the fluid [16].

$$\frac{1}{T_{2S}} = \rho \left(\frac{S}{V}\right)_{pore} \tag{2}$$

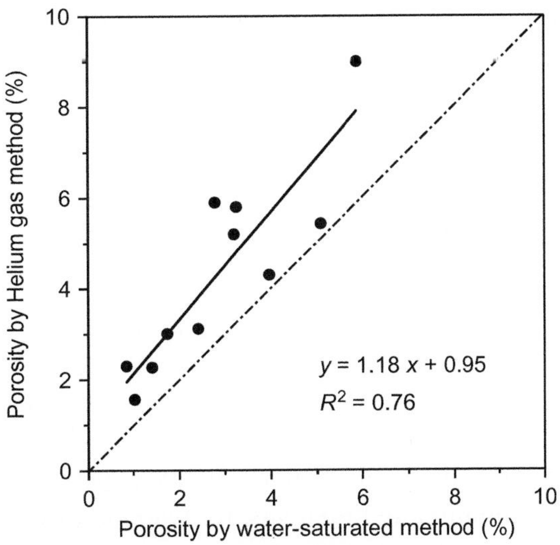

Fig. 1
The relationship between helium porosity and water porosity.

where ρ is the T_2 surface relaxivity (T_2 relaxing strength of the grain surfaces), S represents the surface area of grain, $(S/V)_{pore}$ is the surface area to volume ratio of the pores filled with hydrogen-containing fluid.

Eq. (2) shows that protons in smaller pores relax faster than in larger pores because smaller pores have a larger surface to volume ratio. Consequently, the T_2 distribution in core samples reflects the pore size distribution, with the smallest pores having the shortest relaxation time and the largest pores having the longest relaxation time.

In this chapter, two sets of NMR measurements were performed for all samples, where one measurement was performed under a 100% water saturated condition (S_w) and the second measurement was under an irreducible water condition (S_{ir}). All NMR measurements were conducted using a Rec Core 2500 instrument, which provides a constant magnetic field strength (1200 gauss) and resonance frequency (2.38 MHz). The measurement parameters were set as follows: echo spacing, 0.6 ms; waiting time, 5 s; echo numbers, 2048; scanning numbers, 64. After the measurements, transverse relaxation time (T_2) distributions were computed by multiexponential inversion of the echo data with 51 preset decay times logarithmically spaced from 0.1 ms to 10 s.

After the above experiments, several core plug samples were dried again, and analyzed by MIP. The MIP measurements run up to a pressure of 32 MPa, meaning pore throats as small as 0.02 μm were penetrated [3].

2.3 Theory of Centrifuge Experiment

The centrifuge operation performed by Kenyon [17] was used prior to the second set of NMR experiments. The purpose of centrifuging is to expel the free water within the connective macropores. The centrifuge theory is based on the Washburn equation:

$$r_c = \frac{-2\sigma\cos\theta}{P_c} \tag{3}$$

where P_c is the centrifuging pressure (MPa); r_c is the minimal pore radius (μm) for water to discharge at the pressure P_c; θ is the contact angle (°) between the water molecule and pore surface; and σ is the interfacial tension (N/m) between coal and water.

Although the θ and σ of coal-to-water varies between samples [18], for simplification, they were assigned as 60 degrees and 0.076 N/m, respectively. In this situation, Eq. (3) becomes

$$r_c = \frac{0.14}{P_c} \tag{4}$$

According to Eq. (4), the r_c negatively relates to P_c. In this experiment, the centrifuge pressure of 1.4 MPa indicated the minimum pore throat size for water to discharge was roughly 0.1 μm. In other words, the irreducible water condition for the second set of NMR measurements was where water only remained in pores with a diameter <0.1 μm.

3 T_2 Distributions of the Coal Samples

Using the low-field NMR T_2 measurements, the comparison of T_2 spectra at 100% water saturation and irreducible water condition is shown in Fig. 2. From this data, it seems that the fully saturated T_2 distribution depends on coal ranks. The subbituminous coal with R_o of 0.39% contained two unintermittent peaks at T_2 values of 2.5 and 800 ms, respectively (Fig. 2A). The high volatile bituminous coal with 0.8% R_o revealed a wide unimodal distribution at a T_2 value of ~90 ms (Fig. 2B). Medium volatile bituminous coals showed multimodal distribution, with the dominant peak at a higher T_2 value (Fig. 2C and D). In contrast, low volatile bituminous coals, semianthracite coals, and anthracite coals showed identical T_2 distributions (Fig. 2E–L). Except for sample A3, which was collected adjacent to a dike, all samples with R_o >1.6% had a dominant peak at T_2 of ~1 ms. The magma contact metamorphism may have changed the pore structure of the coal at A3 and thus led to a different T_2 distribution. For lower rank coal, the location of T_2 spectrum peak approached the larger relaxation time area (Fig. 2A–D), whereas the location approached to smaller relaxation time area for higher rank coal (Fig. 2E–L). Fig. 2E and F show that the T_2 distributions of the two subsamples from the same coal sample were very similar in morphology and amplitude, confirming the reproducibility of NMR measurements.

4 Determination of T_2 Cutoff Value

An accurate determination of T_2 cutoff value (T_{2c}) is important when attempting to obtain moveable and irreducible fluid porosities and PSD [19]. T_{2c} is a relaxation time threshold that divides the T_2 spectrum into two parts: the bound water and the free water. The bound water corresponds to small pores that cannot be easily drained by capillary forces, while the free water corresponds to large pores that can be drained. Therefore, the bound water has a characteristic relaxation time $T_2 < T_{2c}$, whereas the free water has $T_2 > T_{2c}$.

An example of the calculation method is illustrated in Fig. 3. Firstly, to calculate T_{2c}, two NMR T_2 distributions were obtained: one corresponding to 100% water-saturated condition (S_w) and the other corresponding to irreducible water condition (S_{ir}). Secondly, the two NMR T_2 distributions were transformed into two accumulative T_2 spectra using the following rule: maximum peak amplitudes (dimensionless) at S_w and at S_{ir} are equal to total porosity and irreducible porosity (IP), respectively. Finally, two projection processes determine the T_{2c}: the first corresponding to a horizontal projection from the cumulative curve at S_{ir}, and the second corresponding to a vertical projection from the cumulative curve at S_w (Fig. 3). The T_2 value at the intersection projected on the T_2 axis is the T_{2c}. The calculated T_{2c} for seven coal samples ranges from 2.5 to 32 ms (Table 1). These values are lower than those from sandstone and carbonate [20].

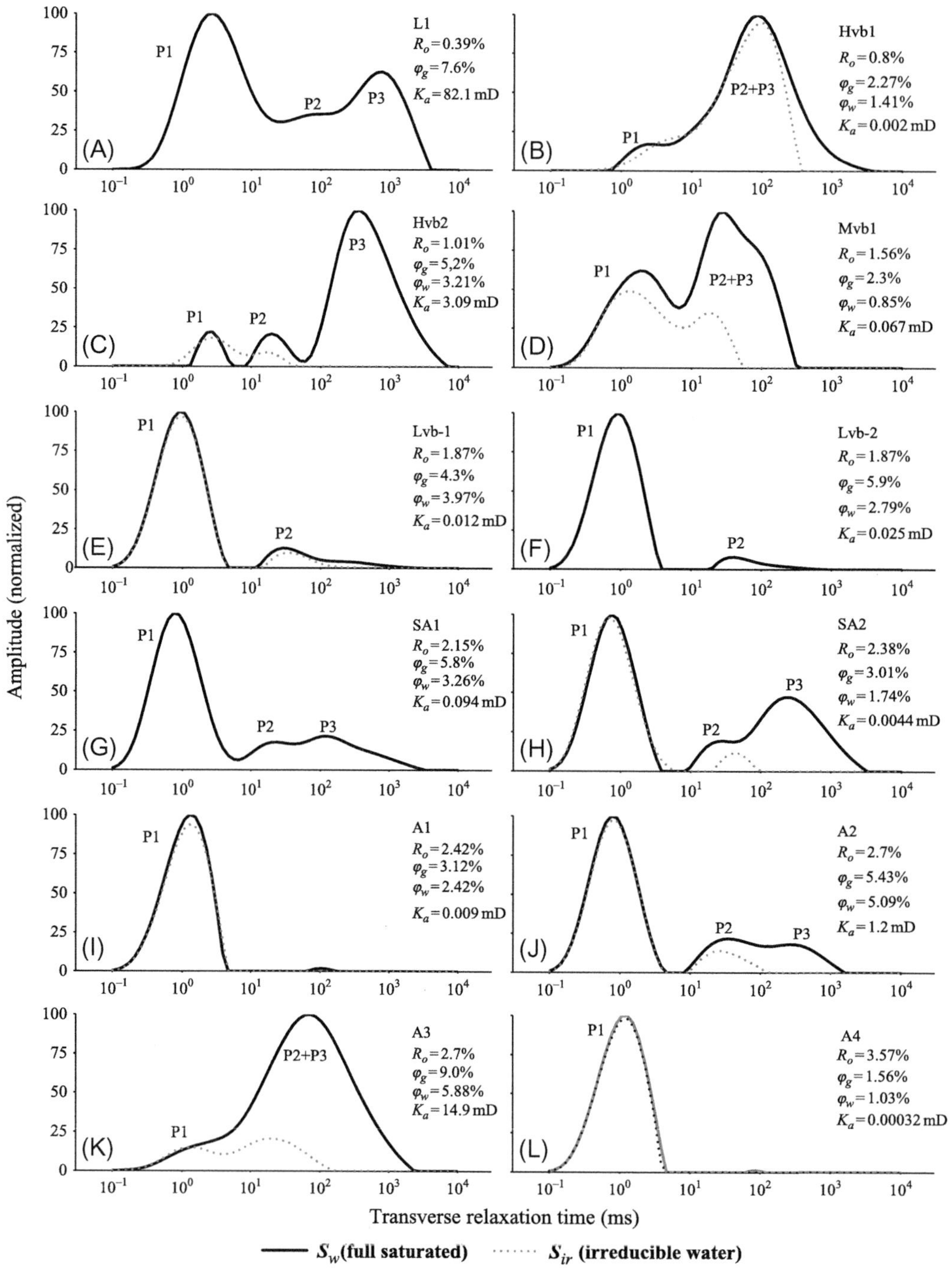

Fig. 2

NMR T_2 distribution of bituminous coal and anthracite samples, (A)–(L) are samples of L1, Hvb1, Hvb2, Mvb1, Lvb-1, Lvb-2, SA1, SA2, A1, A2, A3, and A4, respectively. Sample HC3-5 was broken and failed the φ_w measurement at water saturation. Lvb1 and Lvb2 were subsamples cut from the same coal block. R_o is the maximum vitrinite reflectance in oil. φ_g and φ_w is porosity from helium gas injection and water-saturated methods, respectively. K_a represents air permeability.

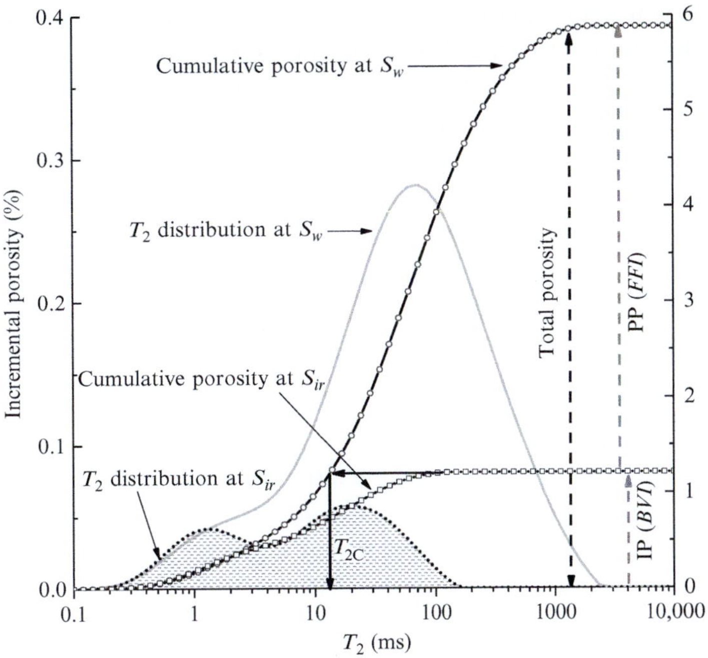

Fig. 3

NMR measurements under 100% water-saturated condition (S_w) and irreducible water condition (S_{ir}) for sample A3 show the method to calculate T_2 cutoff value (T_{2c}).

Table 1 Details about the NMR relaxation results of coal

Sample ID	φ_{NI} (%)[a]	φ_{NP} (%)[a]	*FFI*/*BVI*[a]	T_{2c} (ms)[a]	K_c (mD)[a]	K_p (mD)[a]	k_a (mD)[a]
Hvb2	0.45	2.76	6.13	31	3.32	3.25	3.09
Mvb1	0.36	0.49	1.36	11	0.0008	0.16	0.067
Lvb1	3.63	0.34	0.09	32	0.002	0.09	0.012
SA2	1.03	0.71	0.69	24	0.0036	0.29	0.0044
A2	3.92	1.17	0.3	30	0.049	0.63	1.2
A3	1.21	4.67	3.86	12.5	14.69	15.04	14.9
A4	0.96	0.07	0.07	2.5	0.00049	–	0.00032

[a] φ_{NI} and φ_{NP} are the irreducible and producible porosities based on NMR; *FFI*/*BVI* is the ratio of movable fluid (*FFI*) to bound fluid (*BVI*); T_{2c} is the T_2 cutoff; K_c and K_p are the estimated permeability by Coates model and PP model; K_a is the measured air permeability.

5 Pore Structure Characterization Using NMR

5.1 Pore Classification Based on an NMR Fluid Typing Method

Pores in coal are composed of the primary pore system, consisting of micro- and mesopores (<100 nm), which contain major in situ gases, and the secondary pore system, which includes the macropore (>100 nm), microfractures, cleats and natural fractures and forms the conduit for fluid (gas or water) transfer to the wellbore [7, 21].

Using the NMR theory, water in smaller pores experiences a greater surface relaxation and thus relaxes faster than that in larger pores. A T_2 distribution is similar to a pore size distribution, where each relaxation time corresponds to a pore size: larger pores correspond to longer relaxation time and smaller pores to a shorter relaxation time [3, 22]. Therefore, the T_2 distribution characteristics can imply coal pore types. As shown in Fig. 2, the distributions of T_2 spectra are commonly bimodal, and sometimes unimodal and multimodal based on various pore types. For the unimodal distribution, a narrow distribution represents a single pore type (Fig. 2I), whereas a wide distribution may represent multiple pore types (Fig. 2K). For multiple peaks, the continuity among these peaks can be used to identify the connectivity among pores. For example, the bimodal distribution in L1 (Fig. 2A) suggests that continuous multipores exist in that sample, while the trimodal distribution in Hvb2 (Fig. 2C) may indicate that three disconnected pore types are present in that sample.

In a typical T_2 spectrum, from short to long relaxation times, the adsorption pores, seepage pores and cleats (or fractures) can be recognized sequentially. Adsorption pores are associated with the spectrum peak located at 0.5–2.5 ms, which is always observed in medium to high rank coals (Fig. 2E–J and L). By comparing the NMR measurement in S_w and S_{ir} (Fig. 2E, H–J and L), it was found that adsorption pores were generally saturated with irreducible water that could not be expelled by centrifuging. Seepage pores are commonly associated with the peak located at 20–50 ms, which is observed as the second peak in a spectrum (Fig. 2B–H and J). Cleats (or fractures) are recognized for some samples, such as L1 (Fig. 2A), Mvb1 (Fig. 2D), SA2 (Fig. 2H), A2 (Fig. 2J) and A3 (Fig. 2K). The cleats peak generally has a T_2 value of >100 ms, and can be removed by centrifuging. In addition, an interconnected and continued pore system including adsorption pores, seepage pores, and cleats is seen in A3 (Fig. 2K). Using these relaxation spectra, cleats and fractures generally have connective pores and are meaningful for seepage and movement of gas and water, while the adsorption pores are generally closed and not favored for the movement of gas and water. However, in sample Hvb2 (Fig. 2C), the spectrum amplitude at S_{ir} was higher than that at S_w for some relaxation time area. This was because the poor connectivity of pores in the sample caused some water to be retained at the surface or inside adsorption pores after centrifuging, which led to the increase of relaxation time amplitude [23].

NMR-based pore types agree well with the results obtained from μCT scan of coal samples, demonstrated by five typical coal types with various coal ranks (Fig. 4). For L1, well-developed fractures and seepage pores were shown in the μCT scan image (Fig. 4A), which corresponds to the two peaks (fracture and seepage pore peaks) in the T_2 spectrum (Fig. 4B). This is consistent with the high permeability (82.1 mD) and high porosity of this sample (Fig. 2). For sample Hvb2, the fracture peak in the T_2 spectrum (Fig. 4B) agreed well with the fractures shown in the μCT image (Fig. 4A), suggesting the moderate porosity and permeability of this sample (Fig. 2) were mainly contributed by the fractures in the coal. The T_2 spectrum peaks of samples Mvb1 and A1 were located at <100 ms (Fig. 4B), which means that the cleats and fractures were

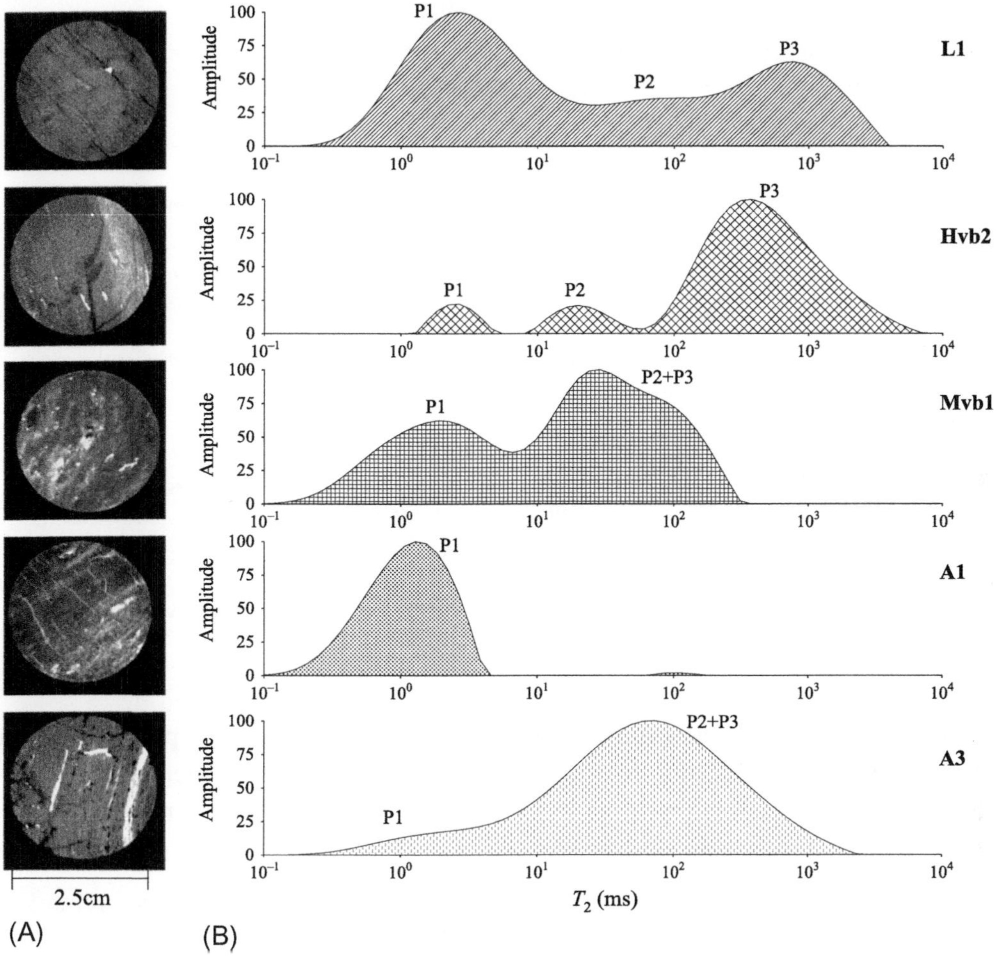

Fig. 4
Porosity in coal types as illustrated by (A) μCT scan image (pores and minerals are represented with black and white colors respectively) and (B) NMR T_2 distribution.

poorly developed. At the same time, the μCT images of these two samples (Fig. 4A) showed that the pores (in Mvb1) or cleats (in A1) were thoroughly filled with minerals, which is the main reason for the low porosity and permeability of these two samples. For A3, it had high porosity and permeability, well-developed seepage pores and cleats (Fig. 4A), and thus a wide T_2 spectrum with a large peak (Fig. 4B).

5.2 Characterization of Pore Size Distribution

The pore types identified by the T_2 spectra and peaks agreed well with the pore structure determined by MIP. Taking SA2 as an example (Fig. 5), the result by MIP showed that the sample had dominant adsorption pore volumes but a few seepage pore volumes, which was

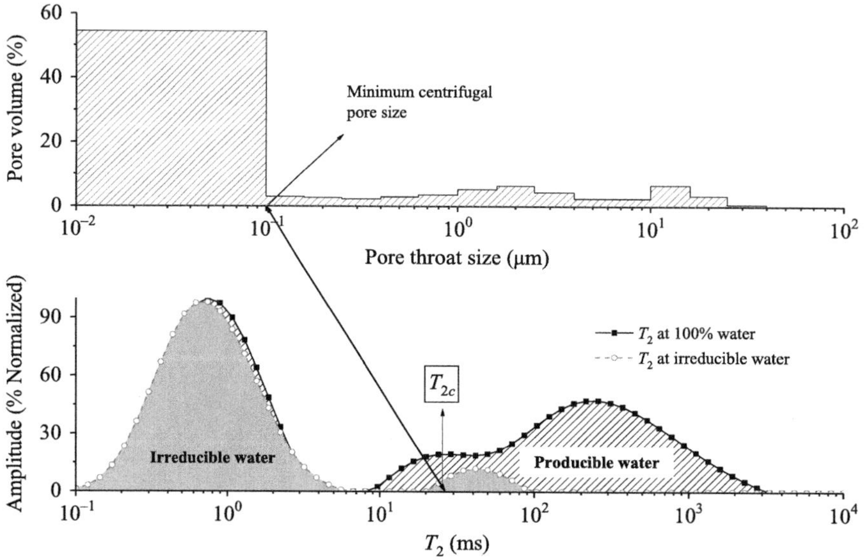

Fig. 5
The similarity of pore size distribution determined by NMR T_2 distribution (lower image) and MIP (upper image) for sample SA2.

consistent with the dominant adsorption pore peak and the small seepage pore peak in the T_2 spectrum. The seepage pores (Fig. 5) were saturated with producible waters that could be expelled by centrifuging. In general, the T_2 spectra distribution accurately represents the pore types and pore size distribution.

As exhibited in Fig. 5, when a shift is applied to the T_2 distribution, it shows a remarkable correlation with the pore size distribution found by MIP. This kind of similarity is universally seen in other coal samples, which means that there could be a relationship between the pore structure by NMR and the pore structure by MIP. If this is the case, the pore structure study by MIP may be substituted by the study based on NMR, and on the other hand, MIP could verify pore structure results found by NMR. For this purpose, accurately scaling the NMR pore structure result by MIP data is discussed further.

According to Eq. (2), when a constant F_s is given [24], the equation can be transformed into

$$\frac{1}{T_2} = \frac{F_s \times \rho_2}{r_c} \tag{5}$$

where F_s is the pore geometry morphologic factor, with $F_s = 3$ for spherical pores; and r_c is the estimated pore size corresponding to a T_2 value. Given a constant C_2, Eq. (5) can be transformed into

$$r_c = C_2 \times T_{2c} \tag{6}$$

Based on Eq. (6), numerous researchers have proposed many transformation methods to obtain the pore structure or capillary pressure curves from NMR data [25, 26]. These methods have been proven suitable or partly suitable for sandstone, but not for the coal types used in this chapter. This is because the methods are not convenient and are difficult to apply to coal and these methods have large estimation errors, especially for coal with multiple peaks in the T_2 spectrum. Here, a new method for scaling the NMR pore structure is proposed.

The method is based on the principle that any relaxation time in the T_2 distribution corresponds to a certain pore size of the coal sample: slow relaxation times correspond to larger pores and vice versa. To quantify the relationship between relaxation time and pore size, a key parameter of T_{2c} is used. Since T_{2c} was determined by a centrifuge experiment and two NMR experiments, the certain pore size (r) corresponding to the certain T_{2c} can be determined. For any relaxation time (T_{2i}) in a T_2 relaxation distribution, the corresponding pore size (r_{ci}) can be determined by Eq. (7).

$$r_{ci} = \frac{r \times T_{2i}}{T_{2c}} \tag{7}$$

where r (μm) is the pore size corresponding to the T_{2c}. For any sample, the r is a constant and corresponds to the minimum pore size that the centrifuge experiment can probe. As discussed in Section 2, r is about 0.1 μm in this chapter. Based on Eq. (7), the pore size distribution by NMR data is determined.

The PSD is scaled for seven coal types, among which two representative samples are shown in Fig. 6. The result shows that the constructive pore size distributions coincide very well with the pore size estimated from MIP, meaning that the pore structure estimated by NMR is reliable.

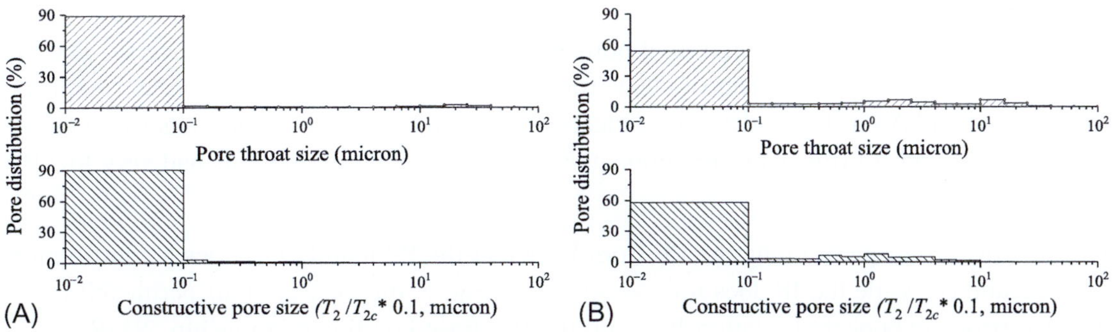

Fig. 6
Comparison of pore size distribution by mercury intrusion porosimetry (MIP) (upper image) and NMR relaxation (lower image) for two coal types: (A) Lvb1 and (B) SA2.

5.3 Characterization of Porosity

Porosity is the volume ratio of pores and fractures in coal. Different to the conventional oil/gas reservoir (e.g., sandstone), the pores and fractures vary greatly in coal, which makes the accurate measurement of coal porosity more difficult.

Here, we use the weighting porosity determined by water saturation method (φ_w) as an approximation of the NMR porosity (φ_N), because theoretically these two porosities should be identical. The φ_w is commonly calculated by estimating the water volume in coal and the φ_N is calculated based on the estimation of the volume of hydrogen liquid NMR signals in coal. In theory, the NMR porosity represents the pore volume fractions occupied by bound (irreducible) and free (producible) water. By definition, bound water is the amount of water in coal that cannot be recovered during CBM production, while free water can be recovered. Thus, the NMR porosity can be divided into the IP corresponding to the bound water fraction, φ_{NI}, and the producible porosity (PP) corresponding to free water fraction, φ_{NP}. The φ_{NI} and φ_{NP} can be determined by comparing the signal of 100% water-saturated sample with that of irreducible water-saturated sample.

The φ_{NI} and φ_{NP} are expressed by

$$\varphi_{NI} = \frac{\varphi_N \times BVI}{BVI + FFI} \tag{8}$$

$$\varphi_{NP} = \frac{\varphi_N \times FFI}{BVI + FFI} \tag{9}$$

where φ_N, φ_{NI} and φ_{NP} are NMR total porosity, IP and PP, respectively; BVI is the bound fluid index (or volume) that can be calculated by the total spectrum area fraction at S_{ir} condition; FFI is the free fluid index (or volume); $BVI+FFI$ represents both the free and bound fluid that can be calculated by the total spectrum area fraction at S_w condition.

The IP and PP were calculated for seven samples (Table 1). The PP values were low, with a range of 0.07%–4.67%, but mostly of 0.1%–2%. The correlation between PP and total porosity was poor. Some samples with high total porosities may have low PP, and vice versa. The PP values related to the development of cleats (or fractures) in the coal. For example, samples Hvb2, Mvb1 and A3 had well-developed cleats (Figs. 2 and 4) and thus had higher PP values; whereas Lvb1 and A4 (Fig. 2) were dominated by adsorption pores and thus had very low PP values.

The significance of partitioning the PP from IP is that the PP mainly provides a contribution to permeability, while the IP does not. As shown in Fig. 7, the measured air permeability shows a strong exponential relationship with the PP. Based on this relationship, NMR permeability can be estimated according to the PP. The estimated NMR permeability results by this model will be discussed below.

6 Characterization of Permeability

Although permeability cannot be directly measured by NMR, it can be estimated based on the porosity and pore size distribution. For sandstone, two commonly used models have been used to estimate permeability using NMR relaxation data: the Free Fluid model and the Mean T_2 model. In this section, these two models are evaluated and a newly presented model will be discussed.

The Free Fluid model (also called Coates model) is based on the Timur-Coates equation [27]:

$$K_c = \frac{\varphi}{C} \times \left(\frac{FFI}{BVI}\right)^2 \tag{10}$$

where φ is NMR porosity, C is a constant, and K_c is the permeability estimated by this equation. The FFI/BVI is determined by the T_{2c} that divides the T_2 distribution into the bound water part (BVI) and free water part (FFI), as shown in Figs. 3 and 5. In well logging applications, a fixed T_{2c} is commonly used. For example, sandstone and carbonate are commonly assigned a T_{2c} value of 33 and 92 ms, respectively [17, 23]. C is an empirically determined constant that relates to the characteristic of rock.

For coal, according to Eq. (10), the C is the linear slope of the correlation between $\phi\sqrt{\frac{FFI}{BVI}}$ and $\sqrt[4]{K_a}$ (K_a, the measured air permeability). Using the linear regression analysis, a goodness-of-fit of 0.99 and a $C=5.9$ are obtained. Thus, the Coates model for coal is defined as:

$$K_c = \frac{\varphi}{5.9} \times \left(\frac{FFI}{BVI}\right)^2 \tag{11}$$

Based on Eq. (11), the K_c for seven coal samples was estimated (Table 1). Although the estimated K_c showed a strongly positive correlation with the measured K_a, the error was significantly large.

The Mean T_2 model, also called Schlumberger-Doll-Research (SDR) model, has many expressions, and is more favorable for the following [23]:

$$K_s = C_1 \times \varphi^4 \times {T_{2g}}^2 \tag{12}$$

where T_{2g} is the geometric mean of the T_2 distribution at water-saturated condition; φ represents NMR porosity; C_1 is a constant relating to the characteristic rock or formation, and K_s is the NMR permeability estimated by this equation.

To estimate the value of C_1, the $\varphi^4T_{2g}^2$ is plotted to K_a. If Eq. (12) is valid for coal, then $\varphi^4T_{2g}^2$ and K_a should be linearly correlated, and the slope of the straight line should be C_1. Unfortunately, our results showed that the measured data could not be correlated, which means that the SDR model is not applicable for coal.

As discussed above, the two classic models both express permeability as a function of porosity. The difference between them is that the Coates model relates to the parameter of free fluid fraction in coal, while the SDR model relates to the parameter of geometric mean of relaxation time distribution. Generally, these two NMR-based computational models for permeability of coal do not work correctly, resulting in the need for a new NMR-based model for coal permeability.

In coal, since most pores refer to the close pores that do not contribute to permeability, the parameter of PP may be much better than the total porosity for permeability estimation. As shown in Fig. 7, air permeability strongly correlates with the PP. At the same time, an exponential regression between air permeability and PP shows a high goodness-of-fit ($R^2=0.997$, Fig. 7). Based on the regression result in Fig. 7, the PP model in this chapter is proposed and defined as:

$$K_p = 0.49 \times e^{\frac{\varphi_p}{1.35}} - 0.54 \tag{13}$$

where φ_p is the PP by NMR analysis and K_p is the estimated NMR permeability.

The estimated NMR permeabilities by PP model are listed in Table 1. When comparing K_p with K_a, it was found that, except for the values for SA2 (with a high error) and A4 (with an unreasonable value <0), the difference between the estimated values and the actual values was within an order of magnitude. The PP model is valid for permeability estimation, especially for highly permeable samples.

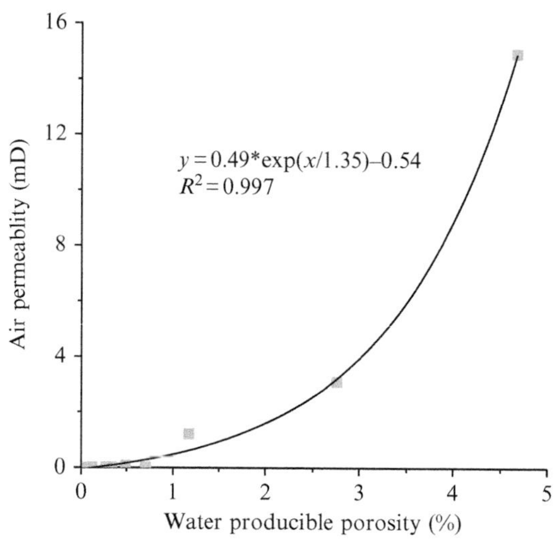

Fig. 7
Exponential relationship between air permeability and the producible porosity.

Both the Coates model and PP model can be used to evaluate the relative size of permeability among coal samples. Comparably, the PP model provides a much better estimation than the Coates model because the permeability of coal relates more closely to the PP, and not the total porosity (integration of PP and IP). Moreover, the PP model is only based on the parameter of PP, and thus it is more convenient than the other models.

7 Conclusions

- The NMR T_2 distributions for 100% water-saturated coal samples are commonly bimodal, unimodal and multimodal, with peaks of adsorption pores (<100 nm) at 0.5–2.5 ms, seepage pores (>100 nm) at 20–50 ms, and the cleats (or fractures) at >100 ms.
- IP and PP are estimated using NMR data of cores at 100% water-saturated and irreducible water conditions. The PP of coals is low, in the range of 0.07%–4.67%, but commonly of 0.1%–2%. The PP strongly relates to the development of cleats (or fractures) and the air permeability values of coal.
- Based on an exponential regression of air permeability and PP, a new NMR-based permeability model is proposed. This model provides significantly better permeability estimations than the classic (Coates and SDR) models originally developed for sandstone. The PSD, pore geometry, and pore connectivity can also be accurately obtained by NMR T_2 values under 100% water-saturated conditions in coal.
- Generally, the NMR technique is a powerful tool to characterize pore structure of coal and has advantages when compared to conventional methods, as it is nondestructive for the testing sample and time saving for operation process.

Acknowledgments

This work was supported by the National Natural Science Foundation of China (41472137) and the National Major Research Program for Science and Technology of China (2016ZX05043-001). We acknowledge Dr. Jianchao Cai for the invitation for this contribution.

References

[1] C.Ö. Karacan, F.A. Ruiz, M. Cotè, et al., Coal mine methane: a review of capture and utilization practices with benefits to mining safety and to greenhouse gas reduction, Int. J. Coal Geol. 86 (2011) 121–156.
[2] T.A. Moore, Coalbed methane: a review, Int. J. Coal Geol. 101 (2012) 36–81.
[3] Y.B. Yao, D.M. Liu, Y. Che, et al., Petrophysical characterization of coals by low-field nuclear magnetic resonance (NMR), Fuel 89 (2010) 1371–1380.
[4] Y.B. Yao, D.M. Liu, Y.D. Cai, et al., Advanced characterization of pores and fractures in coals by nuclear magnetic resonance and X-ray computed tomography, Sci. China Earth Sci. 53 (6) (2010) 854–862.
[5] Y. Zhao, S. Liu, D. Elsworth, et al., Pore structure characterization of coal by synchrotron small-angle X-ray scattering and transmission electron microscopy, Energy Fuel 28 (2014) 3704–3711.

[6] M.M. Mahamud, M.F. Novo, The use of fractal analysis in the textural characterization of coals, Fuel 87 (2008) 222–231.
[7] Y.B. Yao, D.M. Liu, Comparison of low-field NMR and mercury intrusion porosimetry in characterizing pore size distributions of coals, Fuel 95 (2012) 152–158.
[8] J. Hassan, Pore size distribution calculation from 1H NMR signal and N_2 adsorption–desorption techniques, Physica B 407 (2012) 3797–3801.
[9] S.Q. Liu, S.X. Sang, G.F. Wang, et al., FIB-SEM and X-ray CT characterization of interconnected pores in high rank coal formed from regional metamorphism, J. Pet. Sci. Eng. 148 (2017) 21–31.
[10] Y.D. Cai, D.M. Liu, J.P. Mathews, et al., Permeability evolution in fractured coal-combining triaxial confinement with X-ray computed tomography, acoustic emission and ultrasonic techniques, Int. J. Coal Geol. 122 (2014) 91–104.
[11] J. Mitchell, L.F. Gladden, T.C. Chandrasekera, et al., Low-field permanent magnets for industrial process and quality control, Prog. Nucl. Magn. Reson. Spectrosc. 76 (2014) 1–60.
[12] J.Z. Xu, C. Zhai, S.M. Liu, et al., Pore variation of three different metamorphic coals by multiple freezing-thawing cycles of liquid CO_2 injection for coalbed methane recovery, Fuel 28 (2017) 41–51.
[13] J. Liu, Y.B. Yao, D.M. Liu, et al., Experimental evaluation of CO_2 enhanced recovery of adsorbed-gas from shale, Int. J. Coal Geol. 179 (2017) 211–218.
[14] K.K. Darrow, Magnetic resonance: part I—nuclear magnetic resonance, Bell Syst. Tech. J. 32 (1953) 74–99.
[15] C. Straley, D. Rossini, H. Vinegar, et al., Core analysis by low field NMR, Log. Anal. 38 (1997) 84–94.
[16] G.R. Coates, L.Z. Xiao, M.G. Prammer, NMR Logging Principles and Applications, Haliburton Energy Services, Houston, 1999.
[17] W.E. Kenyon, Petrophysical principles of applications of NMR logging, Log. Anal. 38 (1997) 21–43.
[18] F.O. Orumwense, Estimation of the wettability of coal from contact angles using coagulants and flocculants, Fuel 77 (1998) 1107–1111.
[19] X.M. Ge, Y.R. Fan, X.J. Zhu, et al., Determination of nuclear magnetic resonance T_2 cutoff value based on multifractal theory—an application in sandstone with complex pore structure, Geophysics 80 (2015) 11–21.
[20] H. Westphal, I. Surholt, C. Kiesl, et al., NMR measurement in carbonate rocks: problems and an approach to solution, Pure Appl. Geophys. 162 (2005) 549–570.
[21] P. Weniger, J. Franců, P. Hemza, et al., Investigations on the methane and carbon dioxide sorption capacity of coals from the SW Upper Silesian Coal Basin, Czech Republic, Int. J. Coal Geol. 93 (2012) 23–39.
[22] R.L. Kleinberg, C. Straley, W.E. Kenyon, et al., Nuclear magnetic resonance of rocks: T1 vs. T2, in: SPE 26470 Annual Technical Conference and Exhibition, Houston, TX, 3-6, October, 1993.
[23] S.H. Al-Mahrooqi, C.A. Grattoni, A.K. Moss, et al., An investigation of the effect of wettability on NMR characteristics of sandstone rock and fluid systems, J. Pet. Sci. Eng. 39 (2003) 389–398.
[24] R.L. Kleinberg, Utility of NMR T_2 distributions, connection with capillary pressure, clay effect, and determination of the surface relaxivity parameter ρ2, Magn. Reson. Imaging 14 (1996) 761–767.
[25] V. Yakov, A practical approach to obtain primary drainage capillary pressure curves from NMR core and log data, Petrophysics 42 (2001) 334–343.
[26] Y. D. He, Z. Q. Mao, L. Z. Xiao, et al., An improved method of using NMR T_2 distribution to evaluate pore size distribution, Chin. J. Geophys. 48 (2005) 373–378.
[27] A. Timur, Pulsed nuclear magnetic resonance studies of porosity, movable fluid, and permeability of sandstones, J. Pet. Technol. 21 (1969) 775–786.

CHAPTER 3

Characterization of Petrophysical Properties in Tight Sandstone Reservoirs

Fuyong Wang*, **Jianchao Cai**[†]

Research Institute of Enhanced Oil Recovery, China University of Petroleum, Beijing, P.R. China, [†]Hubei Subsurface Multi-scale Imaging Key Laboratory, Institute of Geophysics and Geomatics, China University of Geosciences, Wuhan, P.R. China

Chapter Outline

1 Introduction

Tight sandstone oil and gas reservoirs are a type of unconventional reservoir that is attracting more and more attention for hydrocarbon production. Commercially, developing oil and gas from tight sandstone reservoirs without advanced recovery technologies, such as hydraulic fracturing and horizontal wells, is almost impossible due to the sandstone's ultra-low porosity and permeability. Generally, tight sandstones exhibit the characteristics of small pore-throat size but wide pore-throat size distribution, complex pore structures, and poor

Petrophysical Characterization and Fluids Transport in Unconventional Reservoirs. https://doi.org/10.1016/B978-0-12-816698-7.00003-6

connectivity [1, 2]. Effectively characterizing tight sandstone pore structure is crucial for petrophysical property analysis. There are several techniques that are widely utilized to characterize sandstone pore structures and petrophysical properties, including high-pressure mercury intrusion (HPMI) [3–5], rate-controlled mercury injection (RMI) [6], field emission scanning electron microscope (FE-SEM) [7], nuclear magnetic resonance (NMR) [8, 9], nitrogen gas adsorption (N_2GA) [10], X-ray computer tomography (X-CT) [11, 12], and small angle neutron scattering (SANS) [13, 14]. Comprehensive utilization of various techniques mentioned above is recommended for systematically characterizing tight sandstone pore structures.

The fractal theory has been widely used to characterize pore structures of tight sandstone and model mechanisms of fluid flow [5, 8, 15]. Pore structures of tight sandstone have been demonstrated to have self-similarity characteristics, and the pore size and volume distribution have been proven to have fractal distribution characteristics. Therefore, the fractal theory has been extensively applied to characterize pore structures obtained from HPMI, NMR, SEM, X-CT, N_2GA, and other techniques. Thereafter, the fractal dimension is calculated and used to evaluate pore structure heterogeneity. The petrophysical properties of tight sandstone are quite different from conventional sandstone. The in-situ permeability of tight sandstone is no more than 0.1 mD, with the porosity ranging from 3% to 12% [16]. Due to a significant gas slippage effect and strong liquid-solid interaction, the apparent gas permeability, intrinsic permeability, and liquid permeability are quite different from each other. Tight oil reservoirs exhibit characteristics of low-velocity non-Darcy flow, and gas slippage effect and/or Knudsen diffusion in tight gas reservoirs leads to high gas production compared to the prediction from conventional Darcy flow.

This chapter includes two parts. Firstly, the techniques for pore structure characterization and analysis methods are introduced and analyzed, and then the models for predicting permeability and low-velocity non-Darcy flow are reviewed and discussed.

2 Characterization and Analysis of Tight Sandstone Pore Structures

2.1 Pore Structure Characterization With Thin Section, FE-SEM, and X-CT

Pore type, pore size, and pore geometry can be observed from imaging techniques, such as thin section, FE-SEM, and X-CT. As shown in Fig. 1A, the intragranular dissolved pores, intergranular dissolved pores, and intergranular pores in the tight sandstone can be observed from the thin section, and the pore morphology is highly irregular. Fig. 1B exhibits the FE-SEM image of the tight sandstone with 0.2 μm pixel resolution, and the nanoscale intragranular pores can be clearly identified.

X-CT is a nondestructive technique for pore structure characterization and has been widely utilized for pore analysis [17–19]. The pore size, pore geometry, and pore distributions can be visualized in 3D, and the pore connectivity can be accurately described. Fig. 2 shows 2D images and 3D pore distributions obtained from the X-CT with two different

Fig. 1
Photographs showing microscopic pore structures of tight sandstone obtained from thin section (A) and FE-SEM (B).

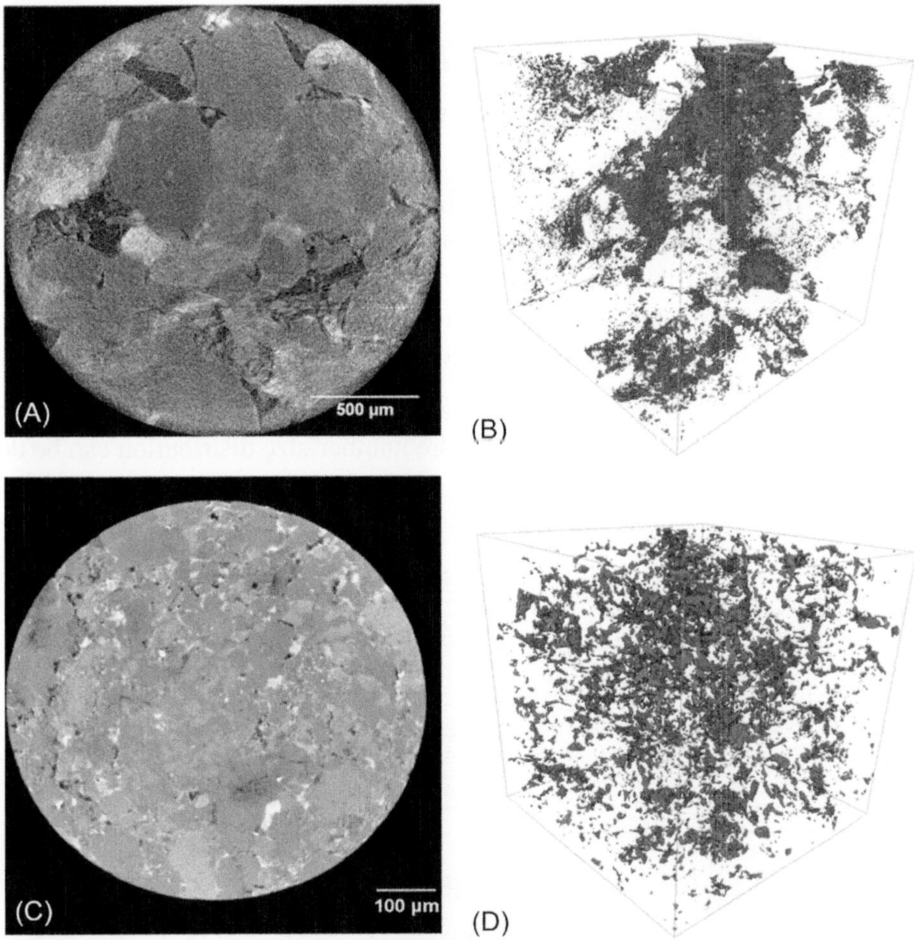

Fig. 2
2D images and 3D pore volume distributions obtained from the X-CT with 2 μm resolution (A and B) and 0.8 μm resolution (C and D).

resolutions from the same core plug. The resolution of Fig. 2A and B is 2 μm and the microscale pores can be identified. However, the connectivity of 3D pore distribution is very poor, and this may be because the pores with a diameter less than 2 μm cannot be identified from X-CT due to the limitation of the pixel resolution. As shown in Fig. 2C and D, another X-CT scanning image with a resolution of 0.8 μm was conducted, and more pores with a pore diameter greater than 0.8 μm were observed. Therefore, considering the wide distribution of tight sandstone pore sizes, multiresolution X-CT scanning is recommended to obtain the entire pore size spectrum and improve the connectivity of the acquired pore network from X-CT.

2.2 Pore Structure Characterization With HPMI

HPMI is a widely applied technique for characterizing pore structure and distribution in tight sandstone. The pore size can be calculated from HPMI by the Washburn equation [20]:

$$P_c = \frac{2\sigma\cos\theta}{r}, \tag{1}$$

where P_c is the capillary pressure, σ is the mercury interfacial tension, r is the pore radius, and θ is the contact angle.

Fig. 3 shows the HPMI capillary curves and pore size distributions of three typical tight sandstone core plugs from the Yanchang Formation of the Ordos Basin in China [8]. Generally, the pore size distributions of tight sandstone are wide, varying from microscales to nanoscales. With the permeability of tight sandstone decreasing, the maximum pore radius r_{max} decreases from several micrometers to hundreds or tens of nanometers.

The fractal theory has been widely used to characterize sandstone pore structure and heterogeneities with the HPMI test [3–5]. The pore number-size distribution can be described with the following power-law correlation [21]:

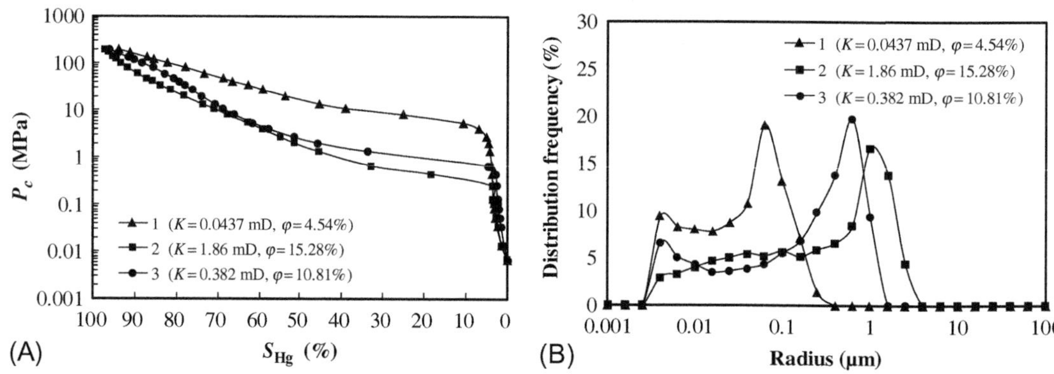

Fig. 3
HPMI capillary curves for three tight sandstone core plugs (A) and the corresponding pore size distributions derived from HPMI (B).

$$N(>r) \propto r^{-D_f}, \tag{2}$$

where D_f is the fractal dimension and $N(>r)$ is the cumulative pore number with pore size greater than r. According to Wang et al. [22], there are generally six different methods of calculating fractal dimensions from HPMI data. As the assumptions of these models are different, the calculated fractal dimensions vary from each other [3, 22].

Assuming the pore space is made up of bundles of capillary tubes, the cumulative tube number with pore radius greater than r can be calculated by:

$$N(>r) = \int_{r_{\max}}^{r} \frac{\Delta V_{\mathrm{Hg}}(r)}{\pi r^2 l} \mathrm{d}r, \tag{3}$$

where $\Delta V_{\mathrm{Hg}}(r)$ is the increased mercury volume when the pore size equals r and l is the core length. The fractal dimension D_f can be calculated from the slope of the $N(>r)-r$ curve in a log-log plot. Taking the HPMI data of the core plug No. 3 in Fig. 3 as an example, Fig. 4A shows the process of fractal dimension calculation using Eq. (3). This model assumes that every capillary tube has the same tortuosity, and in Eq. (3) the tortuosity is set to 1. The calculated fractal dimension using Eq. (3) reflects the pore size distribution in the 2D cross-section of the core plug.

The pores in tight sandstone can be assumed to be made up of spheres with different radii, and then the cumulative pore numbers can be calculated by:

$$N(>r) = \int_{r_{\max}}^{r} \frac{\Delta V_{\mathrm{Hg}}(r)}{\frac{4}{3}\pi r^3} \mathrm{d}r. \tag{4}$$

Fig. 4B shows the calculated fractal dimension using Eq. (4). Generally, the calculated fractal dimension using Eq. (4) is greater than that using Eq. (3) by 1. However, Eq. (4) is not widely used to calculate fractal dimensions, as the pore space in tight sandstone is not often spherically shaped.

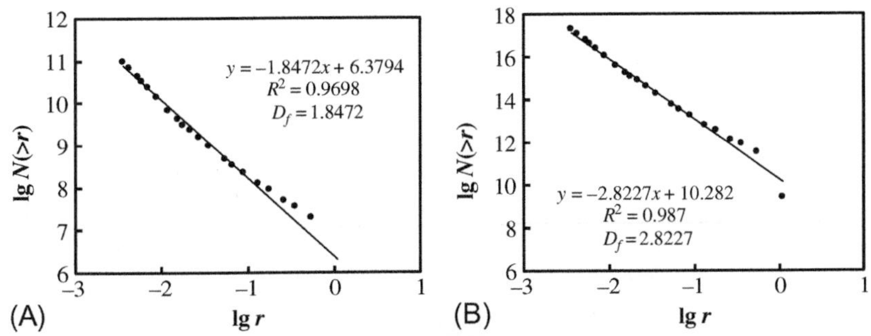

Fig. 4
(A) Fractal dimension calculation using Eq. (3); (B) fractal dimension calculation using Eq. (4).

Li [5] presented a pore size distribution model for calculating the fractal dimension of sandstone pores:

$$N(r) \propto r^{-D_f}, \tag{5}$$

where $N(r)$ is the number of the capillary tubes required to fill the pore space with the capillary tube radius of r, which is different from the definition of the cumulative pore number $N(>r)$ in Eq. (3). $N(r)$ can be calculated with the following equation:

$$N(r) = \frac{V_{Hg}(r)}{\pi r^2 l}, \tag{6}$$

where $V_{Hg}(r)$ is the accumulative mercury volume with the radius equal to r. The mechanism of calculating fractal dimensions using Eqs. (5, 6) is similar to the boxing counting method [23]. Similarly, the fractal dimensions can be obtained from the slope of the $N(r)-r$ curve in a log-log plot, as shown in Fig. 5A. Another version of this method for fractal dimension calculation can be expressed as [5]:

$$S_{Hg} \propto P_c^{-(2-D_f)}, \tag{7}$$

where S_{Hg} is the injected mercury saturation. For the same dataset, the fractal dimensions calculated using Eq. (6) and Eq. (7) are the same. As shown in Fig. 5B, the $S_{Hg}-P_c$ curve usually exhibits two segments in the log-log plot, which can account for the fractal characteristics of small pores and large pores separately. The calculated fractal dimension for the small pores is usually slightly above 2, but the calculated fractal dimension for large pores is usually greater than 3. The split point between the two segments represents the transition between the well-connected large pores and poorly connected small pores [24, 25].

Friesen and Mikula [26] developed an equation to characterize fractal characteristics of coal pore space as:

$$\frac{dS_{Hg}}{dp_c} \propto p_c^{-(4-D_f)}. \tag{8}$$

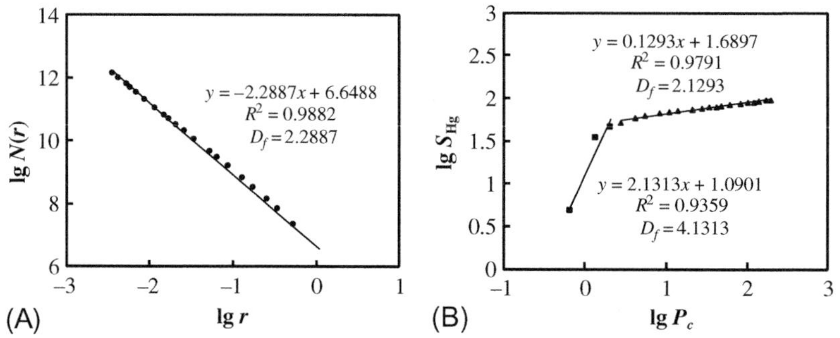

Fig. 5
(A) Fractal dimension calculation using Eq. (6); (B) fractal dimension calculation using Eq. (7).

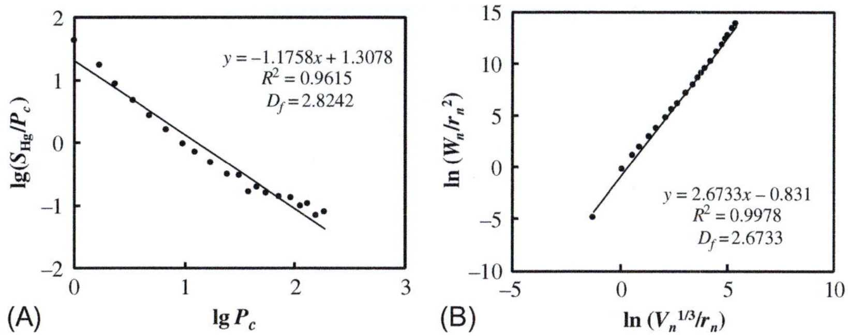

Fig. 6

(A) Fractal dimension calculation using Eq. (8); (B) fractal dimension calculation using Eq. (9).

As shown in Fig. 6A, fractal dimension D_f can be obtained from the slope of the $\frac{dS_{Hg}}{dp_c} - p_c$ curve in a log-log plot. The calculated fractal dimension using Eq. (8) can reflect the heterogeneity in pore geometry.

Zhang and Li [27] proposed a model to determine the surface fractal dimension with the thermodynamic theory, and the equation for fractal dimension calculation can be simplified as [28]:

$$\ln\left(\frac{W_n}{r_n^2}\right) = D_f \ln\left(\frac{V_n^{1/3}}{r_n}\right) + C, \tag{9}$$

where W_n is the surface energy, r_n is the pore radius, and V_n is the injected mercury volume. From Fig. 6B, the fractal dimension can be obtained from the slope of the $\ln\left(\frac{W_n}{r_n^2}\right) - \ln\left(\frac{V_n^{1/3}}{r_n}\right)$ curve. The calculated fractal dimension using Eq. (9) can characterize the roughness and irregularity of the pore surface.

Based on the correlation between the wetting phase saturation S_w and the capillary pressure P_c, He and Hua [29] derived a capillary pressure model:

$$S_w = \left(\frac{P_c}{P_{min}}\right)^{D_f - 3}. \tag{10}$$

Eq. (10) was developed for wetting phase capillary data, but many scholars [3, 30] utilized Eq. (10) to calculate fractal dimension from capillary pressure data measured from mercury through calculating the wetting phase saturation by $S_w = 1 - S_{Hg}$. Fig. 7 presents a case of fractal dimension calculation using Eq. (10). However, as mercury is the nonwetting phase for the rock, the applicability of Eq. (10) to calculate fractal dimension from HPMI data is questionable.

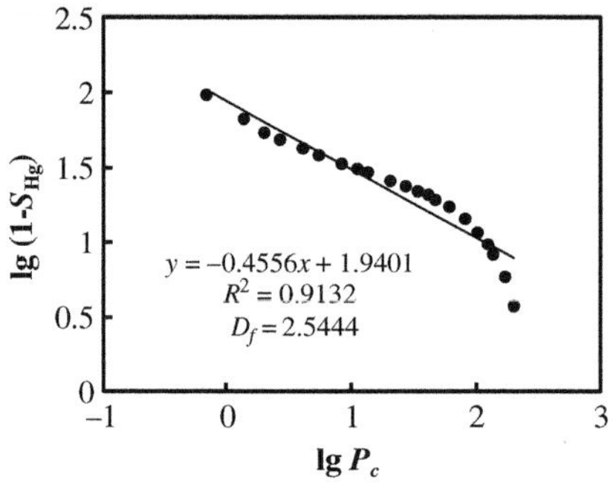

Fig. 7
Fractal dimension calculation using Eq. (10).

Wang et al. [22] found that the fractal dimensions calculated using Eq. (6) and Eq. (10) have strong but opposite correlations with the petrophysical properties, as shown in Fig. 8. Fig. 8A shows that there is a strong and negative correlation between the fractal dimensions calculated using Eq. (6) and the core permeability in a semi-log plot. However, Fig. 8B shows that the fractal dimensions calculated using Eq. (10) are positively correlated with the air permeability, which is contrary to the results presented in Fig. 8A. This might be caused by the mercury capillary pressure rather than the wetting phase capillary pressure that was used to calculate the fractal dimensions using Eq. (10). Therefore, the choice between the different methods should be made according to the purpose of the pore structure analysis. For instance, the fractal dimensions calculated with Eqs. (5, 6) have strong and reasonable correlations with rock petrophysical properties and can be used to predict petrophysical properties from tight sandstone pore structures. Eq. (8) is suitable for analyzing pore geometry characteristics and distributions whilst Eq. (9) is more applicable for characterizing pore surface roughness and irregularity.

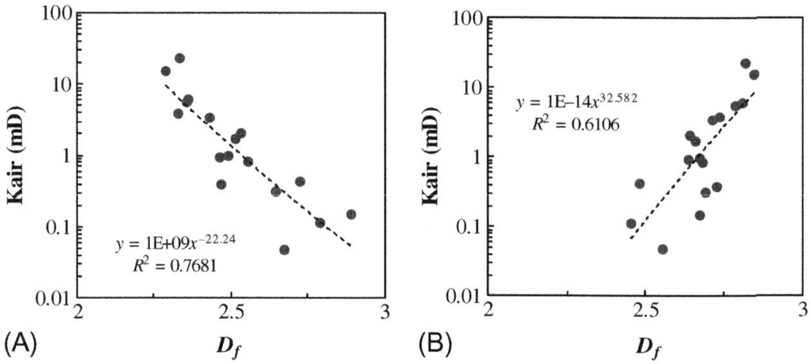

Fig. 8
The correlations between the core permeability and the fractal dimensions calculated from Eq. (6) (A) and Eq. (10) (B), respectively.

2.3 Pore Structure Characterization With RMI

RMI is a different type of mercury intrusion technique for pore structure characterization, compared to HPMI. The mercury is injected with a quasi-static constant and extremely low rate of 0.00005 mL/min, pores are distinguished from throats according to sudden decreases of capillary pressure, and the statistical distributions of pores and throats can be obtained [31]. However, the maximum injected capillary pressure for the current RMI equipment is 900 psi, and therefore the pores and throats smaller than 0.12 μm cannot be identified using the RMI test.

Fig. 9 shows the RMI test results for a tight sandstone core plug. The porosity of the core plug is 15.02%, and the air permeability is 0.168 mD. As show in Fig. 9A, a different capillary pressure curve is obtained when compared to the HPMI data. The three capillary pressure curves represent the mercury intrusion for total pore-throat volumes, pores, and throats respectively. The sum of the mercury saturation of the pores and the throats under a certain capillary pressure equals the mercury saturation of the total pore-throat volume. Due to the limitation

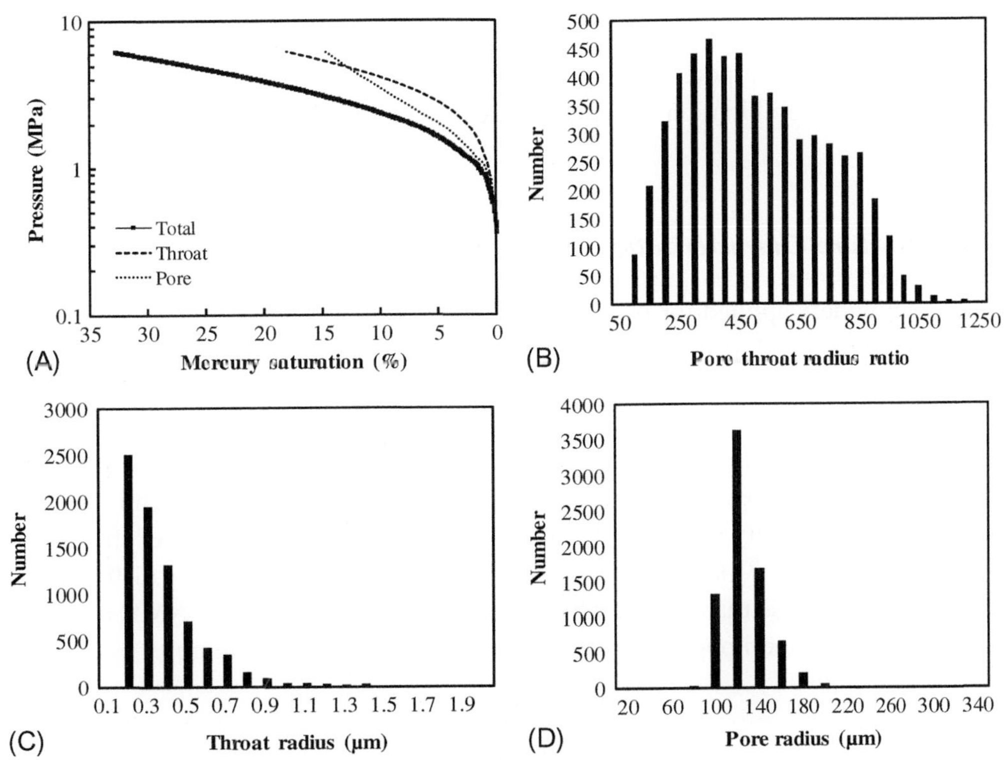

Fig. 9
RMI test results for the core plug with 0.168 mD air permeability and 15.02% porosity (A: the capillary pressure curves of pores, throats and total pore-throat volumes; B: the distribution of pore-throat radius ratio; C: the distribution of throats; D: the distribution of pores).

of the maximum injected capillary pressure (900 psi), the maximum mercury saturation of the total pore-throat volume is less than 35%, which means more than 65% pore volume cannot be characterized using RMI. Fig. 9B demonstrates the weighted mean pore-throat radius ratio to be 509, indicating large variation between pores and throats. This was verified from the porosity and permeability of this core plug. The measured air porosity was more than 15%, but the air permeability was less than 0.2 mD, which indicates that the throats might be very small. Fig. 9C shows the main throat size to be less than 1 μm, and throat number increased with the decreased throat size. However, as the minimum throat radius that can be detected by RMI is only 0.12 μm, more throats less than 0.12 μm cannot be measured. Considering that there is more than 65% of un-injected pore-throat volume, it could be inferred that the throats of this tight sandstone are mainly distributed on the scale of tens of nanometers. In contrast, the pore sizes varied from 80 μm to 300 μm, and the weighted mean pore size was 127.13 μm. The pore volume in this core plug exhibited the characteristics of large pores and narrow throats.

Literatures show that the permeability of tight sandstone is mainly dominated by the size of the throats rather than pores, and with permeability decreasing, the throat size distribution moves from micrometers to nanometers [2, 32]. The above case demonstrates that RMI cannot characterize all the pore-throats of tight sandstone alone and other techniques should be utilized in combination to comprehensively characterize the pore structure of tight sandstone.

2.4 Pore Structure Characterization With NMR

NMR has been widely applied for pore characterization since the 1990s. The advantages of the NMR test are the nondestructive requirement for the tested sample and short measurement time [8, 9]. Fig. 10 shows the measured NMR T_2 spectrum for a tight sandstone core plug with 100% water saturation and irreducible water saturation. NMR T_2 spectrum with 100% water saturation can reflect pore size distribution in tight sandstone, and porosity can also be measured from the NMR test. The measured NMR porosity is 15.38%, which is very close to the measured air porosity of the core plug. After centrifugation, 47.45% of the water is irreducible, and the NMR T_2 spectrum shows that irreducible water is mainly distributed in the pores with a T_2 value less than 60 ms, which represents small pores. The NMR transverse relaxation time T_2 is superimposed by several mechanisms [33] as given by:

$$\frac{1}{T_2} = \frac{1}{T_{2B}} + \frac{1}{T_{2S}} + \frac{D(\gamma G T_E)^2}{12}, \tag{11}$$

where $\frac{1}{T_{2B}}$ is the transverse volume relaxation, $\frac{1}{T_{2S}}$ is the transverse surface relaxation, and $\frac{D(\gamma G T_E)^2}{12}$ is the transverse diffusion relaxation. For a sample in a homogeneous magnetic field with short

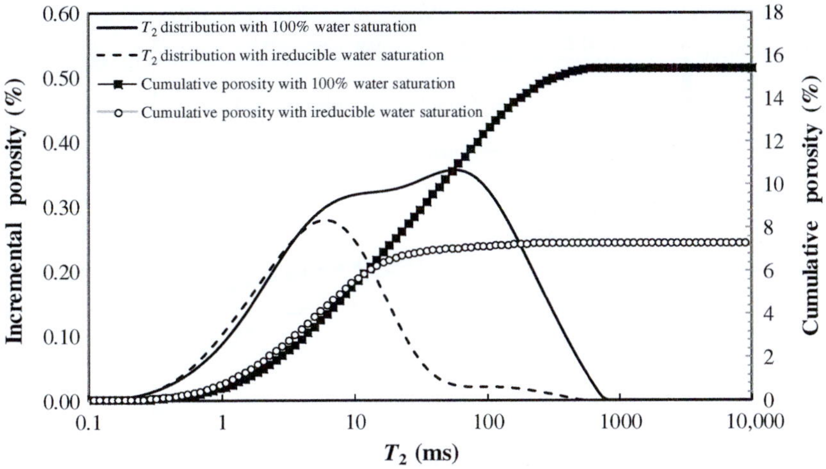

Fig. 10
NMR T_2 spectrum of a tight sandstone core plug with 100% water saturation and irreducible water saturation.

echo interval, $\frac{1}{T_{2B}}$ and $\frac{D(\gamma G T_E)^2}{12}$ can be ignored. However, NMR T_2 is mainly caused by the transverse surface relaxation given by:

$$\frac{1}{T_2} = \frac{1}{T_{2S}} = \rho\left(\frac{S}{V}\right) = F_S\frac{\rho}{r}, \tag{12}$$

where ρ is transverse surface relaxation strength, S is the superficial area of the pore, V is pore volume, F_S is the form factor, and r is pore-throat radius. Rearranging Eq. (12), the linear relationship between transverse relaxation time T_2 and pore radius r yields:

$$T_2 = \frac{r}{F_S\rho}. \tag{13}$$

The power function has also been widely utilized to convert T_2 to r [8, 34–36]:

$$T_2 = \frac{r^n}{F_s\rho}, \tag{14}$$

where n is the exponent.

NMR T_2 spectrum can be converted into pore size distribution with the help of HPMI. Fig. 11 [8] shows the cumulative pore volume distributions derived from NMR and HPMI test data of which T_2 and r corresponding to the same cumulative pore volume can be determined. As presented in Fig. 12 [8], linear regression and power regression were applied to fit the correlations between T_2 and r, and the fitting results of the power regression proved to be satisfactory.

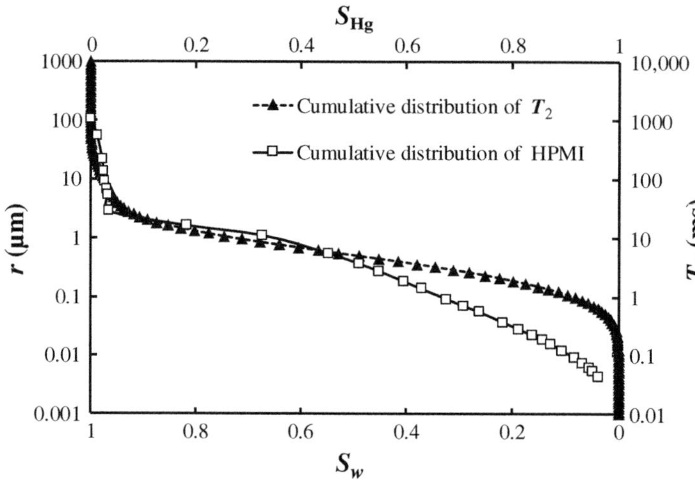

Fig. 11
Cumulative pore volume distributions of NMR and HPMI test data for the same core plug.

Fractal theory also has been widely used to characterize pore structures from NMR test data. When T_2 is assumed to be linearly correlated to r and $r_{min} \ll r_{max}$, the fractal dimension can be calculated from NMR with the following equation as [37]:

$$\lg S_V = (3 - D_f)\lg T_2 + (D_f - 3)\lg T_{2max}. \tag{15}$$

For the power function relationship between T_2 and r, the fractal dimension can be calculated with [8]:

$$\lg S_V = \frac{(3 - D_f)}{n}\lg T_2 + \frac{(D_f - 3)}{n}\lg T_{2max}. \tag{16}$$

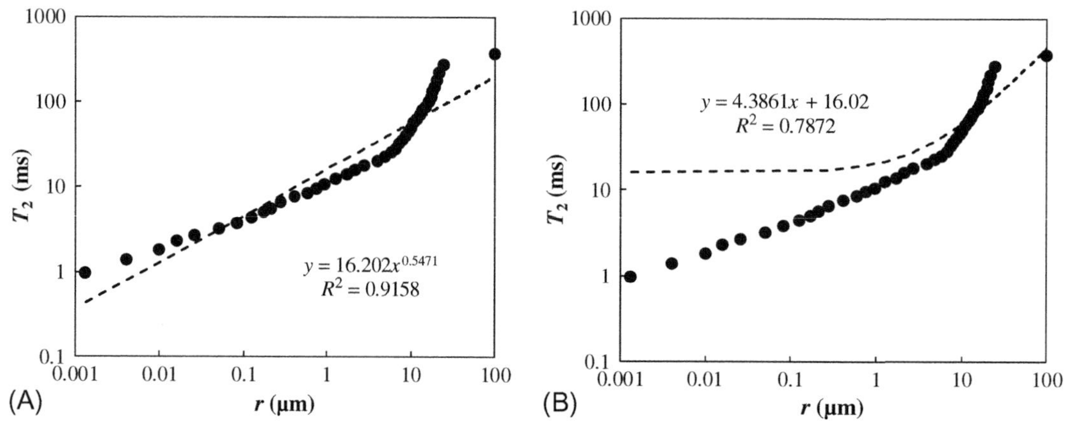

Fig. 12
Regression between transverse relaxation time T_2 and pore radius r (A: power regression; B: linear regression).

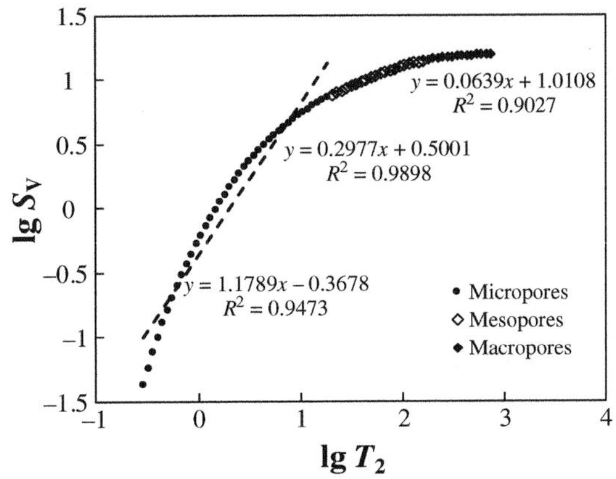

Fig. 13
Fractal dimension calculation for the different pore sizes from the slope of the $S_V - T_2$ curve.

Fractal dimension can be calculated from the slope of the $S_V - T_2$ curve S. For Eq. (15), fractal dimension can be calculated by $D_f = 3 - S$, and for Eq. (16), fractal dimension can be calculated by $D_f = 3 - nS$. As shown in Fig. 13 [8], different pore sizes exhibit different fractal characteristics, and the NMR T_2 spectrum can be divided into three parts. The fractal dimensions for the micropores, mesopores, and macropores can be calculated respectively. As the micropores are occupied by irreducible water, the value of $T_{2\text{cutoff}}$ is generally treated as the split point between the micropores and mesopores, and nine times the value of $T_{2\text{cutoff}}$ is used to separate mesopores from macropores on the $S_V - T_2$ curve [8, 37].

Wang et al. [8] found that the petrophysical properties of tight sandstone are strongly correlated to the fractal dimensions of the mesopores and macropores calculated from the NMR T_2 spectrum but the petrophysical properties have weak correlations with the fractal dimensions of micropores. The increasing fractal dimension means petrophysical properties of tight sandstone become worse, and exponential decreases of air permeability, porosity and median pore sizes with the increasing fractal dimension of mesopores and macropores have been identified.

2.5 Pore Structure Characterization With N_2GA

N_2GA is widely used for characterizing the nanopore structure of unconventional reservoirs. Specific pore volume, specific surface area, pore shape, pore size, and pore size distribution can be obtained from the N_2GA test [38, 39]. Fig. 14 presents results from a low pressure N_2GA test for a tight sandstone core plug. The gas volume under different relative pressures $^{P_0}/_{P}$ was measured. Due to the capillary condensation, there is a hysteresis loop between adsorption and desorption, and the hysteresis loop can be used to understand the morphology of pore shape [40].

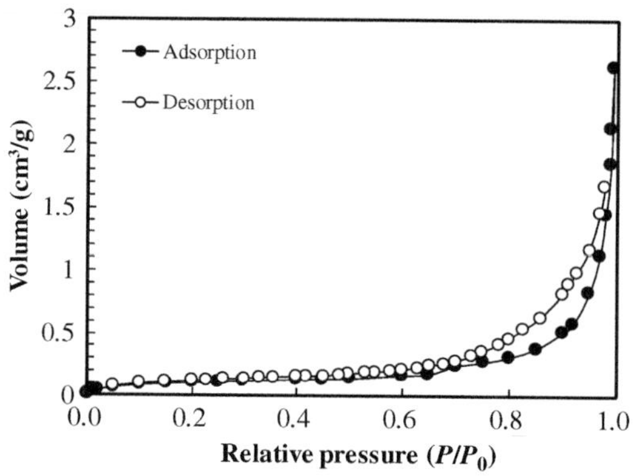

Fig. 14
Low pressure N_2GA test for a tight sandstone core plug.

The multi-point Brunauer-Emmette-Teller (BET) method at the low relative pressure ($0.0098 < {}^{P_0}/_{P} < 0.20$) of N_2GA adsorption branch can be utilized to calculate the specific surface area [41, 42], and the calculated specific surface area is $0.409\,m^2/g$. The pore volume can be calculated from the N_2GA adsorption branch using the Barrette-Joynere-Halenda (BJH) method, and the calculated specific pore volume for this tight sandstone was $0.004\,cm^3/g$. Pore size distribution can be obtained from the N_2GA adsorption branch with the density functional theory (DFT) method [42], and the calculated average pore diameter was 12.55 nm for this tight sandstone.

Fractal characteristics of tight sandstone pore structures have been widely investigated from N_2GA test data, and the fractal Frenkel-Halsey-Hill (FHH) model has been proven to be the most effective approach for describing fractal characteristics of various porous media based on the results of N_2GA test [43–46]. The fractal dimension can be calculated with the following Eq. [47]:

$$\ln\left(\frac{V}{V_0}\right) = A\left[\ln\left(\ln\left(\frac{P_0}{P}\right)\right)\right] + C, \tag{17}$$

where V is the adsorbed gas volume and V_0 is the monolayer coverage volume. For the fractal porous media, $\ln V$ vs $\ln\left(\ln\left(\frac{P_0}{P}\right)\right)$ yields a straight line and fractal dimension can be calculated from the slope A:

$$D_f = A + 3. \tag{18}$$

Fig. 15 presents the process of fractal dimension calculation from the adsorption branch of the N_2GA test. Generally, $\ln V$ vs $\ln\left(\ln\left(\frac{P_0}{P}\right)\right)$ results in two segments with different slopes, and two different fractal dimensions can be obtained from the adsorption under high relative pressure (D_{f1}) and low relative pressure (D_{f2}). The fractal dimension D_{f1} calculated from the adsorption under high relative pressure is widely used to analyze the fractal characteristics of tight sandstone pores. As fractal dimension can reflect the irregularity of pore geometry

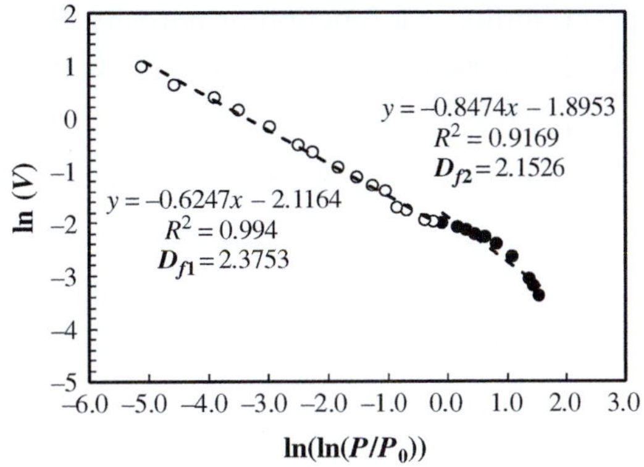

Fig. 15

Plot of $\ln V$ vs $\ln\left(\ln\left(\frac{P_0}{P}\right)\right)$ from the adsorption branch of the N_2GA test.

and roughness of pore surfaces [48, 49], generally with fractal dimension increasing, the specific pore volume and surface area increase whilst average pore diameter decreases, indicating the increase in pore surface roughness and pore geometry complexity.

3 Petrophysical Properties of Tight Sandstone

3.1 Intrinsic Permeability of Tight Sandstone

Permeability is one of the most important petrophysical properties of sedimentary rock. For tight sandstone, the measured apparent permeability with different fluids under different conditions varies greatly. As the pore-throat sizes of tight sandstone are very small, the slippage effect (when permeability is measured with gas) and liquid-solid effect (when permeability is measured with liquid) are significant, and the measured gas permeability and liquid permeability of tight sandstone are different from the intrinsic permeability.

There are several different models for predicting intrinsic permeability of tight sandstone. The most classical approach is the Kozeny-Carman Equation [50, 51]:

$$K = \phi \frac{r_{\text{mh}}^2}{F_s \tau}, \tag{19}$$

where K is the intrinsic permeability, ϕ is porosity, τ is tortuosity, F_s is the shape factor, and r_{mh} is the mean hydraulic radius. r_{mh} can be expressed with surface area S_{gv} and porosity, and Eq. (19) becomes:

$$K = \frac{\phi^3}{(1-\phi)^2}\left(\frac{1}{F_s \tau S_{\text{gv}}^2}\right). \tag{20}$$

As indicated in Eq. (20), permeability is a function of porosity and pore-throat size. Several empirical equations have been proposed for permeability predicting and petrophysical rock type based on the following mathematical model [52]:

$$\log r_i = C_1 + C_2 \log K + C_3 \log \phi, \tag{21}$$

where r_i is the pore-throat size with $i\%$ mercury saturation. C_1, C_2, and C_3 are constants that can be derived from multiple regression analysis. For instance, Winland found the pore-throat sizes corresponding to 35% mercury saturation r_{35} best correlate with the measured porosity and air permeability [53]:

$$\log r_{35} = 0.732 + 0.588 \log K_{\text{air}} - 0.864 \log \phi. \tag{22}$$

For different core samples, the pore-throat size that best fits the measured porosity and permeability differs [54, 55].

Based on the tortuous capillary tube model (as shown in Fig. 16), Yu and Cheng [56] proposed an equation to calculate permeability of the fractal porous media:

$$K = \frac{\pi}{128} \frac{L_0^{1-D_T}}{A} \frac{D_f}{3 + D_T - D_f} \lambda_{\max}^{3+D_T} \left[1 - \left(\frac{\lambda_{\min}}{\lambda_{\max}} \right)^{3+D_T-D_f} \right], \tag{23}$$

where L_0 is the representative length, A is the cross-sectional area, and $A = L_0^2$; D_T is the tortuosity fractal dimension.

Wang et al. [57] derived a similar equation but the coefficient in front of the formula is different:

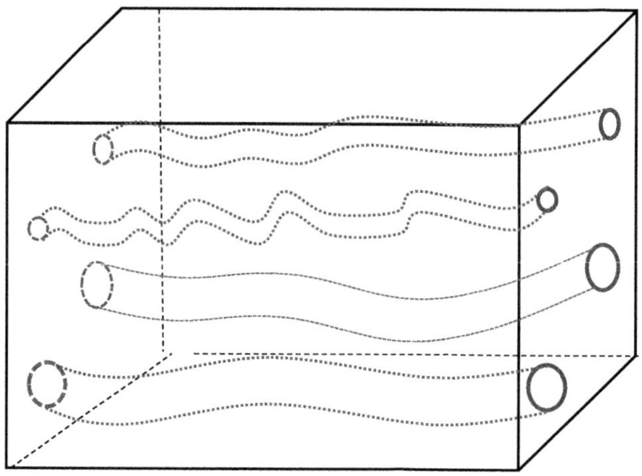

Fig. 16
The tortuous capillary tube model with fractal characteristics.

$$K = \frac{\pi}{32(D_T+3)} \frac{L_0^{1-D_T}}{A} \frac{D_f}{3+D_T-D_f} \lambda_{max}^{3+D_T} \left[1 - \left(\frac{\lambda_{min}}{\lambda_{max}}\right)^{3+D_T-D_f}\right]. \tag{24}$$

The reason is that the tortuosity with fractal characteristics is considered during the process of deriving the flow rate of a tortuous capillary tube [58].

Based on a special fractal space model, Page et al. [59] proposed a different model for permeability prediction with the fractal theory:

$$K = \frac{r_g^2}{8} \frac{\phi}{\tau} \left[\frac{2\phi}{3\tau(1-\phi)}\right]^{\frac{2}{D_f-1}}, \tag{25}$$

where r_g is the average grain radius and D_f is the fractal dimension in three-dimensional space, which is different from that in Eqs. (23, 24). In Eqs. (23, 24), the fractal dimension in two-dimensional space should be utilized.

3.2 Apparent Gas Permeability of Tight Sandstone

Apparent gas permeability K_g is greater than the intrinsic permeability K of tight sandstone due to the gas slippage effect:

$$K_g = K\left(1 + \frac{b}{\bar{p}}\right), \tag{26}$$

where b is the gas slippage factor and $\bar{p}$ is the mean pressure. The mean free path of gas is compatible to the pore-throat size of tight sandstone, which can be evaluated by the Knudsen number K_n [60]:

$$K_n = \frac{l}{\lambda}, \tag{27}$$

where λ is the pore-throat diameter and l is the mean free path of gas, and can be calculated by:

$$l = \frac{\mu}{\bar{p}} \sqrt{\frac{\pi R_g T}{2M}}, \tag{28}$$

where μ is the gas viscosity, T is the temperature, R_g is the gas molecular constant, and M is the gas molecular mass. With the average pore-throat size of tight sandstone, the calculated Knudsen number demonstrates the flow regimes of gas in tight sandstone belong to the slippage flow regime. Based on the tortuous capillary tube model with fractal characteristics, the gas permeability model can be derived. Wang et al. [57] developed a fractal gas permeability model with slippage flow given by:

$$K_g = \frac{\pi}{32(D_T+3)} \frac{L_0^{1-D_T}}{A} \frac{D_f}{3+D_T-D_f} \lambda_{max}^{3+D_T} \left[1 + \frac{4l(3+D_T-D_f)}{\lambda_{max}(2+D_T-D_f)}\right]. \tag{29}$$

The gas slippage factor b can be expressed as [57]:

$$b = \frac{4\mu(3+D_T-D_f)}{\lambda_{max}(2+D_T-D_f)}\sqrt{\frac{\pi R_g T}{2M}}. \tag{30}$$

Eq. (30) is the half of the formula for calculating the gas slippage factor derived by Zheng et al. [61] and is caused by a different definition of the Knudsen number. As shown in Eq. (30), the gas slippage factor b is a function of tight sandstone pore structures and gas properties, and it will increase with the increase of fractal dimension D_f and the decreases of tortuous fractal dimension D_T and maximum pore size λ_{max}. The gas slippage b will increase with the increasing viscosity μ and decreasing gas molecular mass M.

3.3 Liquid Permeability of Tight Sandstone

Generally, the liquid permeability of tight sandstone is less than the intrinsic permeability. One of the widely accepted explanations for this is that the liquid-solid effect in the nanoscale pores is significant, and liquid is assumed to be adsorbed on the surface of nanoscale pores and forms nonflowing boundary layers, which can reduce the effective radius of fluid flow [62–64]. The fractal liquid permeability model for tight sandstone can be expressed as [15]:

$$K_{liquid} = \frac{\pi}{32(D_T+3)}\frac{L_0^{1-D_T}}{A}D_f\lambda_{max}^{D_f}\int_{\lambda_{min}}^{\lambda_{max}}\left[1-\frac{2h}{\lambda}\right]^{(D_T+3)}\lambda^{D_T-D_f+2}d\lambda, \tag{31}$$

where h is the thickness of the nonflowing boundary layer. Tian [65] and Cao et al. [62] developed an empirical equation for estimating the boundary-layer thickness (h) as:

$$h = \begin{cases} r\cdot\beta_1 e^{\beta_2 r}(\nabla p)^{\beta_3}\cdot\mu, & \nabla p < 1\,\mathrm{MPa/m} \\ r\cdot\beta_1 e^{\beta_2 r}\cdot\mu, & \nabla p \geq 1\,\mathrm{MPa/m} \end{cases}, \tag{32}$$

where r is the pore-throat radius; μ is the fluid viscosity; ∇p is the pressure gradient; and β_1, β_2, and β_3 are constants that can be derived from microtube experiments [63, 66, 67]. The above model shows that the effective liquid permeability of tight sandstone is dominated by pore size distributions of tight sandstone and liquid properties. For a porous medium with nanoscale pore-throat sizes, the nonflowing boundary layers can significantly reduce effective liquid permeability of tight sandstone at the low-pressure gradient, and as shown in Fig. 17, the effective permeability will be significantly reduced with fluid viscosity increasing [15].

3.4 Low-Velocity Non-Darcy Flow in Tight Sandstone

Tight oil reservoirs exhibit low-velocity non-Darcy flow. Several mechanisms have been proposed to explain this phenomenon, such as the boundary-layer effect due to liquid-solid

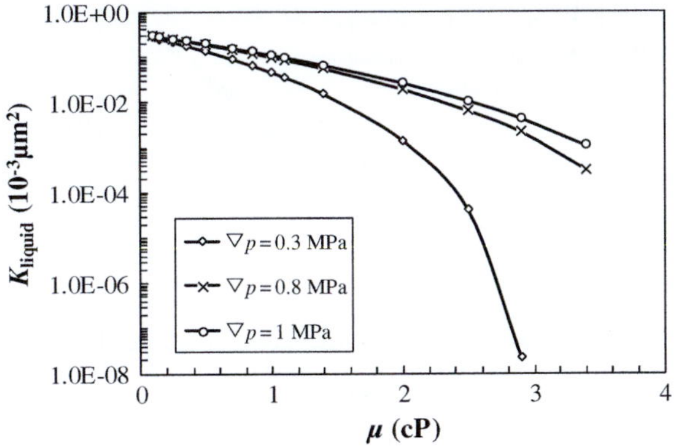

Fig. 17
The correlation between the effective liquid permeability and the fluid viscosity under a different pressure gradient.

interactions [64, 68], particle migration [69], and non-Newtonian flow [70, 71]. The nonflowing boundary-layer assumption is one of the most widely utilized methods to model low-velocity non-Darcy flow. In a single capillary tube with fractal characteristics of tortuosity and a nonflowing boundary layer, the flow rate can be calculated by [58]:

$$q = \frac{\pi}{32(D_T+3)} \frac{(\lambda - 2h)^{D_T+3}}{\mu} \frac{\nabla P}{L_0^{D_T}}. \tag{33}$$

Wang et al. [58] proposed a mathematical model for describing low-velocity non-Darcy flow with fractal approach:

$$V = \frac{Q}{A} = \frac{\pi}{32(D_T+3)} \frac{\nabla p L_0^{1-D_T}}{\mu} \frac{}{A} D_f \lambda_{\max}^{D_f} \int_{\lambda_{\min}}^{\lambda_{\max}} \left[1 - \frac{2h}{\lambda}\right]^{(D_T+3)} \lambda^{D_T - D_f + 2} d\lambda. \tag{34}$$

As the boundary-layer thickness (h) is dependent on pressure gradient ∇p, the relationship between the pressure gradient and flow velocity is nonlinear. With the pressure gradient increasing, the thickness of nonflowing boundary layers decreases, and flow velocity will increase nonlinearly. The boundary-layer thickness can be estimated from the empirical equation proposed by Tian [65] and Cao et al. [62]:

$$h = \frac{1}{2} \lambda \beta_1 e^{\frac{1}{2}\beta_2 \lambda} (\nabla p)^{\beta_3} \mu. \tag{35}$$

Wang et al. assumed that Eq. (30) is valid for $\nabla p \leq 5$ MPa/m to make sure the nonlinear transient flow regime can be entirely described [58]. Fig. 18 shows the flow velocity and apparent liquid permeability under a different pressure gradient calculated with Eq. (29). With a pressure gradient increasing from 0 to 2 MPa/m, the flow velocity increases nonlinearly, and the flow belongs to the nonlinear transient flow regime. When the pressure gradient ∇p is

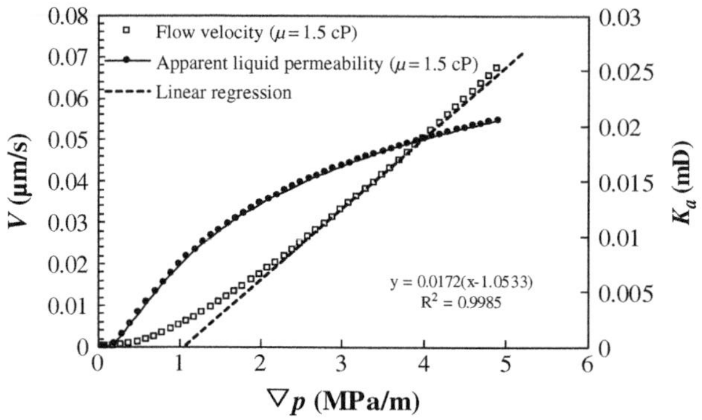

Fig. 18
The flow velocity and apparent liquid permeability under a different pressure gradient.

great than 2 MPa/m, the flow comes into pseudo-linear flow regime, and the relationship between the flow velocity V and pressure gradient ∇p can be described as [58]:

$$V = \frac{K_e}{\mu}(\nabla p - \nabla p_{\mathrm{PTPG}}), \tag{36}$$

where K_e is effective permeability in the pseudo-linear flow regime, and it can be derived from the slope of the $V - \nabla p$ curve; ∇p_{PTPG} is the pseudo threshold pressure gradient (PTPG), which can be calculated from the intercept of the fitted straight line of the $V - \nabla p$ curve. In Fig. 18, the calculated effective permeability is 0.0258 mD, and the calculated PTPG is 1.0533 MPa/m. For the low-velocity non-Darcy flow, apparent permeability K_a will continuously increase and approach effective permeability K_e with the increasing pressure gradient. When the pressure gradient is large enough, the change in boundary-layer thickness is negligible and the apparent permeability can be considered equal to the effective permeability [58].

4 Conclusions

In this chapter, the petrophysical properties of tight sandstone are reviewed and summarized. The techniques for pore structure characterization, such as thin section, FE-SEM, X-CT, HPMI, RMI, NMR, and N_2GA, are briefly introduced and the pore structure analysis methods, especially the analysis methods based on the fractal theory, are reviewed. Comprehensive application of various pore structure characterization techniques is recommended to accurately analyze the tight sandstone pore structures, which have the characteristics of wide pore size distribution, highly complex and irregular pore structures. The models for predicting intrinsic permeability, apparent gas permeability, and liquid permeability of tight sandstone are reviewed, and the behaviors of the low-velocity non-Darcy flow in tight sandstone are also discussed. This chapter would be beneficial to all who analyze the petrophysical properties of tight sandstone.

Acknowledgments

This work was supported by the National Natural Science Foundation of China (Nos. 51604285, 41722403), Scientific Research Foundation of China University of Petroleum, Beijing (No. 2462017BJB11).

References

[1] R. Rezaee, A. Saeedi, B. Clennell, Tight gas sands permeability estimation from mercury injection capillary pressure and nuclear magnetic resonance data, J. Pet. Sci. Eng. 88–89 (2012) 92–99.

[2] H. Wu, Y. Ji, R. Liu, et al., Pore structure and fractal characteristics of a tight gas sandstone: a case study of Sulige area in the Ordos Basin, China, Energ. Explor. Exploit. (2018), https://doi.org/10.1177/0144598718764750.

[3] X. Zhang, C. Wu, T. Li, Comparison analysis of fractal characteristics for tight sandstones using different calculation methods, J. Geophys. Eng. 14 (2016) 120–131.

[4] P. Krakowska, E. Puskarczyk, M. Jędrychowski, et al., Innovative characterization of tight sandstones from Paleozoic basins in Poland using X-ray computed tomography supported by nuclear magnetic resonance and mercury porosimetry, J. Pet. Sci. Eng. 166 (2018) 389–405.

[5] K. Li, Analytical derivation of Brooks-Corey type capillary pressure models using fractal geometry and evaluation of rock heterogeneity, J. Pet. Sci. Eng. 73 (2010) 20–26.

[6] H. Wu, C. Zhang, Y. Ji, et al., Pore throat characteristics of tight sandstone of Yanchang formation in eastern Gansu, Ordos Basin, Petrol. Res. 3 (2018) 33–43.

[7] Z. Zhang, Y. Shi, H. Li, et al., Experimental study on the pore structure characteristics of tight sandstone reservoirs in upper Triassic Ordos Basin China, Energ. Explor. Exploit. 34 (2016) 418–439.

[8] F. Wang, K. Yang, J. Cai, Fractal characterization of tight oil reservoir pore structure using nuclear magnetic resonance and mercury intrusion porosimetry, Fractals 26 (2018). 1840017.

[9] X. Shao, X. Pang, H. Li, et al., Fractal analysis of pore network in tight gas sandstones using NMR method: a case study from the Ordos Basin, China, Energ. Fuel. 31 (2017) 10358–10368.

[10] M. Schmitt, C.P. Fernandes, J.A.B. Da Cunha Neto, et al., Characterization of pore systems in seal rocks using nitrogen gas adsorption combined with mercury injection capillary pressure techniques, Mar. Petrol. Geol. 39 (2013) 138–149.

[11] B. Bai, R. Zhu, S. Wu, et al., Multi-scale method of Nano (micro)-CT study on microscopic pore structure of tight sandstone of Yanchang formation, Ordos Basin, Petrol. Explor. Dev+ 40 (2013) 354–358.

[12] X. Liu, J. Wang, L. Ge, et al., Pore-scale characterization of tight sandstone in Yanchang formation Ordos Basin China using micro-CT and SEM imaging from nm- to cm-scale, Fuel 209 (2017) 254–264.

[13] C.R. Clarkson, M. Freeman, L. He, et al., Characterization of tight gas reservoir pore structure using USANS/SANS and gas adsorption analysis, Fuel 95 (2012) 371–385.

[14] L.M. Anovitz, D.R. Cole, G. Rother, et al., Diagenetic changes in macro- to nano-scale porosity in the St. Peter sandstone: an (ultra) small angle neutron scattering and backscattered electron imaging analysis, Geochim. Cosmoch. Ac. 102 (2013) 280–305.

[15] F. Wang, Z. Liu, L. Jiao, et al., A fractal permeability model coupling boundary-layer effect for tight oil reservoirs, Fractals 25 (2017). 1750042.

[16] C. Zou, R. Zhu, K. Liu, et al., Tight gas sandstone reservoirs in China: characteristics and recognition criteria, J. Petrol. Sci. Eng. 88 (2012) 82–91.

[17] F. Wang, Y. Li, X. Tang, et al., Petrophysical properties analysis of a carbonate reservoir with natural fractures and vugs using X-ray computed tomography, J. Nat. Gas Sci. Eng. 28 (2016) 215–225.

[18] B. Li, X. Tan, F. Wang, et al., Fracture and vug characterization and carbonate rock type automatic classification using X-ray CT images, J. Pet. Sci. Eng. 153 (2017) 88–96.

[19] F. Bai, Y. Sun, Y. Liu, et al., Evaluation of the porous structure of Huadian oil shale during pyrolysis using multiple approaches, Fuel 187 (2017) 1–8.

[20] E.W. Washburn, The dynamics of capillary flow, Phys. Rev. 17 (1921) 273.

[21] B.B. Mandelbrot, The Fractal Geometry of Nature, W.H. Freeman, New York, 1983.

[22] F. Wang, L. Jiao, Z. Liu, et al., Fractal analysis of pore structures in low permeability sandstones using mercury intrusion porosimetry, J. Porous Media 21 (2018) 1097–1119.

[23] H. Liu, J. Zheng, K. Wang, et al., Comments on "comments on "reply to comments on "analytical derivation of Brooks–Corey type capillary pressure models using fractal geometry and evaluation of rock heterogeneity"""", J. Petrol. Sc. Eng. 159 (2017) 614–616.

[24] E. Aliakbardoust, H. Rahimpour-Bonab, Effects of pore geometry and rock properties on water saturation of a carbonate reservoir, J. Pet. Sci. Eng. 112 (2013) 296–309.

[25] B.S. Nabawy, Y. Géraud, P. Rochette, et al., Pore-throat characterization in highly porous and permeable sandstones, AAPG Bull. 93 (2009) 719–739.

[26] W.I. Friesen, R.J. Mikula, Fractal dimensions of coal particles, J. Colloid Interf. Sci. 120 (1987) 263–271.

[27] B. Zhang, S. Li, Determination of the surface fractal dimension for porous media by mercury porosimetry, Ind. Eng. Chem. Res. 34 (1995) 1383–1386.

[28] B. Zhang, W. Liu, X. Liu, Scale-dependent nature of the surface fractal dimension for bi- and multi-disperse porous solids by mercury porosimetry, Appl. Surf. Sci. 253 (2006) 1349–1355.

[29] C.Z. He, M.Q. Hua, Fractal geometry description of reservoir pore structure, Oil Gas Geol. 19 (1998) 15–23.

[30] P. Li, M. Zheng, H. Bi, et al., Pore throat structure and fractal characteristics of tight oil sandstone: a case study in the Ordos Basin, China, J. Pet. Sci. Eng. 149 (2017) 665–674.

[31] H.H. Yuan, B.F. Swanson, Resolving pore-space characteristics by rate-controlled porosimetry, SPE Format. Eval. 4 (1989) 17–24.

[32] H. Zhao, Z. Ning, Q. Wang, et al., Petrophysical characterization of tight oil reservoirs using pressure-controlled porosimetry combined with rate-controlled porosimetry, Fuel 154 (2015) 233–242.

[33] G.R. Coates, L. Xiao, M.G. Prammer, NMR Logging: Principles and Applications, Haliburton Energy Services, Houston, 1999.

[34] A. Li, X. Ren, G. Wang, et al., Characterization of pore structure of low permeability reservoirs using a nuclear magnetic resonance method, J. China Univ. Petrol. (Nat. Sci. Ed.) 39 (2015) 92–98.

[35] Y. Volokitin, W.J. Looyestijn, W. Slijkerman, et al., A practical approach to obtain primary drainage capillary pressure curves from NMR core and log data, Petrophysics 42 (2001) 334–343.

[36] L. Xiao, C. Zou, Z. Mao, et al., A new technique for synthetizing capillary pressure (Pc) curves using NMR logs in tight gas sandstone reservoirs, J. Pet. Sci. Eng. 145 (2016) 493–501.

[37] Z. Zhang, A. Weller, Fractal dimension of pore-space geometry of an Eocene sandstone formation, Geophysics 79 (2014) D377–D387.

[38] H. Gan, S.P. Nandi, P.L. Walker Jr., Nature of the porosity in American coals, Fuel 51 (1972) 272–277.

[39] M.M. Labani, R. Rezaee, A. Saeedi, et al., Evaluation of pore size spectrum of gas shale reservoirs using low pressure nitrogen adsorption, gas expansion and mercury porosimetry: a case study from the Perth and canning basins, Western Australia, J. Pet. Sci. Eng. 112 (2013) 7–16.

[40] K.S.W. Sing, Reporting physisorption data for gas/solid systems—with special reference to the determination of surface area and porosity, Pure Appl. Chem. 57 (1985) 603–619.

[41] S.J. Gregg, K.S.W. Sing, H.W. Salzberg, Adsorption surface area and porosity, J. Electrochem. Soc. 114 (1967) 279C.

[42] Z. Zhang, Z. Yang, Theoretical and practical discussion of measurement accuracy for physisorption with micro- and mesoporous materials, Chin. J. Catal. 34 (2013) 1797–1810.

[43] M. Wang, H. Xue, S. Tian, et al., Fractal characteristics of upper cretaceous lacustrine shale from the Songliao Basin, NE China, Mar. Petrol. Geol. 67 (2015) 144–153.

[44] F. Yang, Z. Ning, H. Liu, Fractal characteristics of shales from a shale gas reservoir in the Sichuan Basin, China, Fuel 115 (2014) 378–384.

[45] M. Schmitt Rahner, M. Halisch, C. Peres Fernandes, et al., Fractal dimensions of pore spaces in unconventional reservoir rocks using X-ray nano- and micro-computed tomography, J. Nat. Gas Sci. Eng. 55 (2018) 298–311.

[46] H. Qi, J. Ma, P. Wong, Adsorption isotherms of fractal surfaces, Colloid. Surf. A 206 (2002) 401–407.

[47] Y. Yao, D. Liu, D. Tang, et al., Fractal characterization of adsorption-pores of coals from North China: an investigation on CH_4 adsorption capacity of coals, Int. J. Coal Geol. 73 (2008) 27–42.
[48] M. Jaroniec, Evaluation of the fractal dimension from a single adsorption isotherm, Langmuir 11 (1995) 2316–2317.
[49] T. Cao, Z. Song, S. Wang, et al., Characterization of pore structure and fractal dimension of Paleozoic shales from the northeastern Sichuan Basin, China, J. Nat. Gas Sci. Eng. 35 (2016) 882–895.
[50] P.C. Carman, Fluid flow through granular beds, Trans. Inst. Chem. Eng. 15 (1937) 150–166.
[51] J. Kozeny, Uber kapillare leitung der wasser in Boden, Royal Academy of Science, Vienna, Proc. Class I 136 (1927) 271–306.
[52] A. Mirzaei-Paiaman, M. Ostadhassan, R. Rezaee, et al., A new approach in petrophysical rock typing, J. Pet. Sci. Eng. 166 (2018) 445–464.
[53] S. Kolodzie, Analysis of pore throat size and use of the Waxman-Smits equation to determine OOIP in Spindle Field, Colorado, in: SPE-9382, Annual Technical Conference and Exhibition, Dallas, Texas, 24 September, 1980.
[54] H.A. Nooruddin, M.E. Hossain, H. Al-Yousef, et al., Comparison of permeability models using mercury injection capillary pressure data on carbonate rock samples, J. Pet. Sci. Eng. 121 (2014) 9–22.
[55] H. Wu, Y. Ji, R. Liu, et al., Insight into the pore structure of tight gas sandstones: a case study in the Ordos Basin, NW China, Energ. Fuel. 31 (2017) 13159–13178.
[56] B. Yu, P. Cheng, A fractal permeability model for bi-dispersed porous media, Int. J. Heat Mass Transf. 45 (2002) 2983–2993.
[57] F. Wang, L. Jiao, P. Lian, et al., Apparent gas permeability, intrinsic permeability and liquid permeability of fractal porous media: carbonate rock study with experiments and mathematical modelling, J. Petrol. Sci. Eng. (2018). https://doi.org/10.1016/j.petrol.2018.10.095.
[58] F. Wang, Z. Liu, J. Cai, et al., A fractal model for low-velocity non-Darcy flow in tight oil reservoirs considering boundary-layer effect, Fractals 26 (2018) 1850077.
[59] H. Pape, C. Clauser, J. Iffland, Variation of permeability with porosity in sandstone diagenesis interpreted with a fractal pore space model, Pure Appl. Geophys. 157 (2000) 603–619.
[60] M. Knudsen, The law of molecular flow and viscosity of gases moving through tubes, Ann. Phys. 28 (1909) 75.
[61] Q. Zheng, B. Yu, Y. Duan, et al., A fractal model for gas slippage factor in porous media in the slip flow regime, Chem. Eng. Sci. 87 (2013) 209–215.
[62] R. Cao, Y. Wang, L. Cheng, et al., A new model for determining the effective permeability of tight formation, Transp. Porous Media 112 (2016) 21–37.
[63] Y. Li, Q. Lei, X. Liu, et al., Characteristics of micro scale nonlinear filtration, Petrol. Explor. Dev. 38 (2011) 336–340.
[64] Y. Xiong, J. Yu, H. Sun, et al., A new non-Darcy flow model for low-velocity multiphase flow in tight reservoirs, Transp. Porous Media 117 (2017) 367–383.
[65] X. Tian, Non-linear Flow Behavior and Mathematical Model in Ultra-Low Permeability Reservoirs and Mathematical Model in Ultra-Low Permeability Reservoirs, China University of Petroleum (Beijing), 2015.
[66] S. Xu, X. Yue, Experimental research on nonlinear flow characteristics at low velocity, J. China Univ. Petrol. (Nat. Sci. Ed.) 31 (2007) 60–63.
[67] J. Wu, L. Cheng, C. Li, et al., Experimental study of nonlinear flow in micropores under low pressure gradient, Transp. Porous Media 119 (2017) 247–265.
[68] X. Wang, J.J. Sheng, Effect of low-velocity non-Darcy flow on well production performance in shale and tight oil reservoirs, Fuel 190 (2017) 41–46.
[69] L. Gavin, Pre-Darcy flow: a missing piece of the improved oil recovery puzzle? in: SPE-89433, SPE/DOE Symposium on Improved Oil Recovery, Tulsa, Oklahoma, 17–21 April, 2004.
[70] M. Yun, B. Yu, J. Cai, A fractal model for the starting pressure gradient for Bingham fluids in porous media, Int. J. Heat Mass Transf. 51 (2008) 1402–1408.
[71] J. Cai, A fractal approach to low velocity non-Darcy flow in a low permeability porous medium, Chin. Phys. B 23 (2014). 044701.

CHAPTER 4

Multifractal Analysis of Pore Structure of Tight Oil Reservoirs Using Low-Field NMR Measurements

Peiqiang Zhao*, Xiangyun Hu*, Xinmin Ge†, Zhenlin Wang‡, Jianchao Cai*

**Hubei Subsurface Multi-scale Imaging Key Laboratory, Institute of Geophysics and Geomatics, China University of Geosciences, Wuhan, P.R. China, †School of Geosciences, China University of Petroleum (East China), Qingdao, P.R. China, ‡Research Institute of Exploration and Development, Xinjiang, Oilfield Company, PetroChina, Karamay, P.R. China*

Chapter Outline

1 Introduction

As typical of unconventional resources, exploration and development of tight oil has drawn attention from many countries. A few of the tight oil reservoirs include the Bakken Formation in the Williston Basin, North America and the Ordos Basin, Junggar Basin, Songliao Basin, and Sichuan Basin in China [1–3]. Compared with conventional reservoirs, the tight oil reservoirs

Petrophysical Characterization and Fluids Transport in Unconventional Reservoirs. https://doi.org/10.1016/B978-0-12-816698-7.00004-8

have low porosity and extremely high permeability, and are always interbedded with or adjacent to source rocks, which require horizontal drilling combined with hydraulic fracturing as the main development techniques [3–5]. A good knowledge of the reservoir quality is necessary for economic production from tight oil formations. Pore structure, which affects the transportation property and reservoir storage capacity, is one of the most critical parameters for determining reservoir quality [6, 7]. Thus, it is important to understand the pore structure characteristics in order to better exploit tight oil reserves.

The pore structure of tight reservoirs and shale samples are commonly determined by two different methods. One is through direct petrophysical analysis on basis of different measurement methods on rocks. The other is conducting the fractal and multifractal analysis based on some of the measurement data. The nano-CT scan, scanning electron microscopy (SEM), field emission scanning electron microscopy (FESEM), and focused ion beam scanning electron microscopy (FIB-SEM) are used to qualitatively observe the pore types and sizes [8-13]. The pore size distributions, pore volumes, and specific surface areas can be characterized through measurements such as low-temperature N_2 and CO_2 adsorption, high-pressure mercury injection (PMI), rate-controlled mercury injection (RMI), and low-field nuclear magnetic resonance (NMR) [14-20]. Low-field NMR is an important method for studying the pore structure of reservoirs as it is a fast method in estimating the porosity, permeability, fluid, and pore size distributions, and it is a nondestructive method [21, 22]. The NMR logs can also provide these petrophysical parameters consecutively, overcoming the discontinuity of core sample analysis, which is the largest advantage over the other pore structure characterization methods.

The fractal and multifractal theory geometry was proposed to investigate the pore space of rocks and other objects by Mandelbrot [23, 24]. The key parameter of fractal dimension is used to reveal the complexity and heterogeneity of pore geometry as it can describe the self-similar features of rock pore volumes [25, 26]. The multifractal theory involves decomposing self-similar measures into intertwined fractal sets, and divides a complex fractal structure into several regions, which are characterized by the singularity strength and generalized fractal dimensions [24, 27]. It has been widely applied in various fields of geosciences [28, 29]. In the field of pore structure characterization, compared to (mono-) fractal analysis, the multifractal analysis provides more information about the pore properties, including the globe and local properties of pores [29, 30]. The NMR measurements can be used to determine multifractal dimensions of the sedimentary rocks in order to better understand the microscopic pore structure and heterogeneity of tight reservoirs or shale [6].

Recently, with the increasing exploration of tight oil in China, the Permian Lucaogou Formation of Jimusaer Sag, Junggar Basin has attracted a large amount of attention. However, the pore structure and its heterogeneity of the Lucaogou tight oil reservoir are not well studied. Zhang et al. [31] used a nano-CT scan and the FIB-SEM method to study the pore structure of the Lucaogou Formation. They concluded that the pore types are mainly intergranular pores, intergranular dissolved pores, intercrystalline pores, and interlayer cracks. Kuang et al. [32] observed from electron microprobe analysis that the dissolved pores, which could be subdivided into intragranular pores, intergranular pores and cracks, were the main pore type, and their diameter ranged from a few micrometers to 30 μm. Zhao et al. [33] reported the median capillary radius of this reservoir ranged from 0.0063 to 0.1482 μm with an average of 0.039 μm. Su et al. [34] found the following pore type: primary mineral pores, dissolved mineral pores, primary organic matter pores, and secondary organic matter pores. In summary, there are a few intensive studies of the pore structure and heterogeneity of the Lucaogou Formation; however, there is still a great amount of further work needed.

In this chapter, we analyze the pore structure of the tight oil rocks from the Permain Lucaogou Formation in Jimusaer Sag, Junggar Basin using NMR measurements with the aid of minerals, porosity, and permeability. Multifractal analysis is applied to quantify the heterogeneity of the pore structures. The relationships of macro petrophysical and multifractal parameters are also investigated. It was observed that several macro petrophysical parameters can capture the heterogeneity of micropore structures.

2 Methodology

2.1 Multifractal Theory

The box-counting method was applied to study the multifractal analysis. For the multifractal property, the target object is segmented into $N(\varepsilon)$ boxes, and the size or scale of the box is $\varepsilon \times \varepsilon$ or ε ($\varepsilon = 2^0, 2^1, 2^2 \ldots$). The ratio of pore volume in the ith box with a size of ε to total pore volume can be taken as mass probability. Thus, the mass probability can be expressed as [29]:

$$P_i(\varepsilon) = \frac{v_i(\varepsilon)}{\sum_{i=1}^{N(\varepsilon)} v_i(\varepsilon)}, \tag{1}$$

where $P_i(\varepsilon)$ is the ratio value of pore volume of the ith box to the total pore volume and $v_i(\varepsilon)$ is the pore volume of the ith box.

For pores with a multifractal property, the relationship of distribution probability $P_i(\varepsilon)$ and scale ε is defined as [35, 36]:

$$P_i(\varepsilon) \propto \varepsilon^{\alpha_i}, \tag{2}$$

where α_i represents the coarse-Holder exponent or singularity strength, and is a noninteger.

The number of boxes with the same or close α value in fractals is denoted as $N_a(\varepsilon)$, which is related to scale ε and can be expressed as:

$$N_\alpha(\varepsilon) \propto \varepsilon^{-f(\alpha)}, \tag{3}$$

where $f(\alpha)$ is the multifractal spectrum or singularity spectrum, indicating the fractal dimension of the boxes with the same or similar α value.

To know the distribution properties, the partition function is defined as:

$$X(q, \varepsilon) = \sum_{i=1}^{N(\varepsilon)} P_i^q(\varepsilon) \propto \varepsilon^{\tau(q)}, \tag{4}$$

where $X(q,\varepsilon)$ is the partition function of q with scale ε; $\tau(q)$ is the mass exponent, which is expressed as:

$$\tau(q) = -\lim \frac{X(q, \varepsilon)}{\log \varepsilon} = -\lim_{\varepsilon \to 0} \frac{\log \sum_{i=1}^{N(\varepsilon)} P_i^q(\varepsilon)}{\log \varepsilon}. \tag{5}$$

The generalized dimension D_q is expressed as:

$$D_q = \frac{1}{q-1} \lim_{\varepsilon \to 0} \frac{\log \sum_{i=1}^{N(\varepsilon)} P_i^q(\varepsilon)}{\log \varepsilon} = \frac{\tau(q)}{q-1}. \tag{6}$$

When $q = 1$,

$$D_1 = \lim_{\varepsilon \to 0} \frac{\log \sum_{i=1}^{N(\varepsilon)} P_i(\varepsilon) * \log P_i(\varepsilon)}{\log \varepsilon}. \tag{7}$$

The $\alpha(q)$ and $f(\alpha)$ can be obtained from:

$$\alpha(q) = \frac{\sum_{i=1}^{N(\varepsilon)} \mu_i(\varepsilon) * \log P_i(\varepsilon)}{\log \varepsilon}, \tag{8}$$

$$f(\alpha) = \frac{\sum_{i=1}^{N(\varepsilon)} \mu_i(\varepsilon) * \log \mu_i(\varepsilon)}{\log \varepsilon}, \tag{9}$$

$$\text{where} \quad \mu_i(\varepsilon) = \frac{P_i(\varepsilon)^q}{\sum_{i=1}^{N(\varepsilon)} P_i(\varepsilon)^q}.$$

They can also be obtained from q and $\tau(q)$ using the Legendre transformation, as [37]:

$$\alpha(q) = \frac{d\tau(q)}{dq}, \tag{10}$$

$$f(\alpha) = q\alpha(q) - \tau(q). \tag{11}$$

The singularity strength range ($\Delta\alpha$), which can represent the heterogeneity of objects, is defined as:

$$\Delta\alpha = \alpha_{\max} - \alpha_{\min}. \tag{12}$$

Generally, the $q - D_q$ and $\alpha - f(\alpha)$ form the basic mathematical language for describing the multifractal properties, and are used for characterizing the heterogeneity of the pore structure.

2.2 Relationship Between NMR T_2 *Time and Pore Size*

According to the NMR theory, the apparent transverse relaxation time T_2 in porous rocks is expressed as [21]:

$$\frac{1}{T_2} = \frac{1}{T_{2b}} + \frac{1}{T_{2s}} + \frac{1}{T_{2d}}, \tag{13}$$

where T_{2b}, T_{2s}, T_{2d} are the bulk, surface, and diffusion transversal relaxation time, respectively.

The bulk transversal relaxation time is an intrinsic property of pore fluids. The T_{2b} of water is commonly between 2000 and 3000ms, which is much larger than the surface transversal relaxation in water wet rocks, thus the bulk transversal relaxation is ignored. The diffusion relaxation is close to zero in a uniform magnetic field with short echo spacing time. The surface relaxation can be expressed as:

$$\frac{1}{T_2} = \rho \frac{S}{V}, \tag{14}$$

where S and V are pore surface and volume, respectively, and ρ is the transversal surface relaxivity.

Thus, Eq. (13) can be expressed as [38, 39]:

$$\frac{1}{T_2} = \rho \frac{S}{V}. \tag{15}$$

The surface/volume ratio of the pore space is related to pore radius and pore shape factor, the transversal relaxation time can also be expressed as [38, 39]:

$$\frac{1}{T_2} = \rho \frac{F_s}{r}, \tag{16}$$

where r is the pore radius, μm; F_s is the pore shape factor, equal to 2 and 3 for cylindrical and spherical pores, respectively.

Based on Eq. (16), the T_2 distribution under 100% water-saturated state can be transformed into a pore size distribution if the surface relaxivity ρ is determined. It is commonly believed that:

$$T_2 \propto r. \tag{17}$$

This equation is the basis for using NMR measurements to study the fractal and multifractal characteristics of porous rocks. Denoting L as the T_2 point number of the 100% water-saturated state sample, L should equal to a power of 2 via inversion or interpolation method, i.e.,

$L = 2^j, j = 1, 2, 3, \ldots$ The scale of T_2 is $\varepsilon = L \times 2^{-k}$ $(k = 0, 1, 2, \ldots, j)$. The T_2 distribution can be segmented in $N(\varepsilon)$ partitions with the scale ε being:

$$N(\varepsilon) = \frac{L}{\varepsilon} = 2^k. \tag{18}$$

Finally, we can conduct the multifractal analysis of NMR T_2 distributions.

3 Samples and Experiments

Samples were drilled from the Permian Lucaogou Formation in Jimusaer Sag, Junggar Basin. The Junggar Basin is the second largest inland basin in China, which is located in the north of the Xinjiang Province, Northwest China. The Jimusaer Sag is located at the southeastern end of the Junggar Basin and is a secondary dustpan-shaped sag in the eastern uplift tectonic unit [40]. It extends about 30 km from the north to the south, about 60 km from the east to the west and is over 1278 km^2. It is bordered by Guxi Uplift to the east, Fukang Fault Zone to the south, Beisantai Uplift to the west, and Shaqi Uplift to the north [40]. The target Lucaogou Formation was developed in the Permain System, which from bottom to top includes Jiangjunmiao, Jingjinggouzi, Lucaogou, and Wutonggou Formations. The Lucaogou Formation consists of two "sweet spot" reservoirs and the shale source rocks between the sweet spots [32]. Four and ten samples were collected from the upper and lower sweet spots in Well Ji174, respectively. Samples to measure porosity, permeability, and NMR properties were cut into a plunger. The parallel powder samples were used to determine the mineralogical compositions. Some additional samples were selected for SEM analysis.

The mineral types and their relative content were determined using X-ray diffraction (XRD) analysis. The XRD analysis was performed on the nonoriented powdered samples using an X-ray diffractometer, which was equipped with a copper X-ray tube that operated at 30 kV and 40 mA. The scan angle ranged from 5–90 degrees at a speed of 2 degrees/min [41]. SEM was performed using a Hitachi S4800 scanning electron microscope with a lowest pixel resolution of 1.2 nm and accelerating voltage of 30 kV, following the standard SY/T 5162-2014 of China.

The cylindrical plugs were made to have a diameter of 2.5 cm (~1.0 in.) and a length of 4 cm (~1.5 in.). The routine helium porosity and permeability measurements were taken on these plunger samples with a helium porosimeter. In order to simulate the reservoir condition, a net confining stress of 5000 psi (34.47 MPa) was exerted during the measuring. The NMR T_2 distributions of the fourteen core samples were obtained using a Geospace2 instrument at a 100% brine saturated and irreducible water (centrifuged) state. All core plugs were subjected to cleaning, drying, vacuuming, and saturating procedure to arrive at the water-saturated state. Water-saturated core plugs were subjected to the NMR and then centrifugal experiments with a centrifuge speed of 10000 r/min to achieve an ideal irreducible water state. The irreducible water state samples were also analyzed through the NMR experiments. The resonant frequency of the Geospace2 instrument was 2 MHZ. The polarization time or waiting time (T_w), the echo spacing (T_E), the number of echoes, and the number of scans was set as 10,000 ms, 0.3 ms, 4096 and 128, respectively.

4 Petrophysical Parameters and Mineralogical Characteristics

The helium porosity (ϕ) and permeability (K) values measured are listed in Table 1. The porosities ranged from 7.38% to 15.92% with an average value of 12.24%. The permeability ranged from 0.002 to 0.641 mD with a logarithmic average of 0.017 mD. The permeability of all samples was lower than 1 mD, and 11 samples showed a permeability of lower than 0.1 mD, demonstrating that the Lucaogou Formation is a tight oil reservoir. Having an average porosity of 12.24% but an extremely low permeability indicates that the pore connectivity of the Lucaogou Formation is poor, which may result from the strong heterogeneity of the pore structures.

In order to better characterize the pore structure and study its multifractal characteristics, we calculated two commonly used parameters associated with the macro physical properties of rocks, the reservoir quality index (RQI) and the flow zone indicator (FZI) [42, 43]. These are the primary parameters of the hydraulic flow unit theory used for formation classification in the presence of strong heterogeneity. The RQI and FZI are, respectively, expressed as [42, 43]:

$$RQI = 0.0314 \times \sqrt{\frac{K}{\phi}}, \tag{19}$$

$$FZI = RQI \times \frac{1-\phi}{\phi}, \tag{20}$$

where RQI is the reservoir quality index; K is permeability; ϕ is porosity; and FZI is flow zone indicator.

Table 1 The porosity, permeability, and other petrophysical properties from NMR measurements

No.	ϕ-Helium (%)	K-Helium (mD)	RQI (μm)	FZI (μm)	ϕ-NMR (%)	S_{wi} (%)	ϕ_{wi} (%)	T_2 Cutoff (ms)	T_{2lm} (ms)	T_{50} (ms)	T_{35} (ms)
1	15.79	0.641	0.063	0.337	16.06	64.26	10.32	103.00	42.63	49.41	94.51
2	12.90	0.044	0.018	0.124	13.30	55.79	7.42	39.00	32.59	26.11	43.55
3	14.22	0.114	0.028	0.170	14.49	70.39	10.20	60.00	21.52	19.59	42.69
4	7.38	0.003	0.007	0.085	7.77	72.84	5.66	93.00	29.24	31.49	54.16
5	8.26	0.002	0.005	0.059	9.35	74.12	6.93	9.47	4.81	4.78	6.90
6	8.28	0.002	0.005	0.058	9.03	78.74	7.11	31.00	13.30	12.73	19.06
7	15.92	0.147	0.030	0.159	16.99	58.27	9.90	18.20	15.00	14.99	20.89
8	13.63	0.027	0.014	0.089	13.67	76.15	10.41	20.28	9.06	8.71	13.45
9	13.63	0.032	0.015	0.097	13.80	85.72	11.83	22.62	6.76	6.02	9.14
10	14.59	0.011	0.009	0.051	14.53	71.16	10.34	4.93	2.99	2.58	3.73
11	10.00	0.002	0.005	0.044	10.56	82.86	8.75	14.86	5.07	4.46	7.24
12	9.30	0.002	0.005	0.047	12.02	78.62	9.45	4.90	2.01	1.78	2.60
13	13.43	0.013	0.010	0.063	12.81	80.72	10.34	22.61	6.86	6.55	10.92
14	14.14	0.020	0.012	0.072	14.27	81.36	11.61	22.61	5.40	5.96	11.20
Average	12.24	0.017	0.016	0.104	12.76	73.64	9.31	33.32	14.09	13.94	24.49

As presented in Table 1, the *RQI* values were extremely low, ranging from 0.005 to 0.0063. The *FZI* also showed similar characteristics, ranging from 0.044 to 0.337 with an average value of 0.104. The T_2 cutoffs were characterized by a large variation range from 4.9 to 103 ms, whereas the irreducible water saturation remained high, ranging from 55.79% to 85.79% with an average of 73.64%, indicating that the majority of the pores were within clay and capillary pores. The T_2 logarithmic mean value (T_{2lm}), which is a sensitive parameter representing the integrated characteristics of T_2 distribution and macro properties of rock pore structure, was calculated in Table 1. The T_{2lm} values were very low, ranging from 2.01 to 42.63 ms with an average of 14.09 ms, suggesting small pore sizes in the tight oil rocks.

On the mercury capillary curve, the Pc_{50}, which equals the pressure when the mercury saturation is 50%, is an important parameter for describing the shape of the curve [33, 44]. The r_{35}, which is the calculated throat size when the mercury saturation is 35%, contributes greatly to the rock permeability in many oil fields [45, 46]. Both of these parameters are associated with the pore structure characteristics. Therefore, we calculated the relaxation times corresponding to the 35% and 50% saturation on the T_2 reverse cumulative curve, as shown in Fig. 1. The values of the two parameters are denoted as T_{50} and T_{35}, respectively. The T_{50} and T_{35} were in the range of 1.78–49.41 ms and 2.60–94.51 ms, respectively. The T_{50} values were less than the corresponding T_{2lm}, while the T_{35} showed greater values than those of T_{2lm}.

From the above discussion, it is apparent that the NMR T_2 distributions at water-saturated condition can be transferred to pore size distribution in a water wet rock. Zhao et al. [33] reported that the wettability of the samples from the Lucaogou Formation after cleaning are water wet, thus long relaxation components of the T_2 spectrum are associated with big or macropores and the short relaxation components are associated with small or micropores. The NMR T_2 distributions of samples at the water saturated state are shown in Fig. 2. The T_2 distributions showed mostly a unimodal (Fig. 2A) shape, with only four distributions being

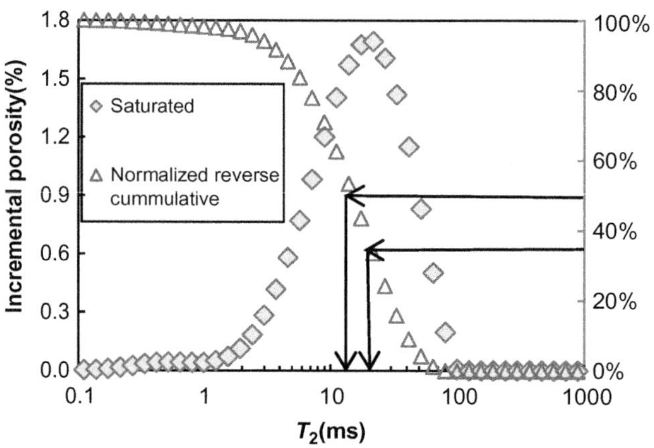

Fig. 1
Schematic diagram showing calculation of T_{35} and T_{50}.

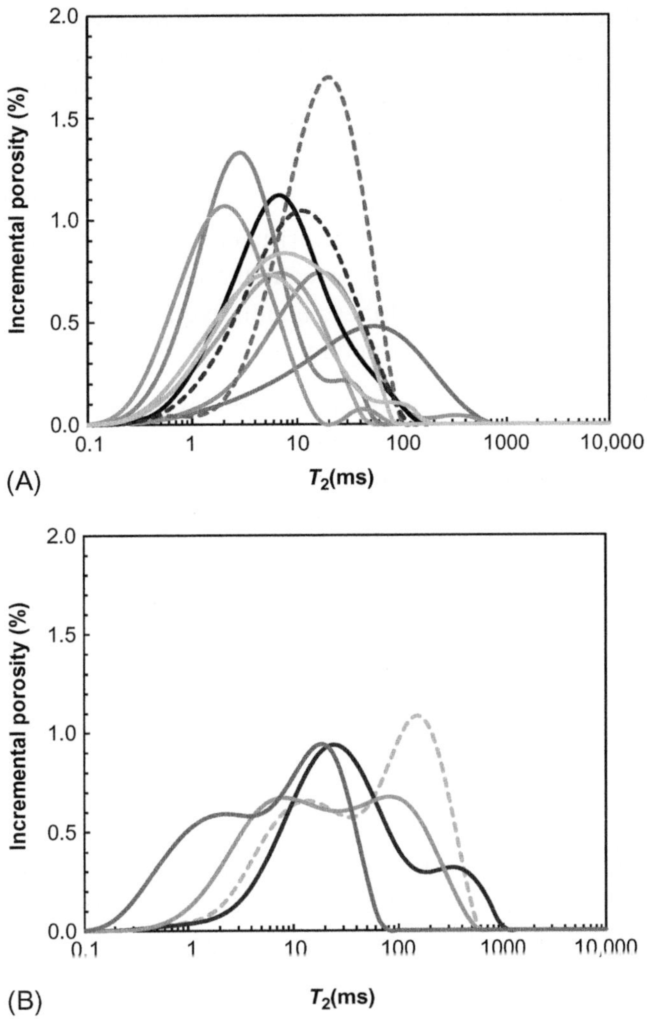

Fig. 2
The NMR T_2 distributions of tight oil rocks, (A) unimodal T_2 distributions, (B) bimodal T_2 distributions.

bimodal (Fig. 2B). As is observed from this figure, there were no pores with a relaxation time greater than 1000 ms, four samples had pores with a relaxation time in the range of 100–1000 ms, while the relaxation time of pores in other samples were all less than 100 ms. The relaxation time of peaks of T_2 distributions without relaxation time greater than the 100 ms varied from 2 to 20 ms, and the minimal time of these T_2 distributions was less than 1 ms, suggesting that the micropores in tight oil rocks develop a lot. It is noted that in rock with a large amount of carbonate mineral, the surface relaxivity is lower than that of clay and quartz, making the T_2 distributions move in the right direction [33, 47]. Therefore, the T_2 distributions may be more on the left if the carbonate contents decrease.

Compared to the T_2 distributions of Longmaxi gas shale in the Jiaoshiba area of the Sichuan Basin [18, 48], the T_2 distributions of the target formation showed more long relaxation components and less short relaxation components, but varied more significantly. Different from the T_2 distributions of conventional reservoir samples, the irreducible water saturation was high even though the T_2 distributions were within a wide range. Fig. 3 shows the NMR T_2 distribution of Sample 11 at the two different conditions. It can be seen that, in spite of a wide variability, the T_2 distribution obtained at the centrifuged state was relatively similar to that of the saturated state sample, demonstrating the irreducible water in large pores occupied a very high proportion of pores. In summary, all these qualitative characteristics of T_2 distributions demonstrated a complex and poor pore structure.

The mineral compositions of the tight oil reservoir samples extracted from the XRD analysis are presented in Table 2. As seen from this table, the mineralogical compositions were dominated by plagioclase, quartz, and dolomite. The plagioclase contents ranged from 13.7% to 49.5% with an average value of 33.2%. The quartz contents ranged from 13.7% to 26.2% with an average value of 19.7% and for dolomite, the contents ranged from 7.7% to 49.4% with an average value of 26.9%. The average content of clay, quartz+feldspar and carbonate mineral (calcite and dolomite) was 8.85%, 56.94%, and 32.77%, respectively. The samples also contained a minor amount of pyrite and siderite. It was observed that the mineral compositions were complex.

5 Multifractal Characteristics

The range of q is set from -10 to 10, where D_{-10} represents D_{min} and D_{10} represents D_{max}. We performed the multifractal analysis for all 14 samples, and the variations of the generalized dimensions (D_q) vs variable q of all samples are shown in Fig. 4.

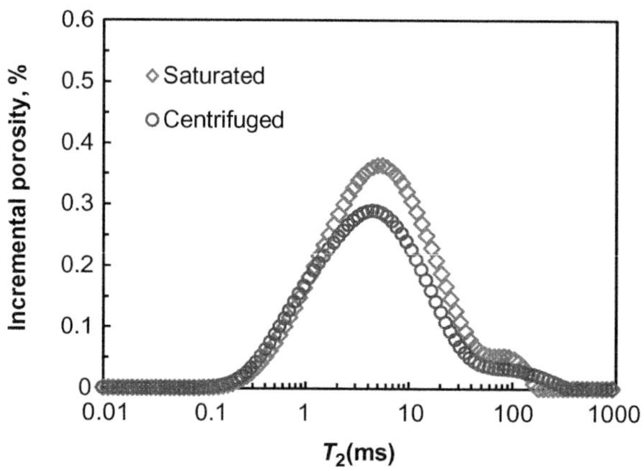

Fig. 3
The NMR T_2 distributions for Sample 11.

Table 2 The studied sample mineral compositions from XRD experiments

No.	Clay (wt.%)	Quartz (wt.%)	*K*-Feldspar (wt.%)	Plagioclase (wt.%)	Calcite (wt.%)	Dolomite (wt.%)	Pyrite (wt.%)	Siderite (wt.%)
1	2.4	18.4	3.2	26.9	0	49.1	0	0
2	11.1	26.2	0	45.6	0	13.6	3.5	0
3	4.2	15.9	2.2	35.3	17.5	24.9	0	0
4	13.9	23.2	3.9	29.4	21.9	7.7	0	0
5	18.2	16.4	4.7	27.7	0	32.5	0.5	0
6	11.6	22.6	2.5	13.7	0	49.4	0	0.2
7	6.3	21.4	7.9	37.5	1.0	18.9	0	7.0
8	6.0	15.6	4.4	25.4	0	48.6	0	0
9	6.9	17.8	5.4	44.4	0	23.6	0	1.9
10	12.2	24.7	4.5	31.1	0	26.5	1.0	0
11	6.6	21.0	1.8	18.2	5.8	41.2	5.4	0
12	7.2	19.9	7.8	49.5	0	15.6	0	0
13	7.5	16.3	5.0	38.4	22.9	9.9	0	0
14	9.8	16.5	3.9	41.0	13.3	15.0	0	0.5
Average	8.9	19.7	4.1	33.2	5.9	27.0	0.7	0.7

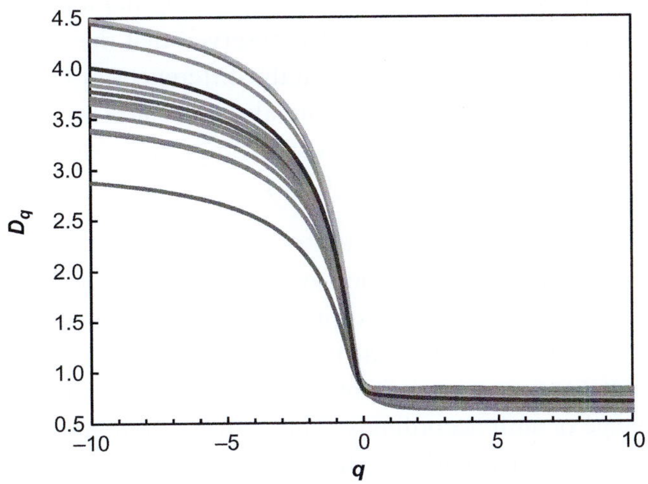

Fig. 4
Generalized dimension (D_q) vs variable q.

Fig. 4 shows that all samples exhibited two variation tendencies of D_q as q increases. D_q showed a significant decrease as q increased when $q < 0$, whereas a slight decrease was observed for D_q as q increased when $q > 0$. The D_q values of the samples presented larger differences for negative values of q, but a smaller difference when $q > 0$. This is different from the multifractal characteristics of rocks based on SEM images, whose D_q values also presented a larger difference when $q > 0$ [7].

Table 3 Multifractal parameters of tight oil rock samples

No.	D_{min}	D_0	D_1	D_2	D_{max}	α_{min}	α_{max}	α_0	$\Delta\alpha$	A
1	2.876	0.847	0.833	0.821	0.803	0.771	3.176	1.030	2.405	0.121
2	3.773	0.887	0.819	0.816	0.776	0.754	4.175	1.190	3.421	0.146
3	3.546	0.863	0.814	0.810	0.821	0.809	3.917	1.149	3.108	0.123
4	3.710	0.873	0.835	0.831	0.805	0.787	4.125	1.118	3.338	0.110
5	4.446	0.810	0.721	0.711	0.698	0.685	4.938	1.230	4.253	0.147
6	3.839	0.873	0.729	0.679	0.633	0.620	4.268	1.220	3.648	0.197
7	3.378	0.836	0.663	0.636	0.616	0.603	3.737	1.282	3.134	0.277
8	3.667	0.836	0.705	0.665	0.597	0.571	4.048	1.228	3.477	0.233
9	3.653	0.858	0.728	0.694	0.652	0.637	4.035	1.295	3.398	0.240
10	4.283	0.816	0.788	0.779	0.725	0.699	4.765	1.133	4.066	0.119
11	4.481	0.866	0.779	0.759	0.739	0.725	4.974	1.222	4.249	0.132
12	3.899	0.827	0.826	0.813	0.755	0.731	4.321	1.058	3.590	0.100
13	3.394	0.823	0.735	0.711	0.684	0.678	3.757	1.128	3.079	0.171
14	4.007	0.827	0.758	0.742	0.695	0.671	4.430	1.153	3.759	0.147
Average	3.782	0.846	0.766	0.748	0.714	0.696	4.190	1.174	3.495	0.162

The important parameters D_0, D_1, and D_2, which are called the capacity dimension, information dimension, and correlation dimension, respectively, are listed in Table 3. For all samples, there was the same relationship, i.e., $D_0 > D_1 > D_2$, thus revealing the multifractal nature of the NMR T_2 distributions for all samples. It was also observed that the D_0 ranged from 0.810 to 0.887 and D_2 ranged from 0.636 to 0.833, and that the differences of D_2 were larger than those of the D_0 and D_1. The values of D_{min} and D_{max} are also presented in Table 3, from which it is seen that D_{min} ranged from 2.876 to 4.481 with an average value of 3.782, with D_{max} ranged from 0.591 to 0.861 with an average value of 0.714.

To further reveal the multifractal characteristics of the micropores of rocks, thė multifractal spectrum, i.e., the relationship between $f(\alpha)$ and $\alpha(q)$, is plotted in Fig. 5. To the left of the crest, corresponding to $q > 0$, $f(\alpha)$ increases as $\alpha(q)$ increases, whereas to the right of the crest, corresponding to $q < 0$, $f(\alpha)$ decreases as $\alpha(q)$ increases. As is observed from this figure, the widths of the multifractal spectrum vary from one sample to another, and the crest of the spectrum can also change or shift accordingly. This proves strong multifractal characteristics, and proves that the pore size distribution or heterogeneity from NMR T_2 distributions of samples saturated with water is different.

The calculated α_{max}, α_{min}, α_0, and $\Delta\alpha$ are presented in Table 3. The high $\Delta\alpha$ values varied from 2.405 to 4.253, with an average value of 3.495, and this indicates that the micropores of all the samples were heterogeneous. The asymmetry of the singularity spectrum denoted as A can be calculated as $(\alpha_0 - \alpha_{min})/(\alpha_{max} - \alpha_0)$, and these values are presented in Table 3. All the asymmetry values were less than 1, demonstrating a slight fluctuation.

The multifractal feature can also be evaluated from the plot of α_0 vs D_0 since the differences in values between α_0 and D_0 is a measure of heterogeneity. The pore structure is homogeneous if

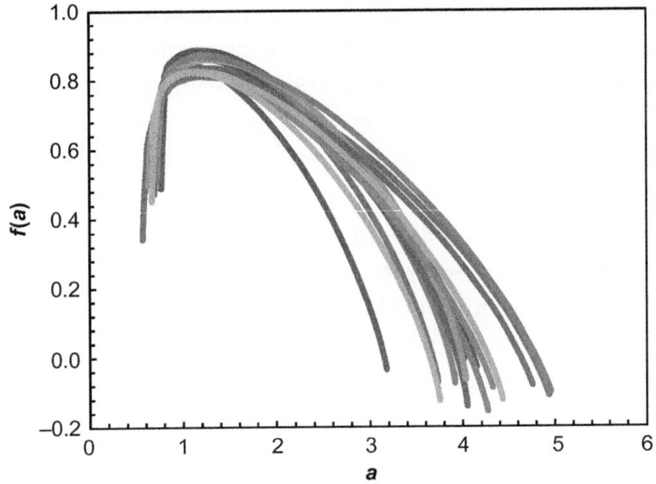

Fig. 5
The multifractal spectrum of the samples.

the data points all fall on the same line. As shown in Fig. 6, the data points are deviated from the same line, demanding that the pore structure should be described by the multifractal spectra rather than a mono-fractal model.

6 Relationship Between Mineral Contents and Multifractal Parameters

The results presented in the preceding section indicate that several multifractal parameters may be used to characterize the heterogeneity of the pore structure. However, the parameter $\Delta\alpha$ has the most sensitive response to the macro petrophysical parameters based on the correlations. Therefore, the discussion in this section is focused on this parameter.

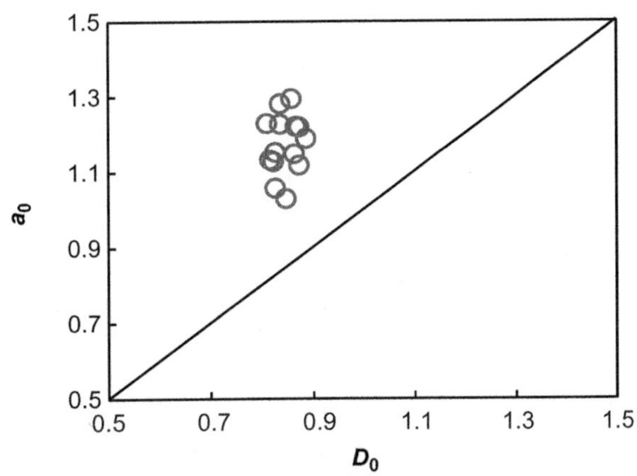

Fig. 6
The cross plot between α_0 and D_0.

The relationships between clay, quartz, feldspar, and carbonate contents vs $\Delta\alpha$ are plotted in Fig. 7. As can be observed from this figure, only the clay mineral contents are positively correlated with $\Delta\alpha$, whereas quartz, feldspar (k-feldspar + plagioclase), and carbonate contents show no meaningful relationship due to poor coefficients of determination (i.e., the square of the coefficient of correlation). The positive correlations between the clay contents and $\Delta\alpha$ demonstrates that the heterogeneity of the micropore structure of rocks becomes stronger as the clay content increases. This observation is in agreement with the previous studies of conventional and unconventional reservoirs concluding that the larger the clay content, the more complex the micropore structure [49]. However, it seems that the other minerals do not contribute to the heterogeneity or the homogeneity according to Fig. 7. According to SEM analysis results, the main pore types of the Lucaogou Formation are the dissolved pores, which can be subdivided into intragranular and intergranular pores. The intragranular dissolved pores frequently develop in albite, which is a type of feldspar, the intergranular dissolved pores often develop in dolomite (Fig. 8). The sizes of the dissolved pores are usually larger than that of

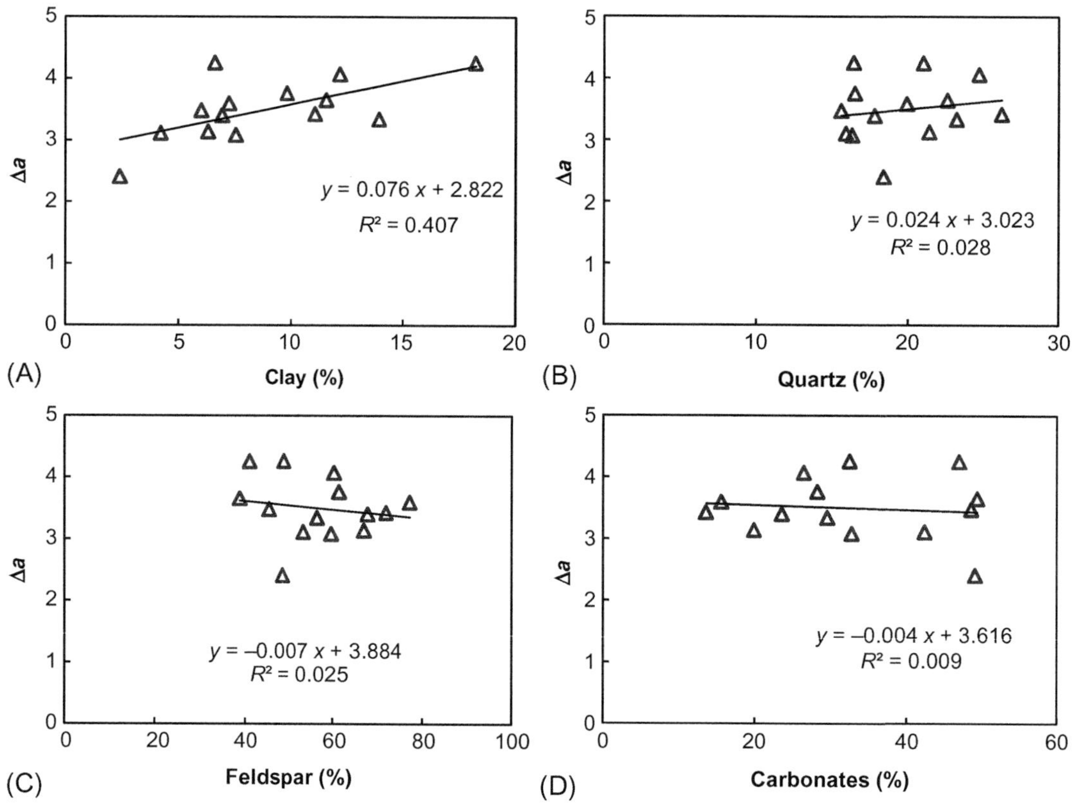

Fig. 7

The cross plots of clay (A), quartz (B), feldspar (C), and carbonates (D) vs $\Delta\alpha$. The clay content has an evident influence on $\Delta\alpha$.

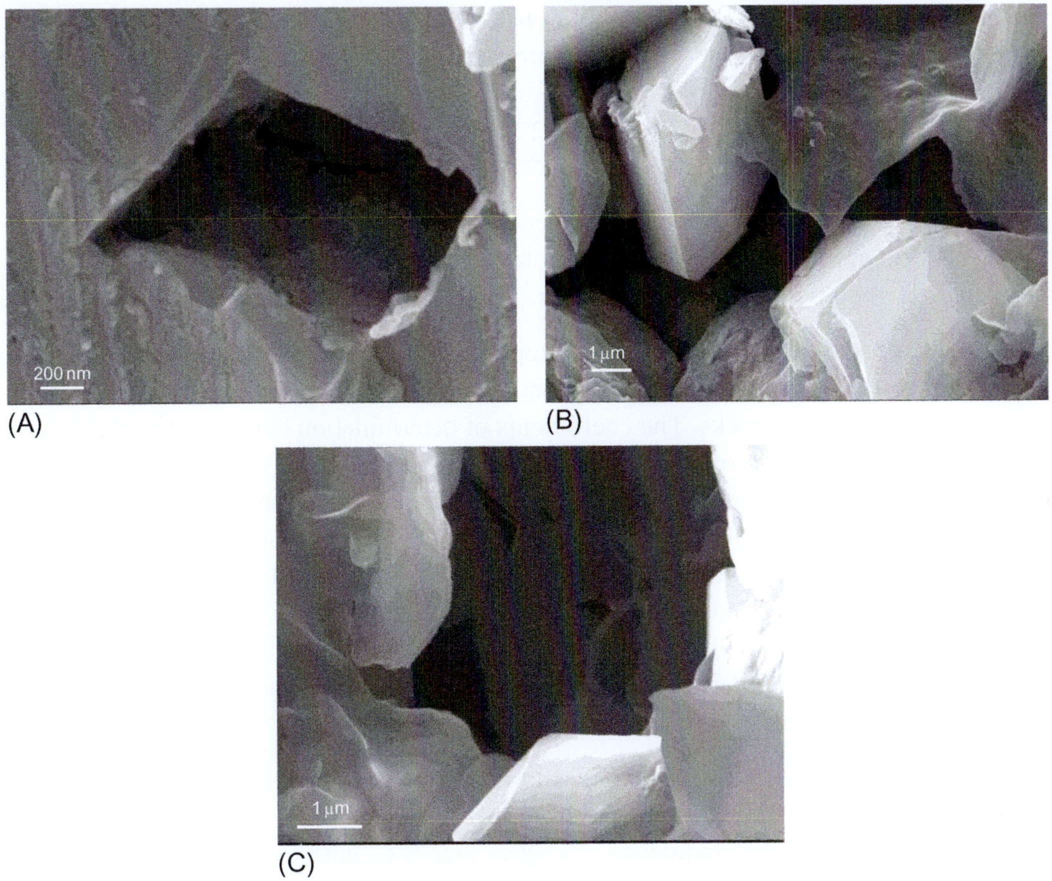

Fig. 8

SEM images of pores in the tight oil reservoirs of the Lucaogou Formation (A) intragranular dissolved pores in feldspar; (B) intergranular dissolved pores in potassium-feldspar and calcite; (C) intergranular dissolved pores in dolomite.

clay pores. Big pores may reduce the heterogeneity. Thus, the relationship between feldspar or carbonates and the multifractal parameters is slightly negative. Nevertheless, the weak correlation cannot ensure that the effect of the main pores on the heterogeneity or homogeneity can be quantified by those minerals.

7 Relationship Between Petrophysical and Multifractal Parameters

The coefficients of determination between the singularity strength range $\Delta\alpha$ and petrophysical properties, in terms of porosity, permeability, *RQI*, *FZI*, irreducible water saturation, irreducible water porosity, T_2 cutoff, T_{2lm}, T_{50}, and T_{35}, are presented in Table 4. As is observed from this table, the permeability, *RQI*, *FZI*, T_2 cutoff, T_{2lm}, T_{50}, and T_{35} correlate well with the

Table 4 The coefficients of determination between $\Delta\alpha$ and petrophysical parameters

ϕ	K	RQI	FZI	S_{wi}	ϕ_{wi}	T_2 Cutoff	T_{2lm}	T_{50}	T_{35}
0.252	0.523	0.629	0.652	0.161	0.061	0.478	0.496	0.539	0.552

singularity strength, compared to porosity, irreducible water saturation, and irreducible water porosity. This suggests that porosity, irreducible water saturation, and irreducible water porosity are not good heterogeneity indicators.

Fig. 9 shows the cross plots between $\Delta\alpha$ and petrophysical parameters. Permeability, RQI and FZI are negatively correlated with $\Delta\alpha$, indicating that the larger those values, the less heterogeneous (or more homogeneous) the rocks. The coefficients of determination between the RQI, FZI, and $\Delta\alpha$ are higher than those between permeability and $\Delta\alpha$, which can be understood easily as RQI and FZI incorporate the porosity and permeability and are representative parameters for the macro properties of reservoirs. In addition, the coefficients of determination between FZI and $\Delta\alpha$ are higher than those between RQI and $\Delta\alpha$. The parameter FZI was defined with the specific surface area per unit grain volume, pore shape factor, and pore tortuosity [43], all of which are good parameters for rock micropore characterization, which may be the reason why the coefficients of determination between FZI and $\Delta\alpha$ are higher. Therefore, FZI appears to be the best out of the three parameters for characterizing the heterogeneity.

As also seen in Fig. 9, the T_2 cutoff, T_{2lm}, T_{50}, and T_{35} are negatively correlated with the $\Delta\alpha$. The T_{2lm} is strongly related to the lithology and pore structure, where the large T_{2lm} means a good pore structure. Therefore, the singularity strength range, i.e., heterogeneity, decreases as the T_{2lm} increases. T_{50} and T_{35} are the relaxation times corresponding to the 35% and 50% saturation on the T_2 reverse accumulative curve. According to the mercury capillary theory, a low pressure indicates a big pore throat and good pore structure. Also, the T_2 reverse accumulative curve can be transformed into a mercury capillary curve, and the pressure is correlated with the reciprocal of T_2. Consequently, a large value of T_{50} and T_{35} would indicate a good pore structure and weak heterogeneity, which is reflected by the negative correlation between the two parameters and $\Delta\alpha$. The T_{35} shows a higher coefficient of determination than the other parameters obtained from NMR measurements, thus, T_{35} is the best heterogeneity indicator among the parameters associated with the T_2 spectrum. It is noted that the $\Delta\alpha$ decreases as the T_2 cutoff increases, which means that the increasing T_2 cutoff reduces the heterogeneity. It seems that the observation is opposite to the knowledge that the T_2 cutoff value is large when pore structure is poor. However, this theory is based on the assumption that T_{2lm} exhibits similar values, which is to say that the correlation between T_2 cutoff and pore structure may not be negative if the T_{2lm} values vary largely. In this work, the T_{2lm} values varied significantly and the T_2 cutoff and T_{2lm} are positively correlated in a random fashion (as shown in Fig. 10). This behavior may be unique to this study and does not represent other reservoir rocks, resulting in a negative correlation between T_2 cutoff values and $\Delta\alpha$.

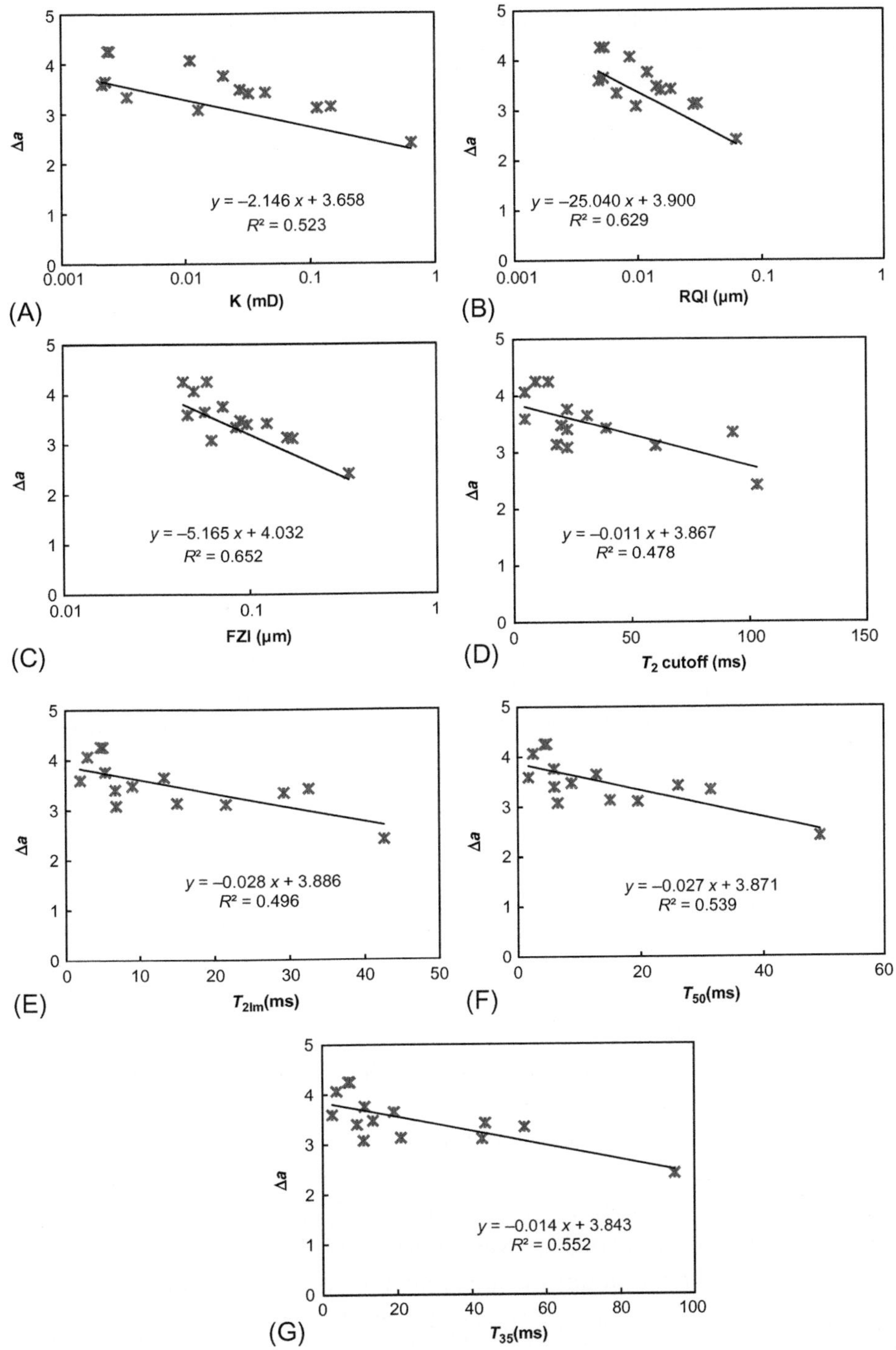

Fig. 9
The cross plots between $\Delta\alpha$ and petrophysical parameters.

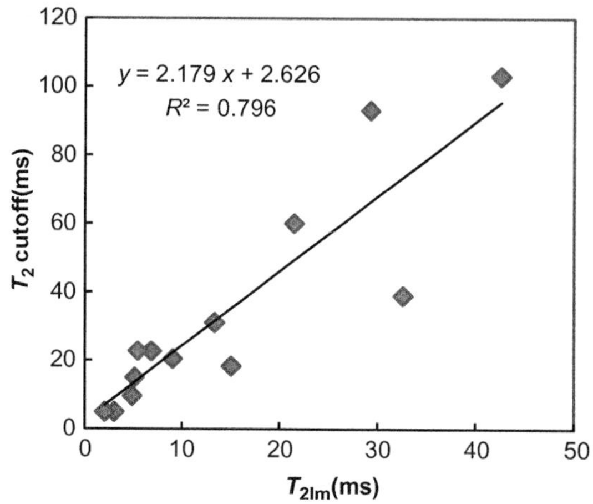

Fig. 10
The cross plot between T_{2lm} and T_2 cutoff.

8 Comparison With Sandstone Reservoirs

Table 5 shows the petrophysical and multifractal parameters of samples of sandstone reservoirs obtained from Ge et al. [29]. The helium porosity and permeability range from 12.26% to 24.36% and from 0.267 to 57.68 mD, respectively. The average for porosity and logarithmic mean for permeability are 17.40% and 2.391 mD, which are larger than that of tight oil samples. The averages for T_{2lm} and S_{wi} are, respectively, 23.921 ms, and 59.91%. Compared with the tight oil reservoir samples, the T_{2lm} is higher, and the S_{wi} is lower. It is noted that the echo time T_E for the sandstone samples was set as 0.1 ms [29], which is lower than the 0.3 ms for the tight oil reservoir samples. Since micropores develop in tight oil reservoirs, the T_{2lm} would be even lower if the T_E was set as 0.1 ms. Thus, when considering porosity and permeability characteristics, this comparison indicated that the pore structure of sandstone reservoirs is better than tight oil reservoirs.

The singularity strength range $\Delta\alpha$ values for sandstone are in the range 1.021–2.231 with an average value of 1.662. These values are lower than those of tight oil reservoir samples, revealing that the heterogeneity of tight oil reservoirs is stronger than that of sandstone reservoirs. Besides, it was mentioned previously that the T_2 cutoff should have a positive correlation with multifractal parameters, however, the observation in the tight oil reservoirs is contradictory because the T_2 cutoff and T_{2lm} are positively correlated. In this case, no correlation between the T_2 cutoff and T_{2lm} was observed (Fig. 11A). Thus, the T_2 cutoff did not have a negative correlation with multifractal parameters, as shown in Fig. 11B.

In summary, the heterogeneity characteristics are consistent with the petrophysical properties for the two types of reservoirs. In other words, good pore structure corresponds to weak heterogeneity, indicating that the multifractal analysis on low-field NMR measurements is effective and reliable.

Table 5 Petrophysical and multifractal parameters of sandstone samples

No.	ϕ-Helium (%)	K-Helium (mD)	T_{2lm} (ms)	T_2 Cutoff (ms)	S_{wi} (%)	D_{min}	D_{max}	D_0	D_1	D_2	Δa
1	19.27	17.975	28.157	37.02	48.72	1.96	0.79	0.907	0.875	0.853	1.669
2	13.58	6.413	5.754	7.243	58.82	1.886	0.904	0.922	0.906	0.912	1.827
3	14.76	2.319	8.941	50.137	79.72	2.468	0.907	0.924	0.908	0.908	1.646
4	13.92	0.267	8.097	59.258	89.11	2.626	0.909	0.983	0.903	0.903	1.077
5	14.82	0.572	10.493	29.804	72.45	1.834	0.897	0.94	0.904	0.9	1.021
6	15.61	0.775	8.902	19.02	64.59	1.855	0.916	0.997	0.908	0.905	1.85
7	15.77	0.833	9.061	18.482	65.1	1.75	0.914	0.924	0.905	0.906	1.452
8	19.88	2.571	34.196	56.268	56.95	2.264	0.78	0.911	0.861	0.83	1.736
9	23.21	30.2	49.241	48.708	41.23	2.483	0.802	0.911	0.869	0.848	1.951
10	16.88	0.578	27.704	69.327	66.25	2.36	0.729	0.922	0.832	0.794	1.895
11	18.36	1.86	41.259	46.434	44.1	2.526	0.8	0.911	0.872	0.852	2.005
12	12.26	0.445	8.397	24.671	71.55	1.963	0.928	0.958	0.919	0.917	1.243
13	20.89	0.492	33.43	65.172	57.4	2.591	0.744	0.922	0.836	0.801	2.131
14	17.36	17.24	34.707	37.931	44.07	2.19	0.801	0.911	0.882	0.862	1.641
15	24.36	57.68	50.475	48.209	38.56	2.31	0.782	0.911	0.87	0.844	1.785
Average	17.40	2.391	23.921	41.179	59.91	2.204	0.840	0.930	0.883	0.869	1.662

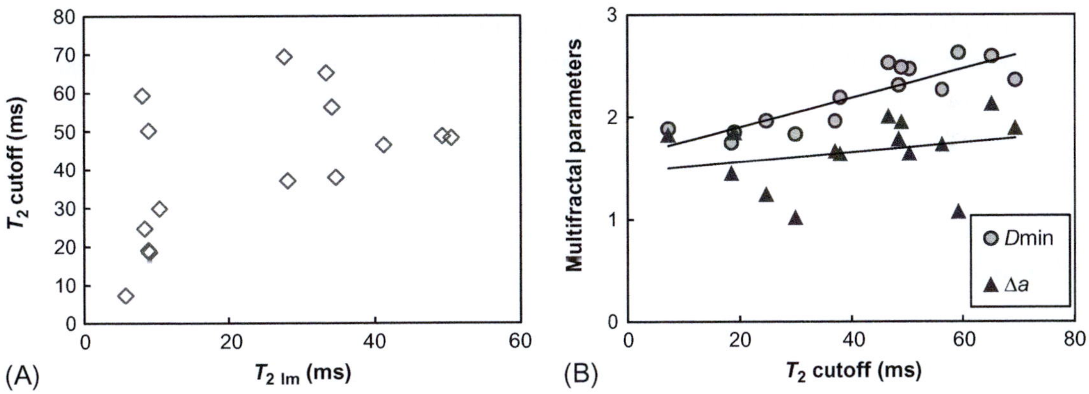

Fig. 11
(A) Cross plot between T_{2lm} and T_2 cutoff; (B) Cross plot between multifractal parameters and T_2 cutoff.

9 Conclusions

In this chapter, the pore structure characteristics of samples from the Permain Lucaogou Formation, Jimusaer Sag, Junggar Basin, were studied. The multifractal characteristics from NMR T_2 distributions were obtained and applied to analyze the heterogeneity of pore structures and the following conclusions were obtained:

(1) Combined with the petrophysical properties, the NMR T_2 distributions reveal that the pore structure of tight oil rocks is complex. Most of the T_2 distributions exhibit a unimodal distribution with no or a short relaxation time. The T_{2lm} values are in the range of 2.01–42.63 ms with an average of 14.09 ms, whereas the irreducible water saturation obtained from the NMR measurements exhibits high values in the range of 55.79% to 85.79% with an average of 73.64%.
(2) The multifractal analysis of the NMR T_2 distributions was conducted. The characteristics of generalized D_q, such as $D_0 > D_1 > D_2$, revealed the multifractal nature of pore structure of tight oil rocks. The singularity strength range $\Delta\alpha$ values are high, varying from 2.405 to 4.253 with an average value of 3.495. The data points deviated from a straight line in the plot of α_0 vs D_0. The large values of singularity strength, the differences in values between α_0 and D_0, revealed that the pore structures of samples were strongly heterogeneous.
(3) The clay contents had an obvious influence on $\Delta\alpha$, an average clay content of 8.85% was not the dominant mineral. The more dominant feldspar, quartz, and dolomite do not have obvious effects on the heterogeneity of rocks.
(4) The *FZI* and T_{35} are good indicators for heterogeneity of pore structure of rocks, as they are well correlated to $\Delta\alpha$. The small *FZI* values indicate a strong heterogeneity. In addition, the T_{35} has a negative linear relationship with the multifractal parameters.
(5) Through comparison with conventional sandstone reservoirs, the heterogeneity characteristics of rock samples were consistent with their petrophysical properties, indicating that the multifractal analysis on low-field NMR measurements is effective and reliable.

Acknowledgment

This work is supported by the National Natural Science Foundation of China (41722403; 41572116) and Fundamental Research Funds for the Central Universities (China University of Geosciences, Wuhan) (CUG170619).

References

[1] B. A. Miller, J. M. Paneitz, M. J. Mullen, et al., The successful application of a compartmental completion technique used to isolate multiple hydraulic-fracture treatments in horizontal Bakken Shale wells in North Dakota, SPE Annual Technical Conference and Exhibition, Denver, Colorado, USA, 21-24 September, 2008.
[2] C.Z. Jia, C.N. Zou, J.Z. Li, et al., Assessment criteria, main types, basic features and resource prospects of the tight oil in China, Acta Pet. Sin. 33 (2012) 343–350.
[3] C.N. Zou, R.K. Zhu, S.T. Wu, et al., Types, characteristics, genesis and prospects of conventional and unconventional hydrocarbon accumulations: taking tight oil and tight gas in China as an instance, Acta Pet. Sin. 33 (2012) 173–187.

[4] C. H. Sondergeld, K. E. Newsham, J. T. Comisky, et al., Petrophysical considerations in evaluating and producing shale gas resources, SPE Unconventional Gas Conference, Pittsburgh, Pennsylvania, USA, 23-25 February, 2010.
[5] P. Zhao, H. Ma, V. Rasouli, et al., An improved model for estimating the TOC in shale formations, Mar. Petrol. Geol. 83 (2017) 174–183.
[6] X. Ge, Y. Fan, J. Li, et al., Pore structure characterization and classification using multifractal theory—an application in Santanghu basin of western China, J. Petrol. Sci. Eng. 127 (2015) 297–304.
[7] K. Liu, M. Ostadhassan, Quantification of the microstructures of Bakken shale reservoirs using multi-fractal and lacunarity analysis, J. Nat. Gas Sci. Eng. 39 (2017) 62–71.
[8] G.R.L. Chalmers, R.M. Bustin, The organic matter distribution and methane capacity of the Lower Cretaceous Strata of northeastern British Columbia, Canada, Int. J. Coal Geol. 70 (2007) 223–239.
[9] Y. Xia, J. Cai, W. Wei, et al., A new method for calculating fractal dimensions of porous media based on pore size distribution, Fractals 26 (2018) 1850006.
[10] K. Jiao, S. Yao, C. Liu, et al., The characterization and quantitative analysis of nanopores in unconventional gas reservoirs utilizing FESEM-FIB and image processing: an example from the lower Silurian Longmaxi Shale, upper Yangtze region, China, Int. J. Coal Geol. 128 (2014) 1–11.
[11] R. Yang, S. He, J. Yi, et al., Nano-scale pore structure and fractal dimension of organic-rich Wufeng-Longmaxi shale from Jiaoshiba area, Sichuan Basin: invstigations using FE-SEM, gas adsorption and helium pycnometry, Mar. Petrol. Geol. 70 (2016) 27–45.
[12] S. Pan, C. Zou, Z. Yang, et al., Methods for shale gas play assessment: a comparison between Silurian Longmaxi shale and Mississippian Barnett shale, J. Earth Sci. 26 (2015) 285–294.
[13] J. Klaver, G. Desbois, R. Littke, et al., BIB-SEM pore characterization of mature and post mature Posidonia shale samples from the Hills area, Germany, Int. J. Coal Geol. 158 (2016) 78–89.
[14] F. Yang, Z. Ning, H. Liu, Fractal characteristics of shales from a shale gas reservoir in the Sichuan Basin, China, Fuel 115 (2014) 378–384.
[15] J. Hu, S. Tang, S. Zhang, Investigation of pore structure and fractal characteristics of the lower Silurian Longmaxi shales in western Hunan and Hubei Provinces in China, J. Nat. Gas Sci. Eng. 28 (2016) 522–535.
[16] H. Giesch, Mercury porosimetry: a general (practical) overview, Part. Part. Syst. Char. 23 (2006) 9–19.
[17] L.M. Novitz, D.R. Cole, Characterization and analysis of porosity and pore structures, Rev. Mineral. Geochem. 80 (2015) 61–164.
[18] L. Wang, Y. Fu, J. Li, et al., Experimental study on the wettability of Longmaxi gas shale from Jiaoshiba gas field, Sichuan Basin, China, J. Petrol. Sci. Eng. 151 (2017) 488–495.
[19] C. Li, M. Ostadhassan, T. Gentzis, et al., Nanomechanical characterization of organic matter in the Bakken formation by microscopy-based method, Mar. Petrol. Geol. 96 (2018) 128–138.
[20] Z. Ouyang, D. Liu, Y. Cai, et al., Fractal analysis on heterogeneity of pore-fractures in middle-high rank coals with NMR, Energ. Fuel. 30 (2016) 5449–5458.
[21] G.R. Coates, L.Z. Xiao, M.G. Prammer, NMR Logging Principles and Applications, Gulf Publishing Company, Texas, 1999.
[22] H. Daigle, A. Johnson, Combining mercury intrusion and nuclear magnetic resonance measurements using percolation theory, Transp. Porous Media. 111 (2016) 669–679.
[23] B.B. Mandelbrot, The Fractal Geometry of Nature, WH Freeman, New York, 1982.
[24] B.B. Mandelbrot, Multifractal measures, especially for the geophysicist, Pure Appl. Geophys. 131 (1989) 5–42.
[25] J. Cai, L. Luo, R. Ye, et al., Recent advances on fractal modeling of permeability for fibrous porous media, Fractals 23 (2015) 1540006.
[26] X. Ge, Y. Fan, S. Deng, et al., An improvement of the fractal theory and its application in pore structure evaluation and permeability estimation, J. Geophys. Res. Solid Earth 121 (2016) 6333–6345.
[27] L. Li, L. Chang, S. Le, et al., Multifractal analysis and lacunarity analysis: a promising method for the automated assessment of muskmelon (Cucumismelo L.) epidermis netting, Comput. Electron. Agr. 88 (2012) 72–84.

[28] J.P. Ferreiro, E.V. Vazquez, Multifractal analysis of Hg pore size distributions in soils with contrasting structural stability, Geoderma 160 (2010) 64–73.
[29] X. Ge, Y. Fan, X. Zhu, et al., Determination of nuclear magnetic resonance T_2 cutoff value based on multifractal theory—an application in sandstone with complex pore structure, Geophysics 80 (2015) D11–D21.
[30] X. Chen, G. Yao, J. Cai, et al., Fractal and multifractal analysis of different hydraulic flow units based on micro-CT images, J. Nat. Gas Sci. Eng. 48 (2017) 145–156.
[31] T. Zhang, G. Fan, Y. Li, et al., Pore-throat characterization of tight oil reservoir in the Lucaogou formation, Jimusar sag (in Chinese with English abstract), CT Theor. Appl. 25 (2016) 425–434.
[32] L.C. Kuang, W.X. Hu, X.L. Wang, et al., Research of the tight oil reservoir in the Lucaogou Formation in Jimusar Sag: analysis of lithology and porosity characteristics (in Chinese with English abstract), Geol. J. Chin. Univ. 19 (2013) 529–535.
[33] P. Zhao, Z. Sun, X. Luo, et al., Study on the response mechanisms of nuclear magnetic resonance (NMR) log in tight oil reservoirs (In Chinese with English abstract), Chinese, J. Geophys. 29 (2016) 1927–1937.
[34] Y. Su, M. Zha, X. Ding, et al., Pore type and pore size distribution of tight reservoirs in the Permian Lucaogou formation of the Jimsar Sag, Junggar Basin, NW China, Mar. Petrol. Geol. 89 (2017) 761–774.
[35] T.C. Halsey, M.H. Hensen, L.P. Kadanoff, et al., Fractal measures and their singularities: the characterization of strange sets, Phys. Rev. A 33 (1986) 1141–1151.
[36] C.G. Gutierrez, F.S. Jose, Multifractal analysis of soil micro and macroporosity using digital images obtained with fluorescent dye, Geophys. Res. Abstr. 8 (2006) 11094.
[37] A. Chhabra, R.V. Jensen, Direction of determination of the f(a) singularity spectrum, Phys. Rev. Lett. 62 (1989) 1327–1330.
[38] Y. Volokitin, W. Looyestijn, Constructing capillary pressure curve from NMR log data in the presence of hydrocarbons, in: SPWLA 40th Annual Logging Symposium, Oslo, Norway, 30 May-3 June, 1999.
[39] Y. Volokitin, W. Looyestijn, A practical approach to obtain primary drainage capillary pressure curves from NMR core and log data, Petrophysics 42 (2001) 334–343.
[40] Z. Cao, G. Liu, B. Xiang, et al., Geochemical characteristics of crude oil from a tight oil reservoir in the Lucaogou formation, Jimusar Sag, Junggar, Basin, AAPG Bull. 101 (2017) 39–72.
[41] Z. Liu, T.J. Algeo, X. Guo, et al., Paleo-environmental cyclicity in the early Silurian Yangtze Sea (South China): tectonic or glacio-eustatic control? Palaeogeogr. Palaeocl. 466 (2017) 59–76.
[42] J.O. Amaefule, M. Altunbay, D. Tiab, et al., Enhanced reservoir description: using core and log data to identify hydraulic (flow) units and predict permeability in uncored intervals/wells, in: SPE Annual Technical Conference and Exhibition, Houston, Texas, 3-6 October, 1993.
[43] M. Abbaszadeh, H. Fujii, F. Fujimoto, Permeability prediction by hydraulic flow units—theory and applications, SPE Form. Eva. 11 (1996) 263–271.
[44] M.R. Rezaee, A. Jafari, E. Kazemzadeh, Relationships between permeability, porosity and pore throat size in carbonate rocks using regression analysis and neural networks, J. Geophys. Eng. 3 (2006) 370–376.
[45] Z. Mao, L. Xiao, Z. Wang, et al., Estimation of permeability by integrating nuclear magnetic resonance (NMR) logs with mercury injection capillary pressure (MICP) data in tight gas sands, Appl. Magn. Reson. 44 (2013) 449–468.
[46] R. Rezaee, A. Saeedi, B. Clennell, Tight gas sands permeability estimation from mercury injection capillary pressure and nuclear magnetic resonance data, J. Petrol. Sci. Eng. 88 (2012) 92–99.
[47] T.S. Ramakrishnan, E.J. Fordham, L. Venkataramanan, et al., New interpretation methodology based on forward models for magnetic resonance in carbonates, in: SPWLA 40th Annual Logging Symposium, Oslo, Norway, 30 May-3 June, 1999.
[48] L. Zhou, Z. Kang, Fractal characterization of pores in shales using NMR: a case study from the lower Cambrian Niutitang formation in the Middle Yangtze Platform, Southwest China, J. Nat. Gas Sci. Eng. 35 (2016) 860–872.
[49] Y. Shi, L. Xiao, Z. Mao, et al., An identification method for diagenetic facies with well logs and its geological significance in low-permeability sandstones: a case study on Chang 8 reservoirs in the Jiyuan region, Ordos Basin, Acta Pet. Sin. 32 (2011) 820–828.

CHAPTER 5

Investigation and Quantitative Evaluation of Organic-Related Pores in Unconventional Reservoirs

Xinmin Ge[*,†], Yiren Fan[*,†], Jianchao Cai[‡], Jier Zhao[*]

[*]*School of Geosciences, China University of Petroleum, Qingdao, Shandong, P.R. China,* [†]*CNPC Key Well Logging Laboratory in China University of Petroleum, Qingdao, Shandong, P.R. China,* [‡]*Hubei Subsurface Multi-scale Imaging Key Laboratory, Institute of Geophysics and Geomatics, China University of Geosciences, Wuhan, P.R. China*

Chapter Outline

1 Introduction

With the rapid increase in demand of fossil energies, unconventional resources, such as tight oil and shale gas, are gaining more and more attention in recent decades. These reservoirs are either a near source accumulation or self-generating and self-preserving. Organic matter plays an important role in unconventional reservoirs since the type and content of organic carbon often controls the reservoir quality and the hydrocarbon potential [1–4]. There has been

Petrophysical Characterization and Fluids Transport in Unconventional Reservoirs. https://doi.org/10.1016/B978-0-12-816698-7.00005-X

increased research to estimate the total organic carbon (TOC) using petrophysical and well logging data [5–11]. However, few reports exist on the characterization and quantitative evaluation of organic-related pores using petrophysical methods.

Generally, the organic-related pores are a result of decomposition of organic material during the hydrocarbon generation process. The diameter of organic-related pore ranges from several to hundreds of nanometers, and can be filled by hydrocarbons [9, 12, 13]. The organic-related porosity, its pore structure and connectivity, can significantly influence the resource storage capacity and mobility. Hence, a comprehensive investigation and evaluation for the organic-related porosity is necessary in the formation evaluation.

In this chapter, we introduce an integrated method to quantify the organic-related pores combining petrophysical, geochemical, and petrographic experiments. We first characterize the organic-related pores using the low-field nuclear magnetic resonance (NMR), scanning electron microscopy (SEM), X-ray diffraction (XRD), and energy dispersive X-ray spectrometer (EDS), and then quantify the organic-related porosity with the help of other data. We investigate relationships between the organic-related porosity and geochemical parameters including TOC, absorbed free gaseous hydrocarbons (S_0), absorbed free liquid hydrocarbons (S_1), residual petroleum potential (S_2), residual carbon (RC), and vitrinite reflectance (Ro) to study the influential factors of the organic-related porosity. At last, we study relationships between the organic-related porosity and petrophysical responses to establish a predicting model for the organic-related porosity.

2 Materials and Methods

2.1 Geological Settings

We performed the experiment on 26 samples drilled from the Luchaogou Formation of Malang Sag in the Santanghu Basin, a typical self-generating and self-preserving shale reservoir. The Santanghu Basin is located in northeastern Xinjiang, China. It is a small intermountain basin sandwiched between Dahafutike-Suhaitu and Moqin Ural Mountains and it covers a total area of $2.3 \times 10^4\,\text{km}^2$. Its first-order structural units can be divided into northern uplift zone, central depression zone, and southern thrust belt. Malang Sag is the main area for oil accumulation in the central depression zone. The sedimentary strata of this sag include Carboniferous, Permian, Triassic, Jurassic, Cretaceous, Paleogene, Neogene, and Quaternary. The Lucaogou Formation of the Middle Permian is the major oil bed in this region, with complex lithology and mineral distribution. Carbonate, clastic, and sedimentary tuffs are unevenly distributed due to unstable sedimentary environments [14]. The location of the Santanghu Basin and Malang Sag and its tectonic units are shown in Fig. 1.

All samples were first analyzed for natural gamma ray and gammy ray spectrums, and divided into three groups: the first group for conventional petrophysical measurements, the second group for organic geochemical tests, and the third group for SEM, EDS, and XRD analysis.

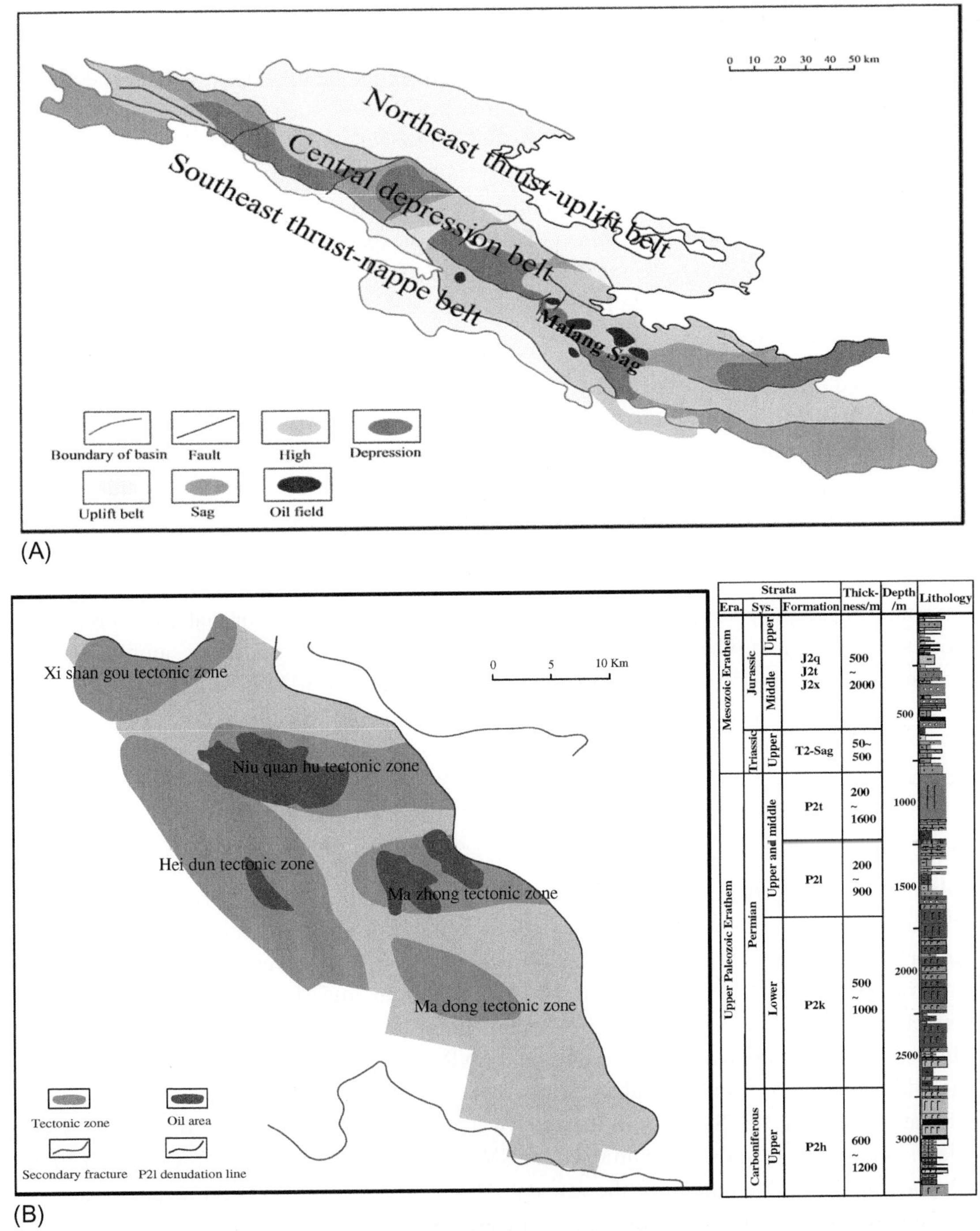

Fig. 1

Geological settings of Santanghu Basin and Malang Sag. (A) Location of Santanghu Basin and Malang Sag and (B) Distribution of tectonic units and the section of Malang Sag.

2.2 Low-Field NMR Test

Low-field NMR has been developed as a sophisticated technique in geophysical prospecting. It can be classified into two categories: (1) one-dimensional NMR such as longitudinal relaxation (T_1) and transversal relaxation (T_2) spectrum and (2) two-dimensional NMR including transversal-diffusion spectrum ($T_2 - D$) and longitudinal-transversal spectrum ($T_1 - T_2$). Although great achievements have been published on characterizing the tight oil and shale gas using two-dimensional NMR spectrums [15–19], the T_2 spectrum is still the most popular technique and serves as a significant method in reservoir exploration [20–22]. Thus, it is essential to determine the valuable information in T_2 spectrum, in order to characterize the unconventional reservoir.

We conducted the NMR T_2 experiments using the MARAN-II ultra-rock spectrometer (Oxford Instrument Incorporation) with a main frequency of approximately 2 MHz. The pulse sequence used was the Carr-Purcell-Meiboom-Gill (CPMG), and the measured data was inversed to T_2 spectrums using the Butler-Reeds-Dawson (BRD) algorithm [23]. Samples were polished to cylinder shapes with a diameter of 2.54 cm and length between 3 and 4 cm, and then heated at a temperature of 100°C to remove the drilling muds and most of the free water, capillary bound water, and light hydrocarbon residing in large pores. NMR signals of this state were measured, representing responses of clay bound water, monolayer water on mineral surfaces, bitumen, and other high viscous hydrocarbons. Next, we put these samples into an auto-saturator container for 72 h, under a confining pressure of 60 MPa to ensure most pores were fully saturated with water, and then measured the NMR signals. And lastly, we used a high velocity centrifugal machine to displace the movable water from the pores, and measure the NMR signals. In summary, each sample was measured at three different states: (1) dry state, (2) full water saturation state, and (3) irreducible water saturation state.

Compared with conventional reservoirs, NMR signals of unconventional reservoirs are weak and difficult to detect. Acquisition parameters should be designed properly, to ensure the quality of NMR data. After many trials, the waiting time (T_W), the echo time (T_E), and the number of echoes (NECH) were set as 6000 ms, 0.1 ms, and 4096, respectively. In addition, the number of scans (NS) per acquisition is also important since it determines the signal to noise ratio (SNR). Five samples under different saturation states were selected randomly to investigate the relationship between NS and SNR (SNR is performed using the software of RINMR version 5.7 automatically). As is shown in Fig. 2, SNR increased nonlinearly with the increase of NS. Samples in the fully water-saturated state reached the largest SNR under a certain NS. Each sample attained an SNR higher than 30 when NS was larger than 512, showing favorable quality of the raw data. Fig. 3 shows the inversion results of a representative water-saturated sample, in a different number of scans. It is observed that there were no obvious variations when the NS was higher than 512. Therefore, the optimal number of NS was chosen as 512 for the NMR measurements.

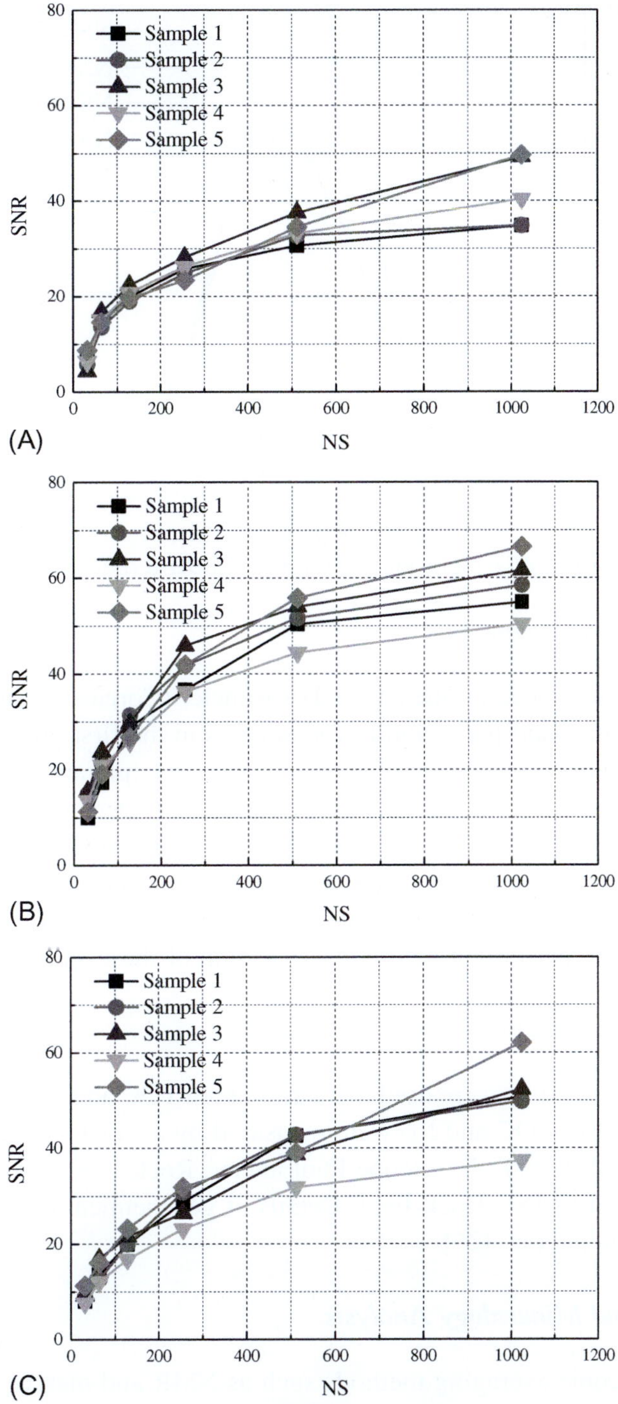

Fig. 2

Relationship between SNR and NS for five randomly selected samples. (A) Dry state, (B) fully water-saturated state, and (C) irreducible water saturation state.

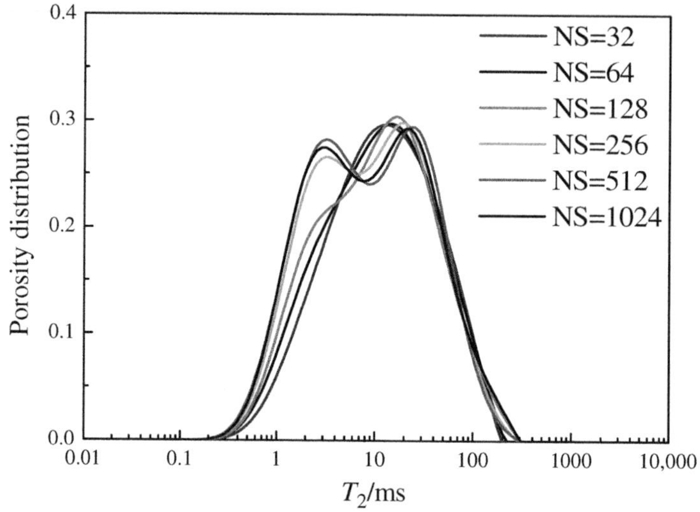

Fig. 3
Inversed T_2 spectrums of a representative sample under a different number of scans.

2.3 Petrophysical Measurements

We used the AP-608 automated Permeameter-Porosimeter (Coretest System Incorporation) to attain the helium porosity and permeability for samples in the first group [24]. Experiments were performed at room temperature and pressure. The principle and flowchart of these measurements were elaborated by Tiab and Donaldson [25]. Density (DEN), resistivity (R), compression velocity (V_p), and shear velocity (V_s) were recorded simultaneously. In addition, natural gamma intensity and the nature gamma ray spectrums were measured to obtain the total natural gamma intensity (GR) and contents of potassium (K), thorium (Th), and uranium (U). These experiments were performed before classification of these samples.

2.4 Geochemical Experiments

Samples from the second group were used to measure both TOC and pyrolysis parameters such as S_0, S_1, S_2, RC, and Ro. TOC and Ro were measured by a LECO CS-200 carbon-sulfur analyzer and TIDAS MSP 200 Microscope Photometer. Rock-Eval pyrolysis measurements were conducted by DELSI INC Rock-Eval. Details of geochemical measurements were referenced from Espitalié et al. [26].

2.5 Petrographic and Mineralogy Analysis

Compared to macroscopic averaging methods such as NMR and mercury injection capillary pressure (MICP), SEM can characterize the microstructure with pore sizes in the nanometer range, and can capture the image of the individual pore directly [27, 28].

We applied the high-resolution backscatter electron (BSE) model of FEI Quanta 250 FEG to collect the images. The BSE detector allows for better contrast due to the high-energy electrons being back scattered by elastic scattering interactions with specimen atoms. All samples were viewed on surfaces oriented perpendicular to the bedding. The lowest and the highest intensity values were colored black and white in the images, whereas the intermediate intensity values were represented by various shades of gray. Therefore, components with different atomic numbers can be identified from the images. Pores developed in organic matter with the lowest atomic number are located in the darkest area of the image. Meanwhile, EDS involving the generation of an X-ray spectrum from the entire scan area of the SEM was captured and analyzed quantitatively to provide specific data on elemental composition. We do not emphasize the extensive principles and experimental procedures for SEM and EDS analysis in this chapter. Detailed information is elaborated on in Goldstein et al. [29] and Legg [30].

After SEM and EDS analysis, we used the sophisticated XRD technique (Rigaku D/max-2600/pc) to quantify the mineral composition and their contents of the third group of samples. Table 1 and Fig. 4 list weight percentages of minerals in the given samples, and their mean values. The mineralogy of samples can vary considerably, but the main minerals were quartz, dolomite, and plagioclase. Clay only accounted for a small portion of these samples, with a mean value of 0.82%.

3 Quantification of the Organic-Related Pores

Fig. 5 shows the BSE images of a representative sample at different magnifications. We observed dissolution pores, micropores, and organic-related pores from the image with a magnification of 4000, and organic micropores from the image with a magnification of 6000. Proportions of elements picked by EDS are shown in Table 2. These organic-related pores and organic micropores were confirmed by the EDS analysis since most signals were contributed to by the carbon (C). With the help of EDS and digital analysis of the BSE images, the organic-related porosity for each sample was conveniently quantified.

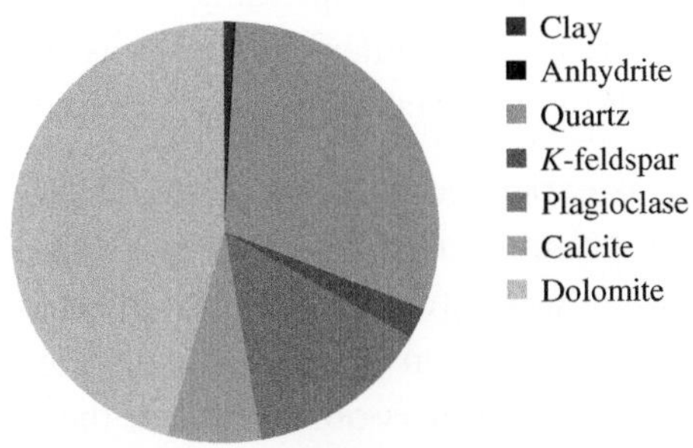

Fig. 4
Pie chart of mineral composition of the given samples.

Table 1 Mineral constitution of the measured samples by XRD

Sample No.	Mineralogical Content (weight %)						
	Clay	Anhydrite	Quartz	K-feldspar	Plagioclase	Calcite	Dolomite
1	0.90	1.00	16.30	0.00	30.10	0.00	51.70
2	0.85	0.00	10.25	0.00	8.10	0.00	80.80
3	0.87	0.00	8.33	2.40	11.00	0.00	77.40
4	0.70	1.10	24.50	0.00	24.10	0.00	49.60
5	1.20	0.00	19.70	0.00	14.00	0.00	65.10
6	0.90	0.00	26.20	0.00	17.00	0.00	55.90
7	1.30	0.00	41.80	0.00	9.10	0.00	47.80
8	0.88	0.00	16.62	0.00	17.40	1.70	63.40
9	0.90	0.00	25.90	5.30	5.80	0.00	62.10
10	0.60	0.00	25.30	4.10	0.00	0.00	70.00
11	0.70	0.00	23.90	7.10	0.00	4.10	64.20
12	0.50	0.00	54.50	3.50	5.40	8.70	27.40
13	0.60	0.00	15.20	0.90	2.00	0.00	81.30
14	0.65	0.00	23.05	2.60	7.00	4.60	62.10
15	0.80	0.00	65.30	3.20	6.30	13.90	10.50
16	0.90	0.00	45.00	3.80	22.20	13.50	14.60
17	0.70	0.00	23.50	0.00	15.50	23.10	37.20
18	0.80	0.00	38.00	6.00	13.80	16.80	24.60
19	0.70	0.00	30.90	8.10	24.10	17.70	18.50
20	0.80	1.30	24.90	0.00	31.10	11.50	30.40
21	0.60	0.00	32.50	0.00	22.20	2.10	42.60
22	0.90	0.00	21.20	0.00	26.00	8.20	43.70
23	0.50	0.00	45.20	7.70	8.00	22.90	15.70
24	1.20	0.00	56.20	0.00	12.80	12.80	17.00
25	0.60	0.00	16.60	2.30	7.30	5.50	67.70
26	1.20	0.00	53.30	4.70	18.90	17.30	4.60

Fig. 6A shows an NMR T_2 decay curve for the representative sample with the same depth under different states. It was observed that the decaying rate was slower and the signal quality was higher when the sample was fully saturated with water. However, the decaying rate became faster when the moveable water was displaced. Oscillations in decay curves were observed for the sample in the dry and irreducible state. Fig. 6B depicts corresponding T_2 spectrums under different saturation states. The spectrum of the dry state is an irregular bimodal-peak distributed with T_2 ranging from 0.1 to 30ms, which is similar to results conducted by Nicot et al. [31] and Kausik et al. [32].

The amplitude and range of T_2 is expanded with the injection of water. The main peak moves right to about 10ms, revealing that intergranular pores of a larger size are occupied with moveable water. After undergoing the centrifugal process, most of the movable water is driven out, leading to a decrease of long relaxation signals. Meanwhile, the main peak moves left to the fast relaxation range (about 3ms). It is observed that the left part of the T_2 spectrum remains unaltered during the centrifugal process, revealing signals of irreducible water.
In addition, the maximal and minimal value of the T_2 spectrum changed between the dry state and the irreducible state, as is shown in Fig. 6B.

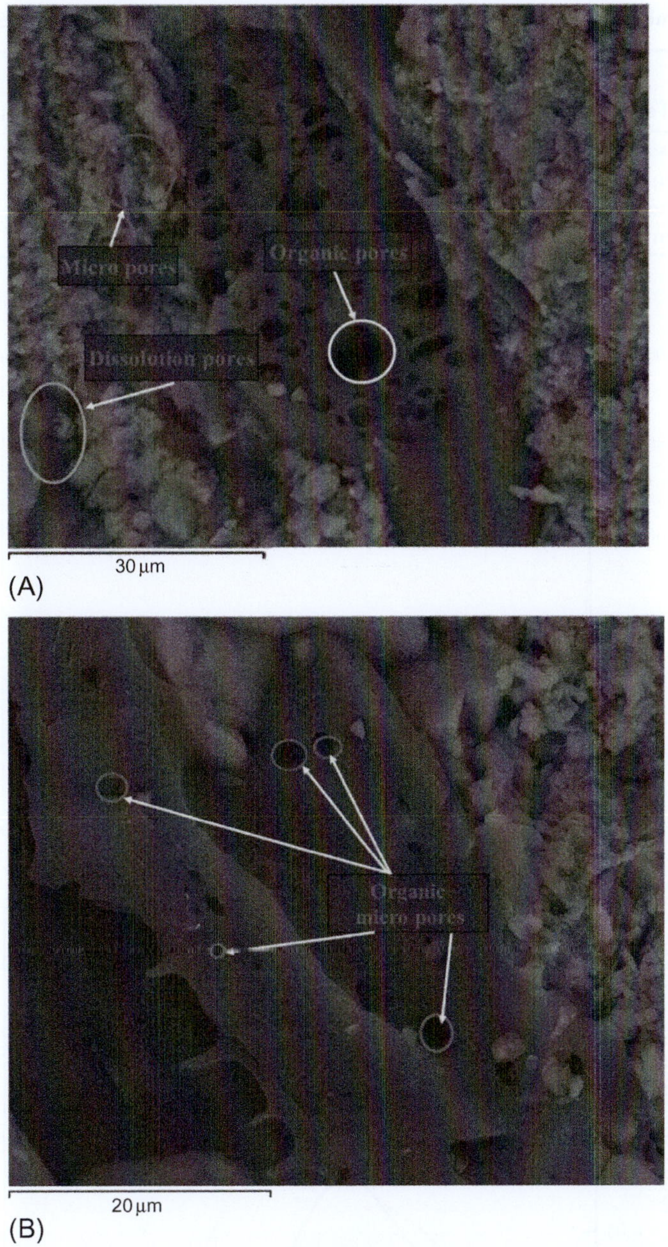

Fig. 5
(A) BSE image with 4000 magnification and (B) BSE image with 6000 magnification.

As mentioned above, most fluids including free water, capillary bound water, drilling mud, and light hydrocarbons are evaporated during the heating process. Thus, the remaining components that contain hydrogen nuclei are mainly composed of bitumen, a monolayer of water on mineral surfaces, and clay bound water. Conventionally, monolayer water on mineral surfaces relaxes too fast to be measured by low-field NMR devices. Signals can be

Table 2 Relative proportions of elements quantified by EDS

Number	Test Result	Elements			Aurum
		Carbon	Oxygen	Silicon	
1	Mass fraction (%)	91.84	4.78	0	3.39
	Atom fraction (%)	96.04	3.75	0	0.22
2	Mass fraction (%)	92.84	0	1.12	6.04
	Atom fraction (%)	99.1	0	0.51	0.39

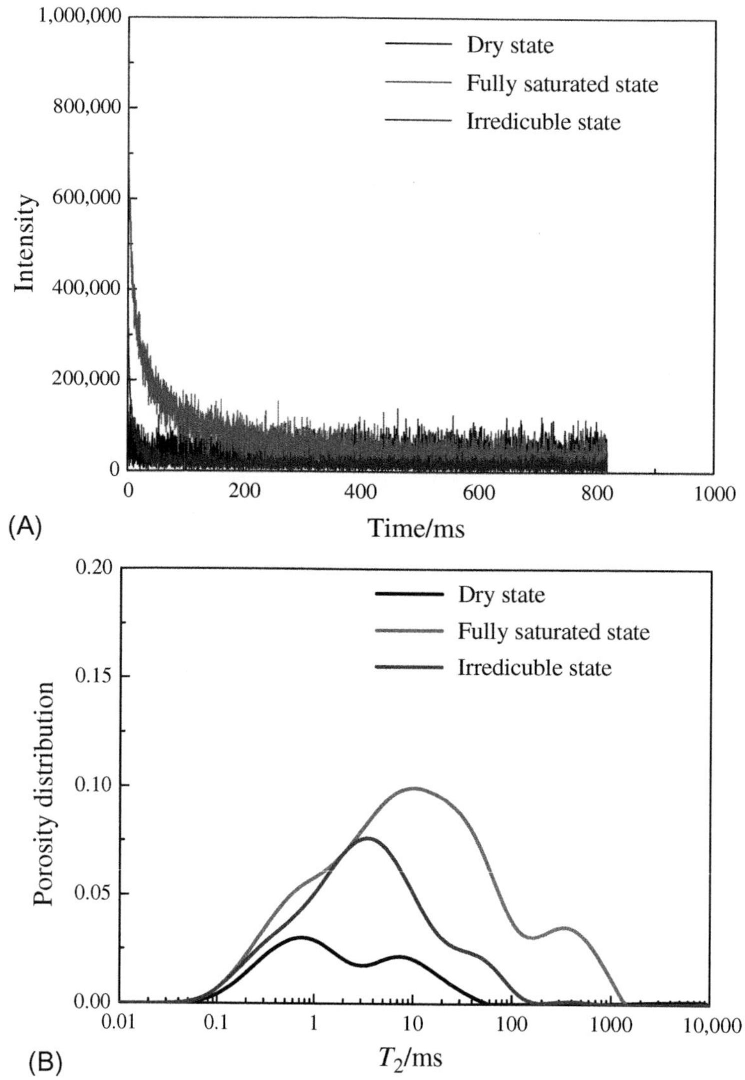

Fig. 6
Measured T_2 decay curves and spectrums of a dolomitic shale under different states. (A) T_2 decay curves and (B) T_2 spectrums.

attributed to the organic-related substances such as bitumen and clay bound water. As is seen from Table 2 and Fig. 4, all of these samples were clean and had a relatively low clay count (less than 1%). The NMR signals of clay bound water are weak and were neglected. With this assumption, we can deduce that NMR signals of samples in the dry state are contributed to by organic-related substances.

From the comparison of T_2 spectrums between dry and irreducible water saturation state, NMR signals of organic-related pores are overlapped with that of capillary bound water. Signals with T_2 smaller than 0.3 ms are almost organic related, since no obvious variations are observed in the three different states. Nevertheless, signals with T_2 located between 0.3 and 30 ms are joint contributions of organic-related substances and capillary bound water, and cannot be easily separated. This proves that the conventional cutoff method used to discriminate the movable and irreducible components from the NMR T_2 spectrum may be invalid when used with unconventional reservoirs [33]. The predicted irreducible water saturation may be higher than the actual value if the organic-related signals are not extracted properly, which could cause potential reservoirs to be missed.

Using the relationship between amplitude and porosity of standard samples [34], we can obtain the apparent porosity at the three different states (dry state, the irreducible water sate, and the fully water-saturated state), which is 1.02%, 2.48%, and 4.55%, respectively. The porosity difference between the fully water-saturated state and the dry state (3.53%) is equivalent to the measured helium porosity (3.6%). Similar results were attained for the remaining samples, as shown in Fig. 7A. These results further explain that NMR relaxation signals in the dry state are mostly organic related.

Considering influences of organic-related substances, the NMR based irreducible water saturation for unconventional resources is modified as,

$$Swi_{NMR} = \frac{\phi_{NMR-irreducible} - \phi_{NMR-organic}}{\phi_{NMR-saturated} - \phi_{NMR-organic}} \times 100\%, \tag{1}$$

where $\phi_{NMR-saturated}$ is the calibrated porosity measured of a sample in a fully water-saturated state, $\phi_{NMR-irreducible}$ is the calibrated porosity of a sample in an irreducible state, and $\phi_{NMR-organic}$ is the calibrated porosity of a sample in a dry state.

Fig. 7B shows the relationship between the irreducible water saturation obtained by Eq. (1) and that of centrifugal experiments. It is seen that the modified irreducible water saturation from NMR data bears good correlation with the conventional centrifugal irreducible water saturation. The comparison supports the theory that the NMR signal of the dry state is mostly contributed to by the organic components, not the clay bound water.

Therefore, the calibrated NMR porosity of samples measured at the dry state is defined as the organic-related porosity. Fig. 7C compares the organic-related porosity of given samples obtained by NMR and SEM-EDS analysis, respectively. It is noticeable that the organic-related porosity obtained by low-field NMR presents a good fit with the organic-related porosity extracted by SEM-EDS analysis.

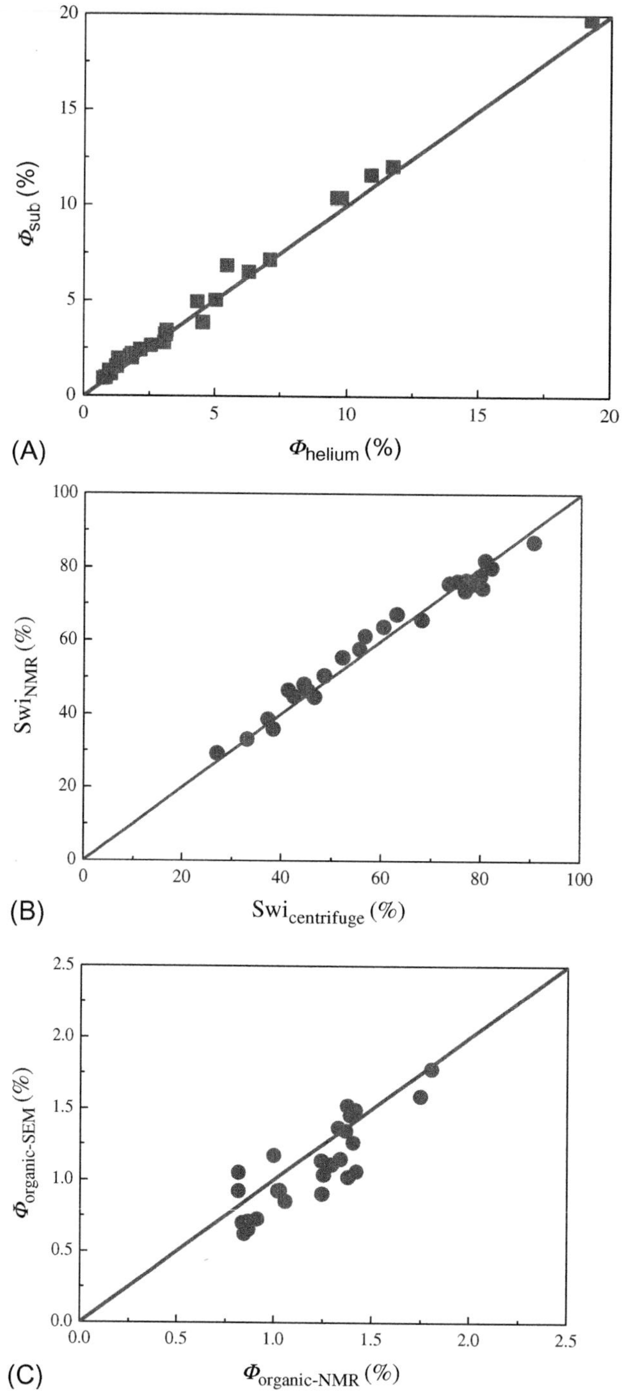

Fig. 7

Comparisons of effective porosity, irreducible water saturation, and organic-related porosity by different methods. (A) Comparison between helium and NMR porosity, (B) comparison between irreducible water saturation of different methods, and (C) comparison between NMR and SEM organic-related porosity.

4 Relationship With Geochemical Parameters

The relationships between the organic-related porosity and geochemical parameters are shown in Fig. 8. The organic-related porosity is positively correlated with TOC, S_0, S_1, S_2, S_1+S_2, and $S_0+S_1+S_2$. NMR measurements are contributed to by high viscous hydrocarbons such as bitumen and kerogen, and the measured S_0, S_1, and S_2 are also positively correlated with the amount of hydrocarbons. TOC describes the quantity of organic carbon including both kerogen and bitumen, and TOC has a good linear relationship with organic-related porosity. Ro reflects the thermal maturity of sedimentary rocks and how it may influence kerogen types and oil-generating windows, but it cannot be used to characterize the hydrocarbon potential. Hence, it is reasonable that the organic-related porosity shows no dependence upon Ro. Fig. 8H depicts the relationship between the organic-related porosity and RC. It is observed that the organic-related porosity is inversely proportional to RC. RC indicates the remaining organic matter after pyrolysis, encountered as background material in sedimentary rocks. Generally, it is an unfavorable factor for the development of organic-related pores, and is negatively correlated with the organic-related porosity.

Table 3 lists the statistical results between the organic-related porosity and the above geochemical parameters, including the residual sum of squares (RSS) and adjusted R squares (Adj.R^2). The organic-related porosity bears the best linear relationship with TOC, then S_1+S_2, and $S_0+S_1+S_2$. Therefore, we can conclude that the organic-related pores may have some similar properties to organic matter.

5 Prediction Using Petrophysical Data

It is vital to investigate the petrophysical responses of organic-related pores since these parameters can be measured directly by well logs. It will be convenient to predict the in-situ organic-related porosity if the relationship between the organic-related porosity and conventional well logging or petrophysical data is established by core calibration.

Fig. 9A–D shows the relationship between the organic-related porosity and radioactive parameters. It is seen that the organic-related porosity has a good correlation with thorium (Th) content and total gamma ray intensity (GR), but a weak correlation with potassium (K) content and uranium (U) content. High K content may be caused by the presence of potassium feldspars or micas, and U is not only naturally found in organic matter, but also affected by the depositional environment. Inversely, Th content is simply correlated with organic-related porosity for samples with low clay content. Hence, the coefficient between the organic-related porosity and GR is lower than that of Th. The results show that U and K are unsuitable to indicate the volume of organic-related pores in this region.

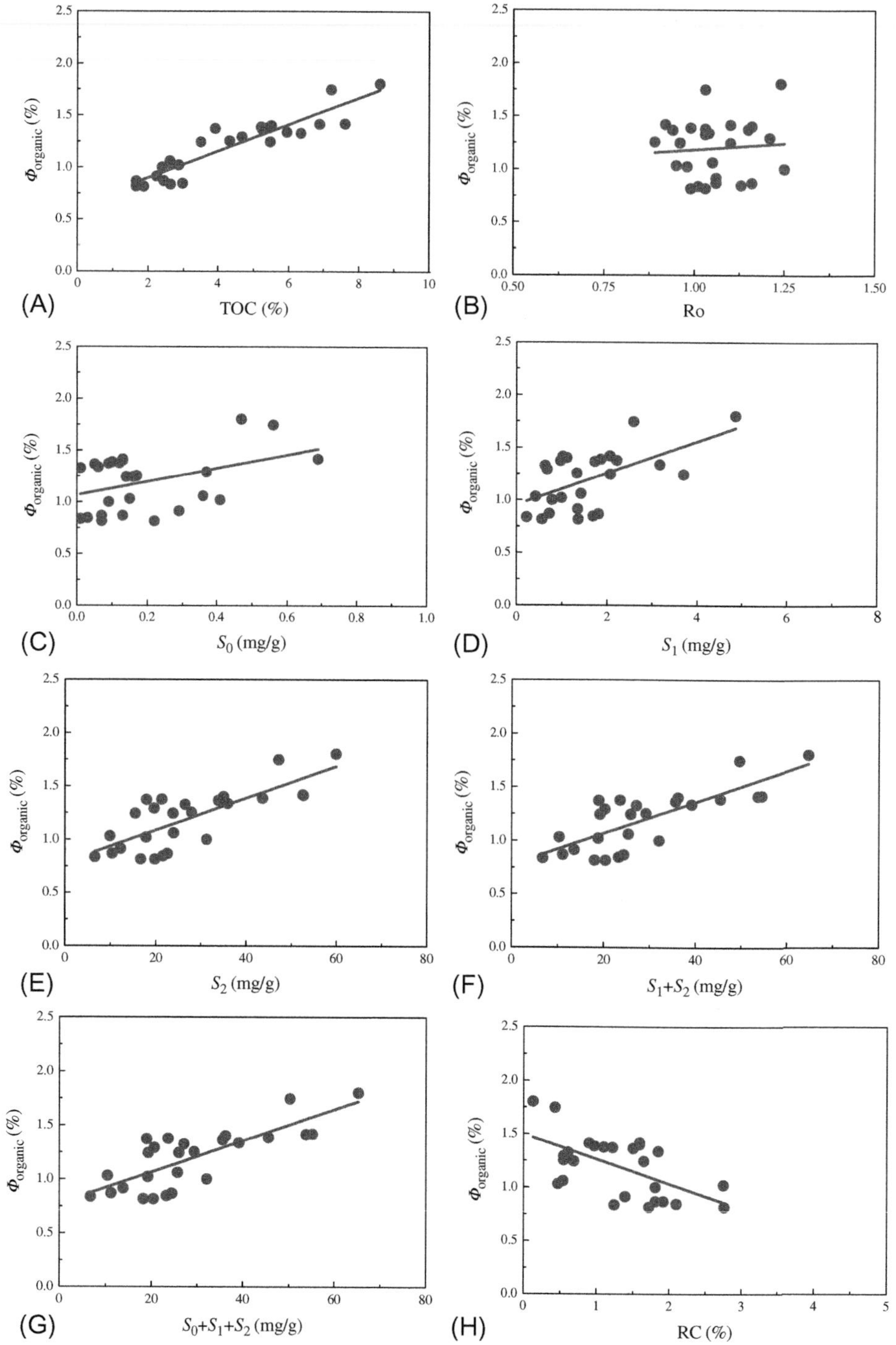

Fig. 8

Relationship between organic-related porosity and geochemical parameters. (A) Organic-related porosity and TOC, (B) organic-related porosity and Ro, (C) organic-related porosity and S_0, (D) organic-related porosity and S_1, (E) organic-related porosity and S_2, (F) organic-related porosity and $S_1 + S_2$, (G) organic-related porosity and $S_0 + S_1 + S_2$, and (H) organic-related porosity and RC.

Table 3 Statistical results of geochemical parameters

Parameter	TOC	S_0	S_1	S_2	S_1+S_2	$S_0+S_1+S_2$	RC
RSS	0.289	1.616	1.298	0.797	0.764	0.762	1.282
Adj.R^2	0.845	0.134	0.304	0.573	0.591	0.592	0.314

The values of DEN, R, V_p, and V_s are plotted against the organic-related porosity, as shown in Fig. 9E–H. Generally, petrophysical responses of the organic-related pores are approximately the same as the organic carbon [4, 11, 35, 36]. Substances residing in the organic-related pores, such as bitumen and kerogen, are nonconductive and low-weighted, yielding high resistivity and low density in petrophysical experiments. Organic-related pores are made up of soft matter showing the characteristics of high acoustic transit time. As expected, an increase in organic-related pores generally produces a decrease in sonic velocity and density, but an increase in resistivity. Furthermore, it was found that for resistivity and organic-related porosity, the fitting result of the exponential function is better than that of the linear function. Though the intrinsic mechanism is not clear, we assume that the phenomenon is similar to the relationship between water saturation and resistivity discovered by Archie [25, 37, 38].

Table 4 summarizes the fitting parameters between organic-related porosity and the above petrophysical responses. The Adj.R^2 of V_p is the highest value, followed by Th. According to these univariate analyses, to predict the organic-related porosity conveniently, the following formula can be obtained by least square regression,

$$\phi_{\text{organic}} = -0.0002V_p + 0.0001V_s - 0.3921\text{DEN} + 0.04\text{Th} + 2.1044R^{0.0396}, \quad (2)$$

where ϕ_{organic} is the organic-related porosity.

It is noted that a positive coefficient of V_s occurred in Eq. (2), which is inconsistent with the univariate analysis. To satisfy the physical meaning, a new regression model, which excluded the V_s data, is presented as,

$$\phi_{\text{organic}} = -0.00014V_p - 0.33495\text{DEN} + 0.04431\text{Th} + 2.0261R^{0.0374}. \quad (3)$$

The correlation coefficients of Eq. (2) and Eq. (3) are 0.91 and 0.89, whereas the mean square errors of Eq. (2) and Eq. (3) are 6.2% and 6.1%, respectively. From the statistical data and Fig. 10, both have similarly favorable fitting results. However, Eq. (3) is superior since it does not need the input of V_s (difficult to obtain by conventional well logging), and all fitting parameters are physically feasible.

To validate the effectiveness of the empirical formula, we tested the remaining 14 samples from the same formation, which were not used for the initial experiment. As shown in Fig. 11, the predicted organic-related porosity was closely related with the experimental data, with a mean square error of 8.37%. The fitting result shows the good performance of our model.

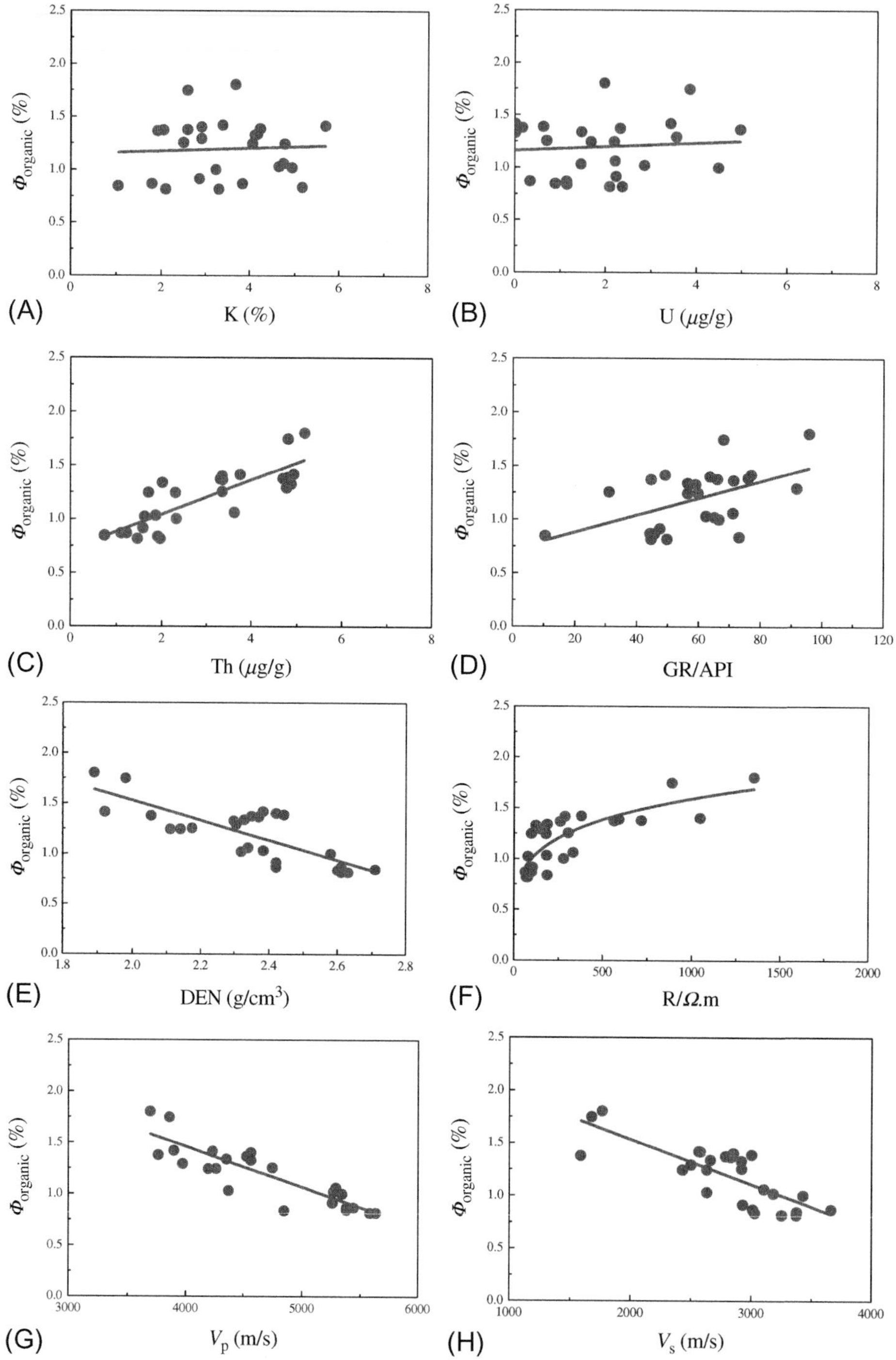

Fig. 9

Relationships between organic-related porosity and petrophysical parameters. (A) Organic-related porosity and K, (B) organic-related porosity and U, (C) organic-related porosity and Th, (D) organic-related porosity and GR, (E) organic-related porosity and DEN, (F) organic-related porosity and R, (G) organic-related porosity and V_p, and (H) organic-related porosity and V_s.

Table 4 Fitting parameters of petrophysical responses

Parameter	K	U	Th	GR	DEN	R	V_p	V_s
RSS	1.938	1.929	0.633	1.438	0.732	0.029	0.496	0.737
Adj.R^2	0.038	0.003	0.661	0.230	0.608	0.629	0.734	0.605

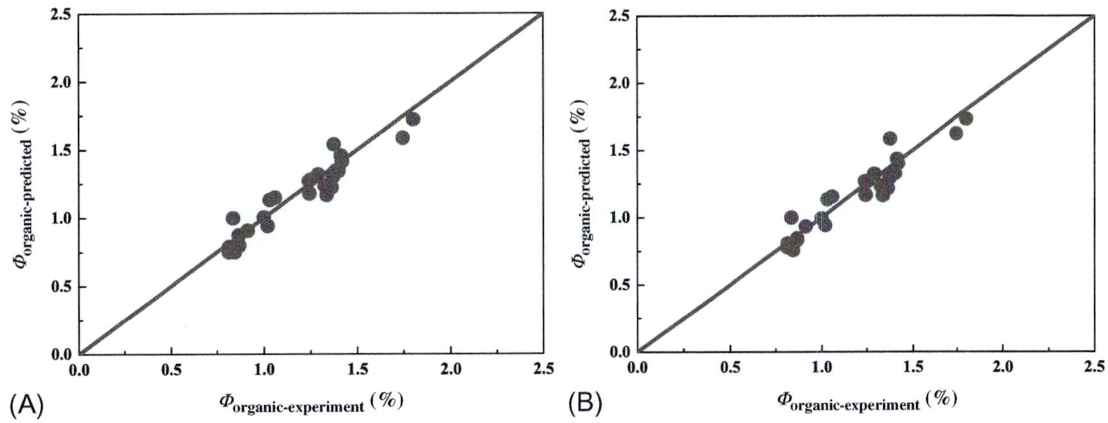

Fig. 10
Relationship between the predicted and measured organic-related porosity. (A) Fitting results by Eq. (2) and (B) fitting results by Eq. (3).

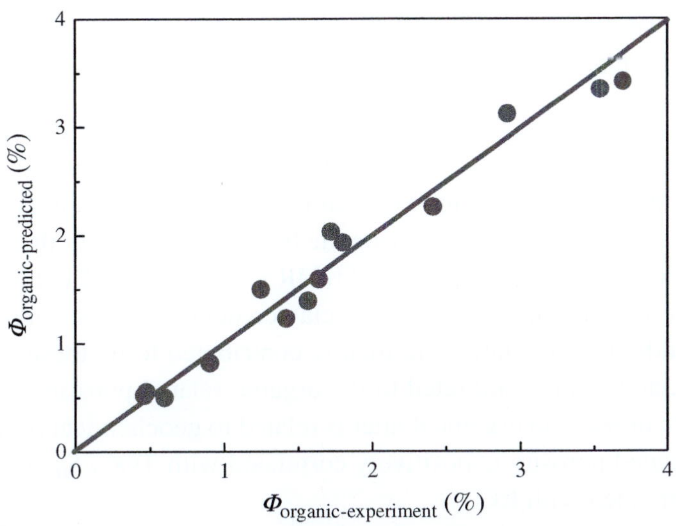

Fig. 11
Model validation using the remaining samples.

Using the result presented above, we can adjust the model to a more general case with the help of core calibration, which is expressed as,

$$\phi_{\text{organic}} = a \times V_p + b \times \text{DEN} + c \times \text{Th} + d \times R^e, \tag{4}$$

where a, b, c, d, and e are empirical fitting parameters, which may vary from region to region, or in lithologies within a given formation.

We can also rewrite the model as the form of an unspecified function,

$$\phi_{\text{organic}} = f\left(V_p, \text{DEN}, \text{Th}, R\right), \tag{5}$$

where $f(\cdot)$ denotes the unspecified function, which can be obtained by experiments.

One of the most important implications for the above analysis is that quantitative information about organic-related pores determined by low-field NMR experiments is in accordance with the value measured by image analysis of SEM and EDS, and can be predicted through petrophysical responses by core calibration.

6 Conclusions

This chapter presents a comprehensive petrographic, geochemical and petrophysical approach to characterizing organic-related pores of tight samples. The organic-related porosity quantified by a low-field NMR experiment was confirmed by other measurements such as helium porosity, BSE and EDS. Geochemical and petrophysical responses of organic-related pores were investigated to determine the intrinsic properties of substances located in these pores. A comprehensive empirical model was established to facilitate the predication of organic-related porosity via petrophysical data. The following conclusions can be summarized based on the above results:

(1) During the heating pretreatment at a temperature of 100°C, most of the free and capillary bound water, light hydrocarbon, and drilling mud are evaporated, whereas high viscous hydrocarbons such as bitumen and kerogen remain in place. Monolayer water on the mineral surfaces cannot be moved during the heating process, but they did not contribute to any relaxation signals in the low-field NMR experiments. Clay bound water was omitted for these clean samples due to the clay content being less than 1%. NMR signals for samples under the dry state were mainly contributed to by these high viscous hydrocarbons, and can be calibrated to the organic-related porosity.

(2) The abundance of pores in organic matter is related to geochemical properties. Generally, the organic-related porosity is positively correlated with TOC, S_0, S_1, and S_2, and negatively correlated with RC.

(3) Petrophysical responses of the organic-related pores were similar to organic carbon. The organic-related porosity increased with the increase of Th, GR, and R, but decreased

with the increase of V_p, V_s, and DEN. Traces of these petrophysical responses can be adopted to predict the in-situ organic-related porosity via core calibration.

The integrated experiments and analysis provided an innovative way to quantify the organic-related pores of tight samples with low clay content, which can be used as a useful tool for geologists and geophysicists. The geological and sedimentary environment is far more complex and can vary sharply in different regions and formation, causing the established model to be futile in other cases. But we still recommend it as an important method to characterize the organic-related pores if relationships between the organic-related porosity and petrophysical parameters are established.

Acknowledgments

This work was supported by the National Natural Science Foundation of China (41404086; 41722403), China Postdoctoral Science Foundation (2015T80759), Major National Science and Technology Projects of China (2011ZX05020-008), National Key Foundation for Exploring Scientific Instrument of China (2013YQ170463) and CNPC Science and Technology Project (2012E-34-12).

References

[1] G.R.L. Chalmers, R.M. Bustin, The organic matter distribution and methane capacity of the lower cretaceous strata of northeastern British Columbia, Canada, Int. J. Coal Geol. 70 (2007) 223–239.

[2] D.J.K. Ross, R.M. Bustin, The importance of shale composition and pore structure upon gas storage potential of shale gas reservoirs, Mar. Petrol. Geol. 26 (2009) 916–927.

[3] Q.R. Passey, K. Bohacs, W.L. Esch, et al., From oil-prone source rock to gas-producing shale reservoir-geologic and petrophysical characterization of unconventional shale gas reservoirs, in: SPE-131350, International Oil and Gas Conference and Exhibition in China, Beijing, China, 8–10 June, 2010.

[4] M. Tan, Q. Liu, S. Zhang, A dynamic adaptive radial basis function approach for total organic carbon content prediction in organic shale, Geophysics 78 (2013) D445–D459.

[5] Q.R. Passey, S. Creaney, J.B. Kulla, et al., A practical model for organic richness from porosity and resistivity logs, AAPG Bull. 74 (1990) 1777–1794.

[6] C.D. Patterson, C.A. Quarles, J.A. Breyer, Possible new well-logging tool using positron Doppler broadening to detect total organic carbon (TOC) in hydrocarbon source rocks, Radiat. Phys. Chem. 68 (2003) 523–526.

[7] M. Khoshnoodkia, H. Mohseni, O. Rahmani, et al., TOC determination of Gadvan formation in south pars gas field, using artificial intelligent systems and geochemical data, J. Pet. Sci. Eng. 78 (2011) 119–130.

[8] A. Kumar, S.K. Sinha, Total organic carbon prediction for shale gas exploration using statistical clustering, multiple regression analysis, in: 10th Biennial International Conference and Exposition, Rae Bareli, India, 2013.

[9] M.F. Romero-Sarmiento, M. Ducros, B. Carpentier, et al., Quantitative evaluation of TOC, organic porosity and gas retention distribution in a gas shale play using petroleum system modeling: application to the Mississippian Barnett shale, Mar. Petrol. Geol. 45 (2013) 315–330.

[10] M. Tan, Y. Zou, X. Wang, et al., A novel wireline logs based approach for assessment of mineral concentrations in organic shale, Oil Shale 31 (2014) 132–146.

[11] X. Ge, Y. Wang, Y. Fan, et al., Determination of total organic carbon (TOC) in tight reservoir using empirical mode decomposition-support vector regression (EMD-SVR): A case study from XX-1 basin, Western China, ASEG Extended Abstr. 2015 (2015) 1–10.

[12] V. Kuchinskiy, Porosity development within an organic matter of Lower Silurian and Ordovician source rocks of Poland shale trend, in: AAPG Annual Convention & Exhibition, Pittsburgh, PA, May 19–22, 2013.

[13] U. Kuila, D.K. McCarty, A. Derkowski, et al., Nano-scale texture and porosity of organic matter and clay minerals in organic-rich mudrocks, Fuel 135 (2014) 359–373.

[14] X. Ge, Y. Fan, J. Li, et al., Pore structure characterization and classification using multifractal theory—an application in Santanghu basin of western China, J. Pet. Sci. Eng. 127 (2015) 297–304.

[15] K.E. Washburn, J.E. Birdwell, Updated methodology for nuclear magnetic resonance characterization of shales, J. Magn. Reson. 233 (2013) 17–28.

[16] R. Lewis, P. Singer, T. Jiang, et al., NMR T2 distributions in the Eagle Ford shale: Reflections on pore size, SPE-164554, in: SPE Unconventional Resources Conference-USA, The Woodlands, Texas, USA, April 10–12, 2013.

[17] A.E. Ozen, R.F. Sigal, T1/T2 NMR surface relaxation ratio for hydrocarbons and brines in contact with mature organic-shale reservoir rocks, Petrophysics 54 (2013) 11–19.

[18] R. Kausik, K. Fellah, E. Rylander, et al., NMR petrophysics for tight oil shale enabled by core resaturation, in: SCA2014-073, International Symposium of the Society of Core Analysts, Avignon, France, September 8–11, 2014.

[19] M. Saidian, T. Rasmussen, M. Nasser, et al., Qualitative and quantitative reservoir bitumen characterization: a core to log correlation methodology, Interpretation 3 (2015) SA143–SA158.

[20] E. Odusina, R.F. Sigal, Laboratory NMR measurements on methane saturated Barnett shale samples, Petrophysics 52 (2011) 32–49.

[21] N. Musharfi, A. Almarzooq, M. Eid, et al., Combining Wireline Geochemical, NMR, and dielectric data for formation evaluation and characterization of shale reservoirs, in: SPWLA 53rd Annual Logging Symposium, Cartagena, Colombia, 2012.

[22] J.H. Chen, J. Zhang, G. Jin, et al., Capillary condensation and NMR relaxation time in unconventional shale hydrocarbon resources, in: SPWLA-2012-186, SPWLA 53rd Annual Symposium, Cartagena, Colombia, June 16–20, 2012.

[23] J.P. Butler, J.A. Reeds, S.V. Dawson, Estimating solutions of first kind integral equations with nonnegative constraints and optimal smoothing, SIAM J. Numer. Anal. 18 (1981) 381–397.

[24] D. Tiab, E.C. Donaldson, Petrophysics: Theory and Practice of Measuring Reservoir Rock and Fluid Transport Properties, Gulf Professional Publishing, 2015.

[25] J. Cai, W. Wei, X. Hu, et al., Electrical conductivity models in saturated porous media: a review, Earth-Sci. Rev. 171 (2017) 419–433.

[26] J. Espitalié, G. Deroo, F. Marquis, La pyrolyse Rock-Eval et ses applications, Oil. Gas. Sci. Technol. 40 (1985) 755–784.

[27] C.H. Sondergeld, R.J. Ambrose, C.S. Rai, et al., Microstructural studies of gas shales, in: SPE-131771, SPE Unconventional Gas Conference, Pittsburgh, Pennsylvania, USA, February 23–25, 2010.

[28] M.E. Curtis, C.H. Sondergeld, R.J. Ambrose, et al., Microstructural investigation of gas shales in two and three dimensions using nanometer-scale resolution imaging, AAPG Bull. 96 (2012) 665–677.

[29] J.I. Goldstein, D.E. Newbury, D.C. Joy, et al., Scanning electron microscopy and x-ray microanalysis, third ed., Springer Science and Business Media, 2003.

[30] J.A. Legg*, R. Donahoe, R. Martens, Assessment of the petroleum generation potential of the neal shale in the black warrior basin, Alabama, in: Unconventional Resources Technology Conference, Denver, Colorado, August 25–27, 2014.

[31] B. Nicot, N. Vorapalawut, B. Rousseau, et al., Estimating saturation in organic shales using [31D] NMR, in: SCA2015-024, International Symposium of the Society of Core Analysts, St. John's Newfoundland and Labrador, Canada, August 16–21, 2015.

[32] R. Kausik, K. Fellah, E. Rylander, et al., NMR relaxometry in shale and implications for logging, Petrophysics 57 (2016) 339–350.

[33] G.R. Coates, L. Xiao, M.G. Primmer, NMR Logging Principles and Applications, Haliburton Energy Services, Houston, 1999.

[34] K.J. Dunn, D.J. Bergman, G.A. Latorraca, Nuclear Magnetic Resonance: Petrophysical and Logging Application, Elsevier Science Ltd., 2002.
[35] M.S. El Sharawy, G.R. Gaafar, Application of well log analysis for source rock evaluation in the Duwi formation, southern gulf of Suez, Egypt, J. Appl. Geophys. 80 (2012) 129–143.
[36] N. Aadil, M.H. Tayyab, A.M. Naji, Source rock evaluation with interpretation of wireline logs: a case study of lower Indus basin, Pakistan, Nucleus 51 (2014) 139–145.
[37] G.E. Archie, The electrical resistivity log as an aid in determining some reservoir characteristics, Trans. AIME 146 (1942) 54–62.
[38] W. Wei, J. Cai, X. Hu, et al., An electrical conductivity model for fractal porous media, Geophys. Res. Lett. 42 (2015) 4833–4840.

CHAPTER 6

Permeability of Fractured Shale and Two-Phase Relative Permeability in Fractures

Harpreet Singh*, Jianchao Cai†

National Energy Technology Laboratory, Morgantown, WV, United States, †Hubei Subsurface Multi-Scale Imaging Key Laboratory, Institute of Geophysics and Geomatics, China University of Geosciences, Wuhan, P.R. China

Chapter Outline

1 Introduction

Darcy's law is well established as a method to compute permeability of conventional reservoir rocks. However, it cannot be used to estimate permeability of ultra-tight organic-rich mudrocks (also commonly referred to as shale) because Darcy's law highly underestimates the

Petrophysical Characterization and Fluids Transport in Unconventional Reservoirs. https://doi.org/10.1016/B978-0-12-816698-7.00006-1

flow rate when compared to field observations. Specifically, Darcy's law assumes that the flow of a Newtonian fluid in a rock is only a function of its pressure difference, and that the permeability of rock is a constant parameter that does not change with time. Shale reservoir formations contain an abundance of dispersed, sedimentary organic matter (*OM*) with fluid-like properties (e.g., density and hydrogen index) [1] and laboratory studies over the years have shown that these sites are a source of additional transport mechanisms such as desorption and diffusion, which contribute significantly to the gas produced from the shale matrix. The permeability of shale matrix derived from these additional transport mechanisms leads to an expression that is a sensitive function of pressure, especially at low reservoir pressure when the flow is diffusive, besides other secondary variables.

Predicting upscaled permeability of fractured shale rock requires accounting for major permeable features, besides the ultra-tight matrix, such as *OM*, microfractures, and other natural fractures at the scale of interest. The scale at which shale permeability is predicted is important because the permeable features that allow flow of fluids, and hence impact permeability, vary at different scales as shown in Fig. 1.

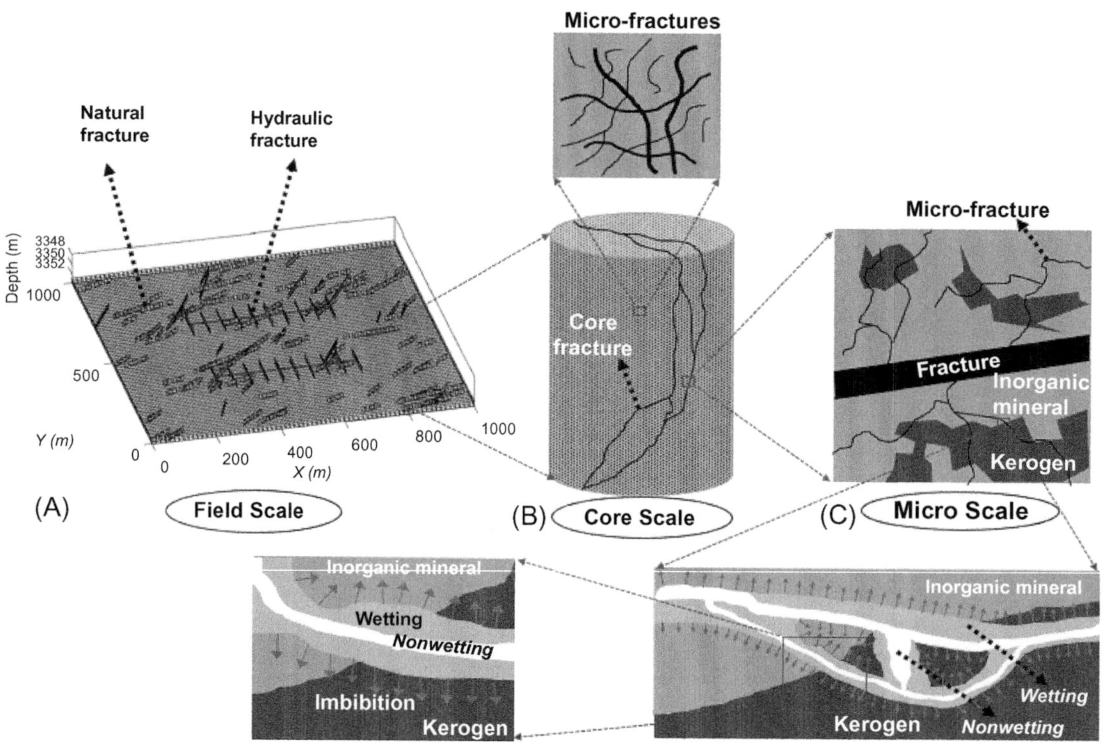

Fig. 1
A schematic diagram illustrating the ubiquitous presence of distinct permeable features in shale at different scales.

This chapter critically discusses the permeability of shale from experimental and modeling perspectives, including a discussion on the transport processes in shale that contribute to shale permeability at different scales. This chapter also presents a nonempirical model for two-phase relative permeability in shale fractures, derived using conservation of momentum, which does not require any experimental data or curve fitting. Further, this chapter discusses recent progress made in history-matching techniques to predict field-scale permeability of fractured shale in terms of its spatially varying distinct features (fractures, matrix and organic matter), and the use of well logs in predicting field-scale shale permeability.

2 Experimental Observations

2.1 Transport Processes in Shale

Gas produced from shale reservoirs is contributed to by convective flux coming through natural and hydraulic fractures (by Darcy flow), desorbed gas from clays and organic matter (by surface diffusion) and diffusive flux from inorganic matrix (by Knudsen diffusion) as shown in Fig. 2. The gas and liquid produced initially during the flowback period has a high production rate [2] that is primarily the result of depletion of compressed fluids accumulated in large-scale fractures by convective flow (function of pressure gradient), and it can decrease by more than half of its maximum rate within the first year [3]. Gas produced subsequent to the high production rate is the result of flow contributed by surface diffusion [4, 5] from kerogen/clay minerals (function of pore pressure), which is a rate-limiting transport. Additionally, the gas produced from the matrix is also contributed to by Knudsen diffusion [6] from extremely small pores in inorganic minerals of shale (function of pore sizes and pore pressure) that do not store any measurable amount of liquid and this tends to attenuate diffusive flow. Although the production rate of gas from the matrix is considerably lower than the

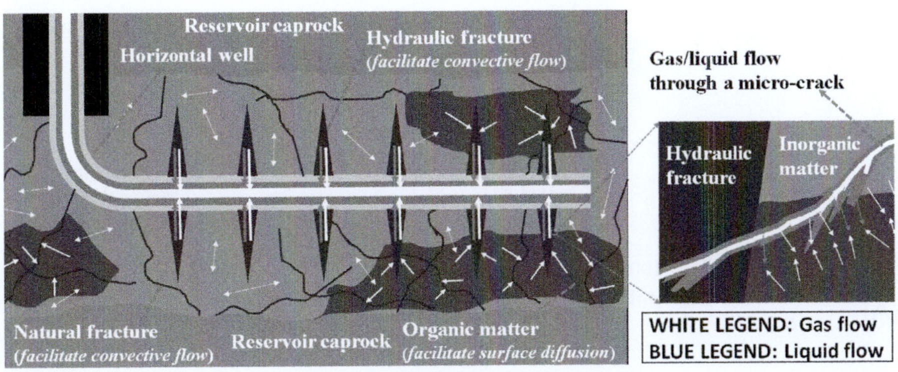

Fig. 2
A schematic diagram illustrating the transport processes in a shale reservoir facilitated by distinct permeable features.

initial rate from fractures, this lower rate can be sustained for a long time as the matrix could hold a significantly larger amount of gas than the fractures.

Gas produced by the three mechanisms (convection, surface diffusion and Knudsen diffusion) leads to a permeability of shale gas that changes with the contribution of each mechanism to gas flow with time. The contribution of each of these three mechanisms is strongly tied to pore pressure and pore sizes in the medium, and the combined effect of these two parameters is used to define flow regimes by a dimensionless parameter called *Knudsen number* (Kn) that gives an idea of which flow mechanism is dominating in the reservoir [7]. For example, convective flow would be the dominant mechanism if the flow regime is continuum ($0 \leq Kn < 10^{-2}$), while the contribution of surface diffusion and Knudsen diffusion increases when the flow regime moves towards the slip flow regime ($10^{-2} \leq Kn < 10^{-1}$) as pore pressure is depleted. The Kn is estimated for pore pressures ranging from 0.1 to 50 MPa and pore sizes ranging from 1 to 1000 nm. The corresponding flow regimes are plotted in Fig. 3.

Besides the three primary mechanisms of fluid flow in shale discussed above, two other mechanisms may be contributing to the production of gas, which are (i) imbibition of slick frack water into the matrix by osmosis that pushes gas out towards fractures by a process known as *counter-imbibition*, and (ii) gas dissolved in organic minerals that adds to the gas adsorbed on its surface. However, the significance of these two mechanisms in gas production has been investigated by only a few studies, so their role in the literature is not widely discussed when compared to the other three mechanisms discussed above.

2.2 Shale Permeability Measurement Techniques

Permeability of a shale rock sample can be measured using three different techniques: (i) steady-state, (ii) unsteady-state, and (iii) other alternative techniques [8]. Unlike steady-state method that is less popular among experimentalists because of the large time it takes to achieve

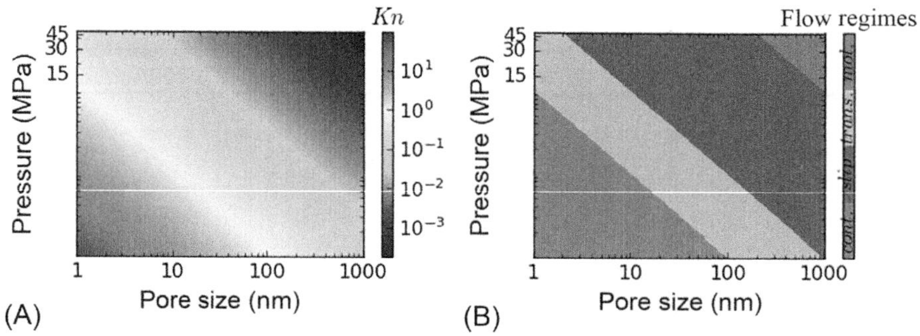

Fig. 3

(A) Kn value as a function of channel height and pore pressure, and (B) graphical representation of gas flow regimes (continuum/convection lies in the range $[0 \leq Kn < 10^{-2}]$, slip lies in the range $[10^{-2} \leq Kn < 10^{-1}]$, transition lies in the range $[10^{-1} \leq Kn < 1]$, and molecular lies in the range $[Kn \geq 1]$) based on the value of Kn.

steady-state, unsteady-state method [9] to measure shale permeability has been modified into several variations that include pressure pulse-decay, oscillating pressure, and the GRI method (applicable to both crushed samples and intact cores). Other alternative permeability techniques include indirect permeability measurement (e.g., mercury porosimetry, nuclear magnetic resonance), canister desorption test, and probe technique. These measurement methods, along with their advantages and disadvantages, are discussed in Table 1.

Table 1 Experimental permeability measurement techniques for shale

Method		Description	Advantage	Disadvantage
Steady-state [10,11]		Core is pressurized and allowed to equilibrate either by constant flow rate or differential pressure	(i) Simple setup (ii) Simple analysis of data using Darcy's law expression	(i) Takes a long time (ii) Accuracy is greatly affected with decreasing permeability
Unsteady-state	Pulse-decay [12]	Similar to steady-state except the flow requires gas storage reservoirs at upstream and downstream ends. Pressure decay and build up on the upstream and downstream ends are measured	(i) Can measure both fracture and matrix permeability with early-time and late-time pressures (ii) Shorter measurement time	(i) Interpretation of data is complicated (ii) Very sensitive to temperature fluctuations, porosity, and system volume
	Oscillating pressure [13,14]	Experimental setup is similar to pulse-decay, except that a sinusoidal pressure is transmitted through the upstream end	(i) Can simultaneously determine permeability, diffusivity and pore volume (ii) More accurate than steady-state method	(i) Complicated and expensive experimental setup (ii) More complicated data interpretation than pulse-decay
	GRI [15,16]	Can be applied to both crushed and intact core, where crushed sample is analyzed using an adsorption cell, while a core is analyzed using a method that mimics pulse-decay, but with only one storage reservoir	(i) Less complicated than pulse-decay (ii) Can incorporate in situ stress conditions with core sample	(i) Lack of standard data interpretation method (ii) Cannot measure fracture permeability

Continued

Table 1 Experimental permeability measurement techniques for shale—cont'd

Method		Description	Advantage	Disadvantage
Alternative techniques	Indirect measurement [17–19]	Measurements from mercury porosimetry or nuclear magnetic resonance are used in theoretical models or partial least-square regression	(i) Measured data can be obtained easily without complications of a core flood	(i) Can only provide estimate of permeability, not accurate measurement
	Canister desorption test [20]	Cores are held in canister and stored in an incubator at reservoir temperature, where gas desorption at ambient pressure is measured	(i) Apart from measuring permeability, this method can provide gas content in a formation as well as estimate of diffusion coefficient	(i) Tests are suggested to be performed on-site to avoid losing gas content of the core due to desorption
	Probe technique [21]	Gas is injected into the sample by a small injection tip (probe) by controlling either the flow rate or the pressure. Can measure permeability at different local regions	(i) Can measure heterogeneity in the sample (ii) Cost effective means of gathering large quantity of data	(i) Cannot provide in situ estimate

Out of all the methods discussed in Table 1, pulse-decay is the popular choice to measure shale permeability. The model for evolution of this pressure pulse through an extremely tight-rock was first proposed by Brace et al. [12], although their method did not consider the effect of fluid compressibility that is essential for compressible fluids such as gas (e.g., methane, CO_2). A numerical model to describe the flow of compressible gas in an extremely tight and adsorptive media [22, 23] should account for these two important effects as shown below.

$$\nabla^2 p = c\frac{\partial p}{\partial t} \tag{1}$$

$$c = \frac{\mu}{k}[\phi\beta + \beta_b - \beta_s(1+\phi)] \tag{2}$$

where β_b, β_s, and β are the compressibility of the bulk sample, the sample matrix, and the fluid, respectively.

3 Permeability Modeling

Permeability models generally describe the relationship between the fluid flux and the parameters that control the flux, for example. pressure gradient, gas properties. The conventional approach to modeling permeability is based on Darcy's law that assumes a constant permeability of a rock, which may not be an appropriate assumption for unconventional shale rocks. Therefore, to account for the dynamic permeability of a shale rock, many permeability models have been developed that assume a range of parameters and processes to describe the dynamic permeability. There is no conditional dependence between the type of experimental method and the permeability model, so the data used to validate a permeability model can come from any experimental method as long as the experimental operational conditions are honored by the model parameters.

The permeability of shale can be defined primarily in terms of its matrix and fracture, where shale matrix is further divided into organic matter (*OM*) and inorganic matter (*iOM*).

3.1 Matrix Permeability

A typical characteristic of shale matrix that makes it different from conventional reservoir rocks is its extremely tight pores, which are finely dispersed and heterogeneous in shape as shown by Fig. 4 (from Ref. [7]). The gas transport from these characteristic pores of shale matrix cannot be predicted by Darcy law, which has prompted several investigations on the nature of gas storage and transport in shale matrix and led to the development of various models that are summarized in Table 2 with their advantages and disadvantages.

3.2 Fracture Permeability

Experiments on effective permeability of fractures, $k_{f,\,eff}$, in tight rocks have shown that it varies as a function of stress, strain, and damage, as follows.

$$k_{f,\,eff} = k_f . f(\text{stress, strain, damage}) \tag{3}$$

where k_f is the permeability at initial equilibrium condition, and f is any function of parameters affecting the fracture permeability. Although many relationships have been proposed for the function f, to account for variation in fracture permeability [40], most of these expressions adopt a segmented approach to model the variation in permeability for undamaged or damaged rock. One of the hurdles in estimating permeability of a damaged fracture is an estimate of the damage parameter, which is difficult to obtain. Shi et al. [41] proposed a novel stress-strain deformation relationship that avoids the use of the damage parameter as follows:

$$k_{f,\,eff} = k_f \times \exp\left(-a\varepsilon_1 + br_d^3\right) \tag{4}$$

Moreover, fracture permeability in a highly porous medium is different to an ideal permeability of slit because of the nonideal nature of fractures in porous medium, such as

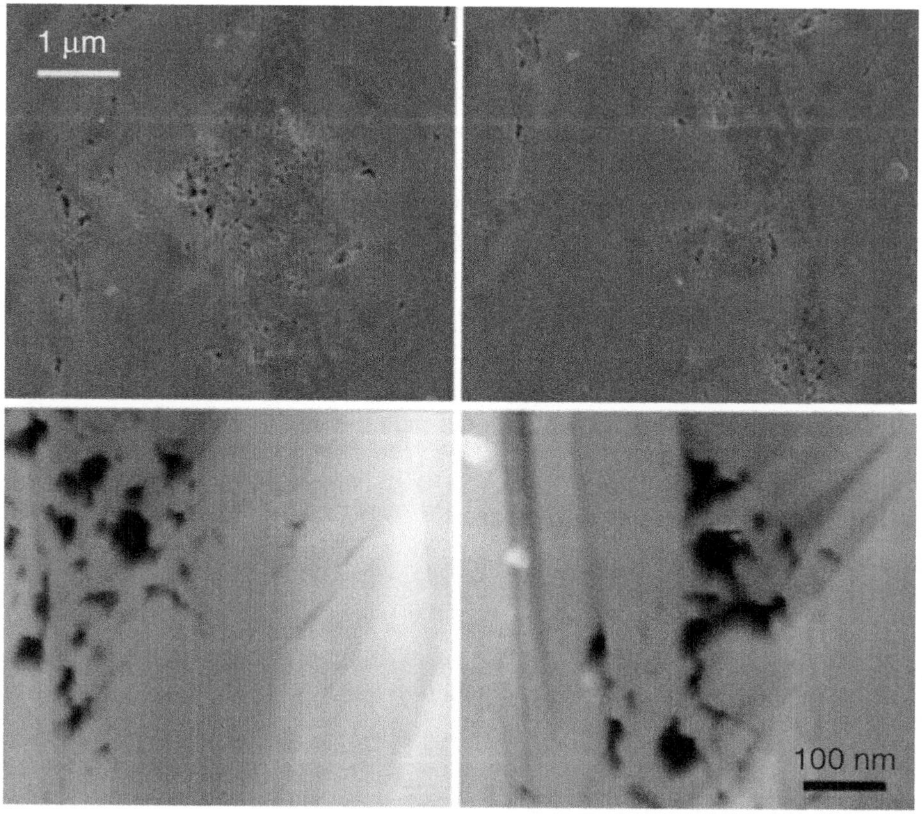

Fig. 4
SEM images (top row) and AFM topography images (bottom row) showing pores and their respective scales in shale matrix.

Table 2 Summary of various permeability models developed for shale matrix

Model	Description	Advantages	Disadvantages
Javadpour [24]	Model developed using slip flow assumption, represented by Maxwell theory. Accounts for Knudsen diffusion. Modeled only for straight capillary tubes	Simple	Limited to straight tubes. Ignores desorption, and stress-strain effect
Civan [25]	Model developed using slip flow assumption, represented by simplified second-order slip model. Contains several empirical parameters	Higher-order slip flow	Several empirical parameters. Ignores stress-strain effect
Darabi et al. [26]	Model developed using slip flow assumption, represented by Maxwell theory. Accounts for surface roughness and Knudsen diffusion in a porous medium	Includes tortuosity and pore surface roughness	Needs TMAC values Ignores desorption, and stress-strain effect

Table 2 Summary of various permeability models developed for shale matrix—cont'd

Model		Description	Advantages	Disadvantages
Sakhaee-Pour and Bryant [27]		Model developed using slip flow assumption, represented by Maxwell theory. Accounts for Knudsen diffusion	Geometry of porous media included	Needs TMAC values. Does not explicitly account for desorption. Ignores stress-strain effect
Singh et al. [7]		Model developed using Navier-Stokes equation and kinetic theory (no slip flow assumption). Accounts for Knudsen diffusion, porous medium and sorption	Simple. No empirical coefficient	Ignores slip flow, and stress-strain effect
Wu et al. [28]		Model developed by coupling slip flow and Knudsen diffusion	Accounts for different pore shapes	Ignores desorption. Ignores stress-strain effect
Naraghi and Javadpour [29]		Model developed by stochastically characterizing organic and inorganic pores. Accounts for slip flow, Knudsen diffusion, surface roughness and desorption	Distinguishes different pore systems in organic and inorganic matter	Needs additional information from SEM images. Needs TMAC values. Ignores stress-strain effect
Singh and Javadpour[30]		Model developed using the Langmuir slip condition and it does not carry several shortcomings associated with the use of Maxwell slip. Reliably predict apparent permeability in shale	Simple and analytic. Gets slip coefficient from sorption data	Ignores local heterogeneity, and stress-strain effect
Cao et al. [31]		Model developed by coupling viscous flow, Knudsen diffusion, and surface diffusion that also accounts for stress-strain effect	Distinguishes different pore systems in organic and inorganic matter	Ignores local heterogeneity, and slip flow
Wang et al. [32]		Model developed by coupling viscous flow, and Knudsen diffusion that also accounts for stress-strain effect	Accounts for stress-strain effects	Ignores local heterogeneity, slip flow, and desorption
Singh and Cai [33]		Developed by extending Singh et al. [7] model to include spatial distribution of organic matter	Accounts for spatial distribution of *OM*.	Ignores slip flow, and stress-strain effect
Based on fractal porous media	Sheng et al. [34]	Model adapted from Beskok and Karniadakis [35] and based on a fractal tortuous capillary tubes	Accounts for system heterogeneity	Complex numerical model and ignores desorption. Ignores stress-strain effect
	Yuan et al. [36]	Model adapted from Beskok and Karniadakis [35] and based on a fractal tortuous capillary tubes that also accounts for surface diffusion	Accounts for system heterogeneity	Ignores stress-strain effect. Several empirical parameters

Continued

Table 2 Summary of various permeability models developed for shale matrix—cont'd

Model		Description	Advantages	Disadvantages
	Wang et al. [37]	Model developed using viscous flow, and molecular diffusion based on a bundle of fractal tortuous capillary tubes	Accounts for system heterogeneity	Ignores slip flow, desorption, and stress-strain effect
	Geng et al. [38]	Model developed using viscous flow, diffusion, and surface diffusion based on a bundle of fractal tortuous capillary tubes	Accounts for system heterogeneity	Complex numerical model. Ignores stress-strain effect
	Song et al. [39]	Model developed using bulk gas and surface diffusion based on a bundle of fractal tortuous capillary tubes	Accounts for system heterogeneity	Complex numerical model. Ignores stress-strain effect. Several empirical parameters

length, porosity, and tortuosity, which affect the permeability. The transition zone between fracture and matrix also imbibes some fluid from the fracture walls into the matrix, which affects the permeability of the fracture as shown by Fig. 5.

Singh and Cai [33] proposed the method to calculate permeability of a fracture that combines the effect of stress mechanics, and the nonideal nature of the fracture, including the imbibition of fluid from fracture walls into the matrix, which is given as follows:

$$k_{f,pm} = \begin{cases} \left[k_f.\exp\left\{-c_f\left(p_c - \chi p_p\right)\right\}\right].\dfrac{\phi_f}{\tau_f} \text{ without damage} \\ \left[k_f.\left\{\exp\left(-a\varepsilon_1\right) + br_d^3\right\}\right].\dfrac{\phi_f}{\tau_f} \text{ with damage} \end{cases} \tag{5}$$

where

$$k_f = k_m + k_{slit} + \sqrt{3k_{slit}k_m}.\tan\text{h}(3) \tag{6}$$

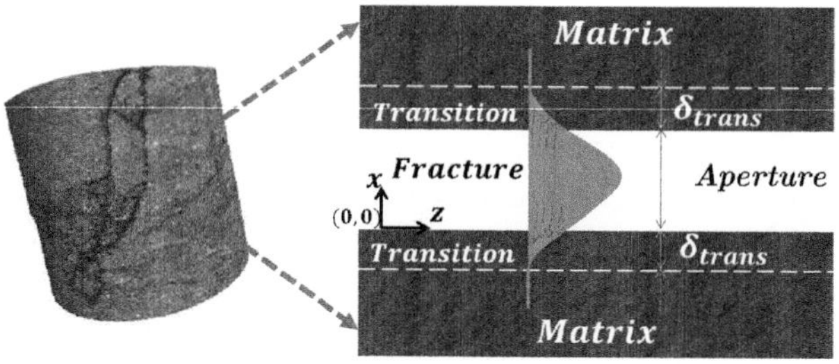

Fig. 5
A conceptual sketch used to illustrate how fracture permeability is affected by fluid imbibition into the adjacent matrix.

where k_m is the matrix permeability in the vicinity of the fracture, k_{slit} is the permeability of the fracture assuming it is an ideal slit, p_c is the confining pressure, p_p is the pore pressure, χ is the coefficient of pore fluid pressure, c_f is the coupling parameter that reflects the sensitivity of stress on the permeability of fracture, ε_1 is the compressive deformation, $r_d\left(=\frac{\varepsilon_3-\varepsilon_3'}{\varepsilon_3}\right)$ is a ratio that represents the weight of the dilatant deformation in the σ_3 direction, and a, b are the rock material-specific parameters that reflect the effects of ε_1 and r_d on the permeability.

3.3 Combined Fracture-Matrix System Permeability

One of the drawbacks in current shale permeability models is their inability to predict an upscaled permeability of a fractured rock with diverse features (e.g., matrix, microfractures, laminations, fractures) where each feature will be significantly different from each other in terms of their permeabilities. This variation in permeability of different features in shale is observed at all scales of a fractured tight-rock, which is composed of minerals with significant differences in their permeability (e.g., organic, inorganic), and other discrete features at the microscale (e.g., microfractures) to larger scales (e.g., fractures) that impact the permeability of shale; this can be visualized graphically in Fig. 1, which shows various features impacting shale permeability that are not possible to account for reasonably by permeability models summarized above that model either matrix or fracture permeability. Singh and Cai [33] proposed a method that accounts for distinct permeable features within a shale rock, including inorganic matrix (*iOM*), organic matter (*OM*), and fractures that are summarized in Fig. 6, with the method to develop finite difference grids for fractures and matrix described below:

3.3.1 Finite-difference grid for fractures

The procedure to discretize the fractures is based on approximation of actual fractures on a finite-difference grid [33, 42] as shown schematically in Fig. 7. Each cell through which the approximate (discretized) fracture passes is assigned discrete fracture properties such as length, aperture, orientation, porosity and tortuosity.

3.3.2 Finite-difference grid for organic and inorganic matter

1. Quantify the probability of finding an *OM* $\left(\propto \frac{1}{\rho_b \phi}\right)$ using bulk density and porosity as follows:

$$P_{OM} = \left[\frac{\left(\frac{1}{\rho_b\phi}\right) - \left(\frac{1}{\rho_b\phi}\right)_{\min}}{\left(\frac{1}{\rho_b\phi}\right)_{\max} - \left(\frac{1}{\rho_b\phi}\right)_{\min}}\right] \tag{7}$$

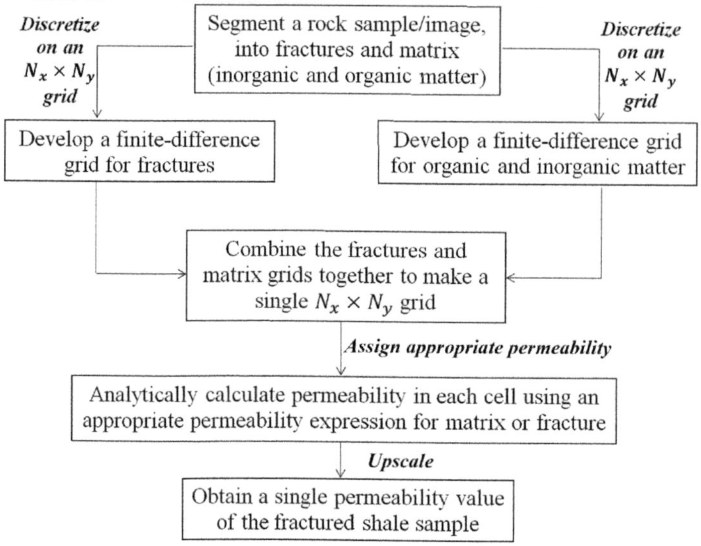

Fig. 6
A flowchart summarizing the method to predict permeability of shale with fractures and matrix (inorganic and organic) in shale rock.

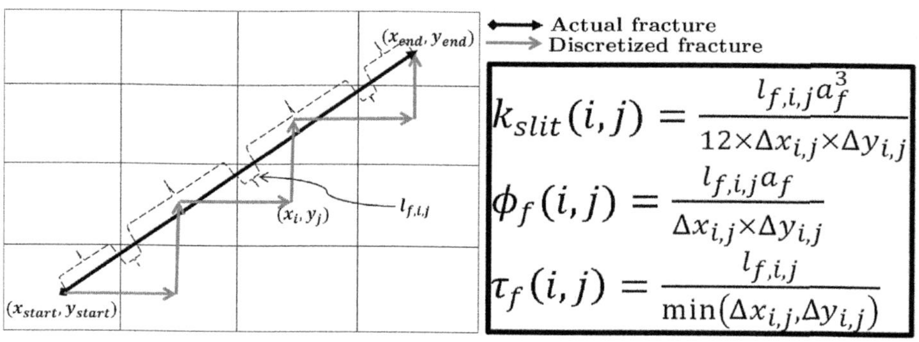

Fig. 7
Schematic diagram to illustrate fracture discretization on a grid.

2. Spatially locate cells on a grid occupied by *OM* using a Bernoulli distribution that uses the probability map calculated in step 1 as an input.
3. Calculate the total volume of the cells occupied by *OM* as calculated in step 2. If the total volume of the cells is equal to the observed *total organic carbon* (TOC) for this rock then save the map, otherwise go to step 2 and repeat the process until the observed volumetric TOC content is satisfied.

Fig. 8 shows an example that illustrates the method of developing a grid for organic and inorganic matter.

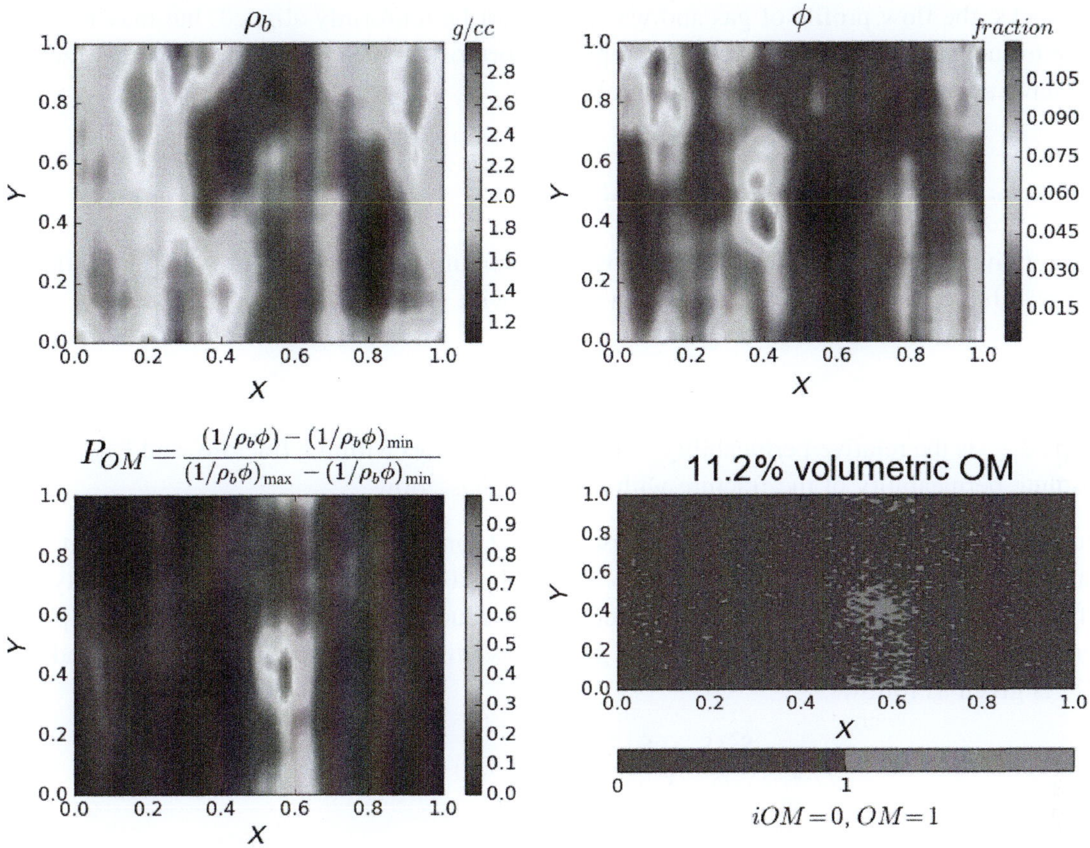

Fig. 8
Illustration of the algorithm to spatially sample *OM* on a grid representing shale matrix.

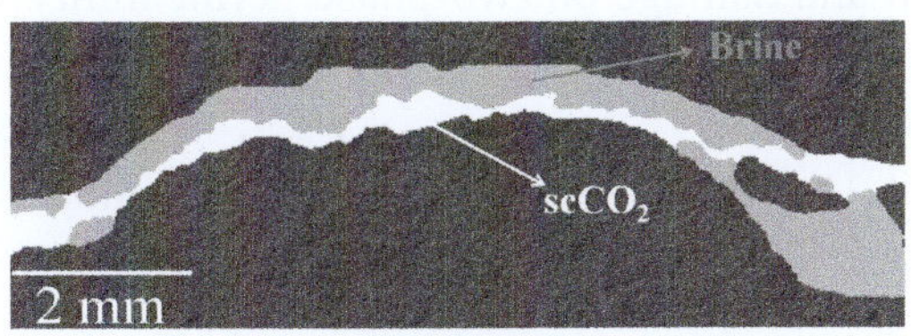

Fig. 9
Processed image of $scCO_2$ and brine flowing through a fracture as captured in a micromodel experiment.

4 Relative Permeability in Shale Fractures

Unlike an extremely tight matrix, fractures in shale can easily carry two-phase flow, which is shown in Fig. 9 for a sample of Utica shale containing brine and supercritical carbon dioxide ($scCO_2$) in the fracture (adapted from Ref. [43]). For flow through rough fractures in real rocks, the flow profile of gas and water may not be uniformly aligned, but may be more tortuous as shown in Fig. 9, so the impact of tortuosity on each fluid phase in the fracture can have a significant effect on its permeability.

4.1 Two-Phase Flow Relative Permeability

The relative permeability of two-phase flow in a fracture is the ratio of the two-phase flow permeability to the absolute fracture permeability with single phase flow as follows:

$$k_{ri,f} = \frac{k_{f,i}}{k_{f,pm}}, i = w, g \tag{8}$$

where $k_{ri,f}$ is the relative permeability of the fluid i (=w, g) within a fracture, and $k_{f,pm}$ is the absolute permeability of the fracture with a single-phase flow.

Singh and Cai [33] proposed a physics-based model for two-phase flow permeability in fractures that accounts for the tortuosity of each fluid phase in the fracture, fracture porosity, the effect of stress-mechanics, capillary pressure and irreducible water saturation. The two-phase permeability of water (wetting phase) and gas (nonwetting phase) proposed by Singh and Cai [33] is given as follows:

$$k_{f,w} = \frac{S_w^2(S_w - S_{wr})}{1 - S_{wr}}\left[2(S_w - S_{wr}) + 3S_g\right].\left(\frac{H^2}{24}\right) \tag{9}$$

$$k_{f,g} = \frac{S_g^2}{1 - S_{wr}}\left[2S_g^2 + 3\alpha(S_w - S_{wr})^2 + 6\alpha(S_w - S_{wr})S_g\right].\left(\frac{H^2}{24}\right) \tag{10}$$

The impact of stress, strain and damage on two-phase permeability within fractures [33] is given as follows:

$$k_{f,i,pm} = \begin{cases} \left[k_{f,i}.\exp\left\{-c_f\left(p_c - \chi p_p\right)\right\}\right].\dfrac{\phi_f}{\tau_{f,i}} \text{ without damage} \\ \left[k_{f,i}.\left\{\exp(-a\varepsilon_1) + br_d^3\right\}\right].\dfrac{\phi_f}{\tau_{f,i}} \text{ with damage} \end{cases} \tag{11}$$

where i depicts the phase of the fluid (=g, w), S_{wi} is the saturation of phase i, S_{wr} is the residual saturation of the wetting phase, α is the ratio of the gas to water (nonwetting to wetting) phase viscosity, H is fracture thickness (aperture).

Here, $\tau_{f,i}$ is the tortuosity of each phase inside the fracture, which is empirically derived using an experimental data model for gas and water tortuosity in a rough channel from various studies [44] and is as follows:

$$\tau(S_w) = \frac{1}{0.74 S_w^{norm} + 0.26} \tag{12}$$

$$\tau(S_g) = \frac{1}{0.43 S_g^2 + 0.38 S_g + 0.19} \tag{13}$$

$$S_w^{norm} = \frac{S_w - S_{wr}}{1 - S_{wr}} \tag{14}$$

The above two-phase permeability model does not differentiate between the type of fluid except for the position of fluids that depict their wetting (close to walls) and nonwetting (away from walls) behavior. Therefore, these expressions are equally applicable for oil and gas phases or for oil and water phases.

4.2 Relative Permeability of Gas and Water in Utica Shale Sample

Fig. 10A shows an image of a fracture etched on a Utica shale sample with two phases present ($scCO_2$ and brine) inside the fracture (adapted from Ref. [43]), and Fig. 10B shows the permeability of the sample (matrix and fracture) estimated using the method of Singh and Cai [33]. Fig. 10B shows the predicted value of absolute permeability of the fracture (without any matrix) is ~30 mD.

The two-phase relative permeability of $scCO_2$ ($k_{rg,\,f}$) and brine ($k_{rw,\,f}$) in the fracture is calculated using the model of Singh and Cai [33] for four different assumed values of residual brine saturation (S_{wr}=0, 0.1, 0.2, 0.4).

Fig. 11 shows that the relative permeability of gas is significantly reduced when the brine saturation increases beyond 0.2, and it becomes almost zero as brine saturation (S_w) approaches 0.7. One interesting observation from Fig. 11 is that, although the relative permeability model does not account for any residual saturation of the gas (nonwetting) phase, "effective" residual gas saturation (based on $k_{rg,\,f} \approx 0$) in this sample suggests $S_{gr} \approx 0.30$.

5 Field-Scale Permeability

5.1 Permeability of Shale Using History-Matching

5.1.1 Drawback of conventional history-matching methods

Different approaches exist to predict field-scale permeability by history-matching [45, 46], such as training image-based method [47], ensemble-based methods [48, 49], etc., but they usually require some reservoir scale information (e.g., reservoir geology, seismic) to constrain

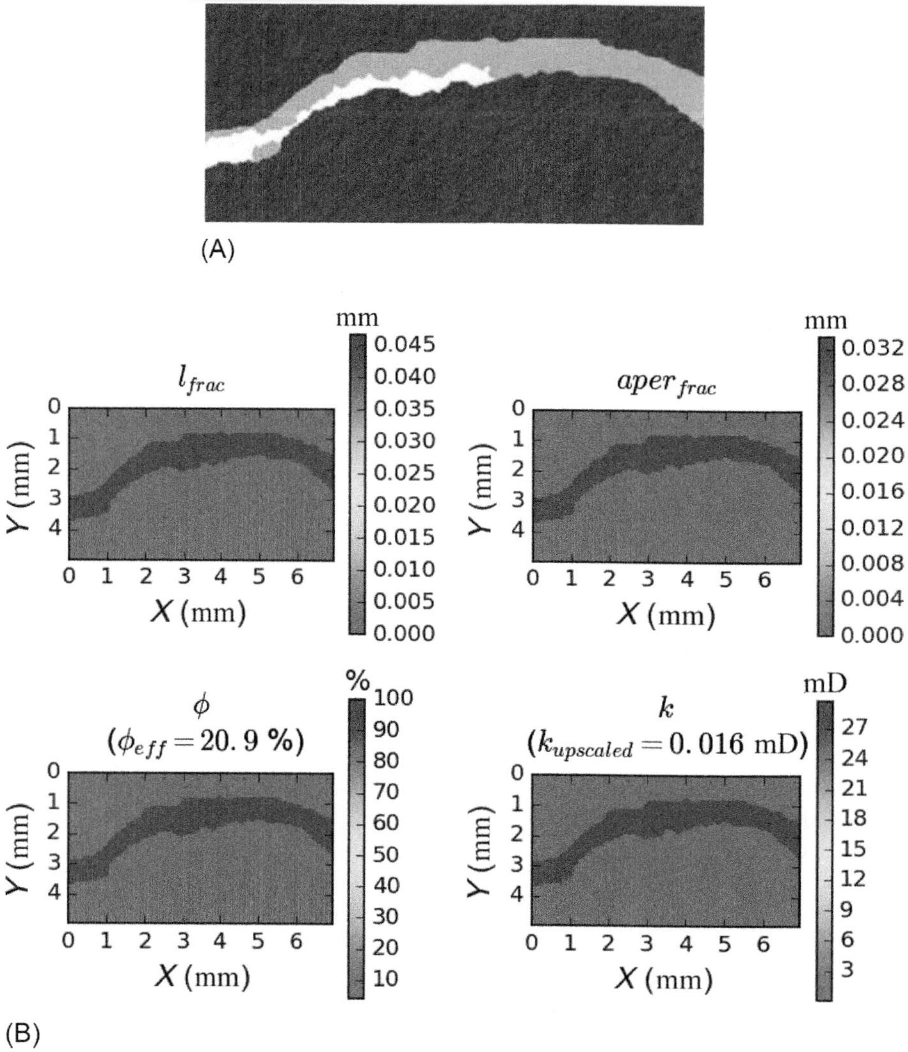

Fig. 10

(A) Image of $scCO_2$ displacing water while passing through a fracture etched on a Utica shale sample and (B) discretized sample with predicted properties (porosity and permeability) using the method proposed by Singh and Cai [33]. The predicted value of absolute permeability of the fracture (without any matrix) is ~30 mD.

the optimization process. However, most of these approaches work for conventional reservoirs without any fractures.

Some methods exist to history-match fractured reservoirs [50–52], but they work by optimizing the upscaled permeability of the fractured reservoir, which does not ensure consistency between the underlying fracture distribution and optimally-matched fractured reservoir as shown in Fig. 12.

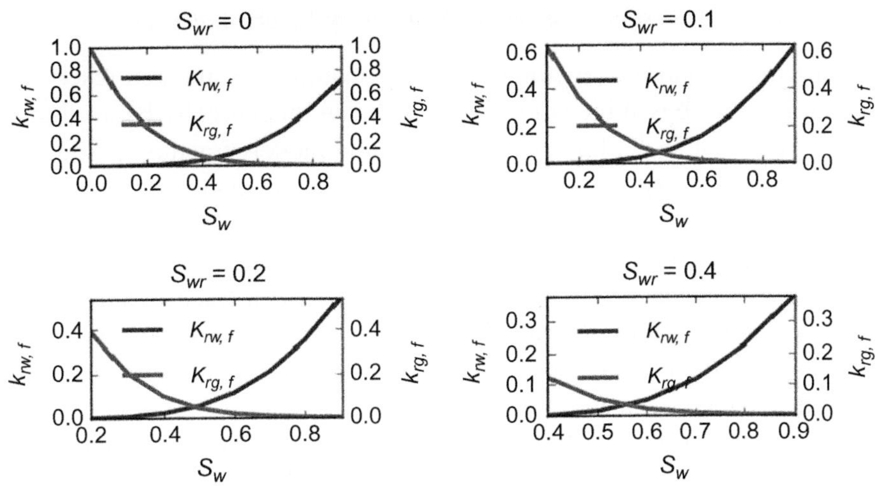

Fig. 11
Two-phase relative permeability for water and $scCO_2$ within the fracture at four different residual saturation values (S_{wr} = 0, 0.1, 0.2, 0.4; shown by four different subplots). The permeability of the fracture with single phase flow is ~30 mD.

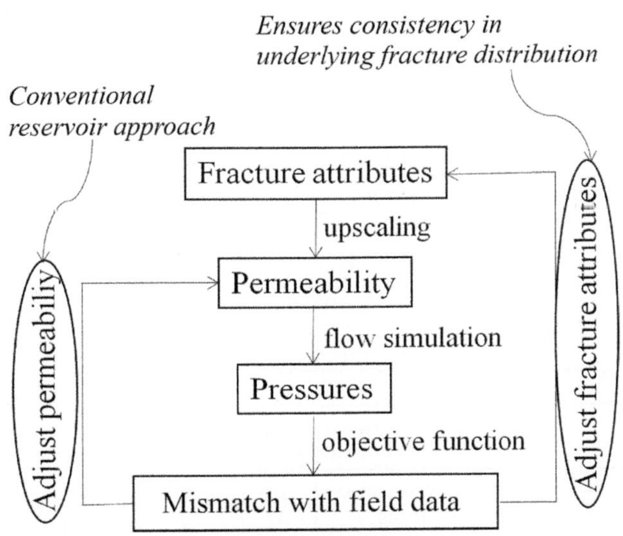

Fig. 12
A flowchart summarizing the optimization approach used in this study. This study adopts a history-matching approach that ensures consistency between the underlying fracture distributions and the best-fitting permeability model.

5.1.2 Prospective methods to estimate shale permeability by history-matching

Chang and Zhang [53] proposed a method to history-match the simulated reservoir volume of shale using an iterative ensemble smoother, which can theoretically be used to predict field-scale permeability of shale; however, their method models fractures as dual-permeability/dual-porosity such that it is difficult to verify the obtained fracture network without their length and aperture distributions, and the authors acknowledge the fact that microseismic data may not be reliable. Chai et al. [54] proposed a method to history-match a fractured shale reservoir that ensures the consistency between fracture lengths, but it does not ensure consistency for fracture aperture sizes in addition to ignoring the random distribution of natural fractures and their abundance. Singh and Cai [55] proposed a method to history-match a fractured shale reservoir that ensures consistency between the underlying fracture distribution and optimally-matched fracture lengths and their apertures, in addition to accounting for random distribution of fractures and their abundance. The summary of the method proposed by Singh and Cai [55] to predict field-scale permeability of a shale reservoir is shown in Fig. 13.

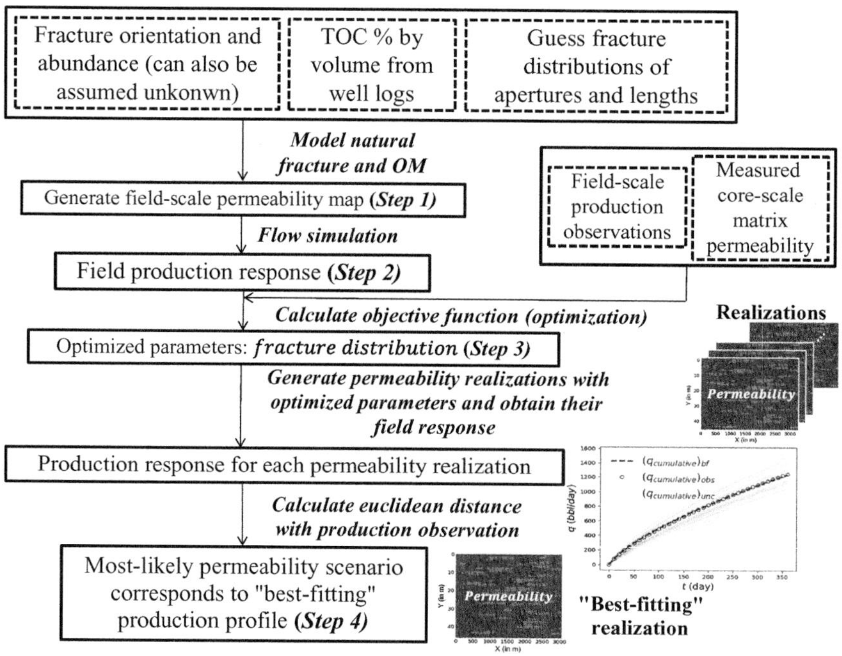

Fig. 13

A flowchart summarizing the four steps to predict field-scale permeability using the lab-scale permeability estimate, fracture orientation and abundance (from an outcrop), and field production history in a history-matching optimization procedure that minimizes an objective function composed of permeability and field production.

The objective function used by Singh and Cai [55] in their study is shown as follows:

$$obj\begin{cases} \min\left[\frac{1}{2}\left\{\sum_{i=1}^{N_w}(q_i - q_{i,obs})^2\right\} + \frac{1}{2}\left\{k_{core} - (k_{field})_{upscaled}\right\}^2\right] \\ \text{subject to } l_{\min} = [0, \max(L_x, L_y)] \\ \text{subject to } l_{\max} = [0, \max(L_x, L_y)] \\ \text{subject to } l_{\min} < l_{\max} \end{cases} \tag{15}$$

Here, k_{field}, $q_i = f(l_{\min}, l_{\max}, \alpha_l, F_a, \kappa_a)$, i.e. k_{field} and q_i are functions of these five unknown parameters $(l_{\min}, l_{\max}, \alpha_l, F_a, \kappa_a)$ of the natural fracture distributions. Where $l_{\min}$, $l_{\max}$ are the minimum and maximum length of the fracture length distribution, L_x, L_y are the dimensions of the field along the X, Y directions, respectively, α_l is a parameter in the power law describing the fracture length distribution, F_a, κ_a are parameters in the power law describing the fracture aperture distribution, k_{core} is the averaged value of laboratory measured permeabilities of various rock samples, with and without fractures, from the desired reservoir, $(k_{field})_{upscaled}$ is the upscaled (averaged) value of the estimated permeability field, q_i, $q_{i,\ obs}$ are the estimated and observed production rates from well i, and N_w are the total number of wells in the field from where observed production rates are available.

5.1.3 An example of shale permeability prediction using history-matching

An example from Singh and Cai [55] is presented here to illustrate the application of history-matching in predicting field-scale permeability using only readily-available data.

In the absence of a real field production history, production history, $q_{obs}(t)$, is simulated numerically for a synthetic field. This synthetic field is generated using all known parameters described in the flowchart shown in Fig. 13, where it is assumed that the upscaled (averaged) permeability is equal to the observed permeability (k_{core}~2.90 μD) as measured in the laboratory, while the remaining parameters are arbitrarily chosen. The permeability field is generated using the method as described in Figs. 6–8 earlier. The synthetic field generated using this method and shown in Fig. 14 is used as a reference truth (true solution) that is used to verify the accuracy of the predicted permeability field.

The parameter values of the fracture distribution (length and aperture distributions) obtained after a history-match and those used in developing the reference truth (Fig. 14) are as follows:

$$l_{\min}, l_{\max}, \alpha_l, F_a, \kappa_a = \begin{cases} 76.2\,\text{m}, 762.0\,\text{m}, 0.10, 4.6 \times 10^{-5}, 0.40 & \text{reference truth} \\ 81.94\,\text{m}, 1275.84\,\text{m}, 0.183, 7 \times 10^{-5}, 0.3113 & \text{optimum} \end{cases} \tag{16}$$

The optimized parameters are used to generate various realizations of permeability that depict the uncertainty remaining after optimizing the unknown parameters for fracture distribution. Realizations of permeability obtained after optimization are used to pick a single "best-fitting"

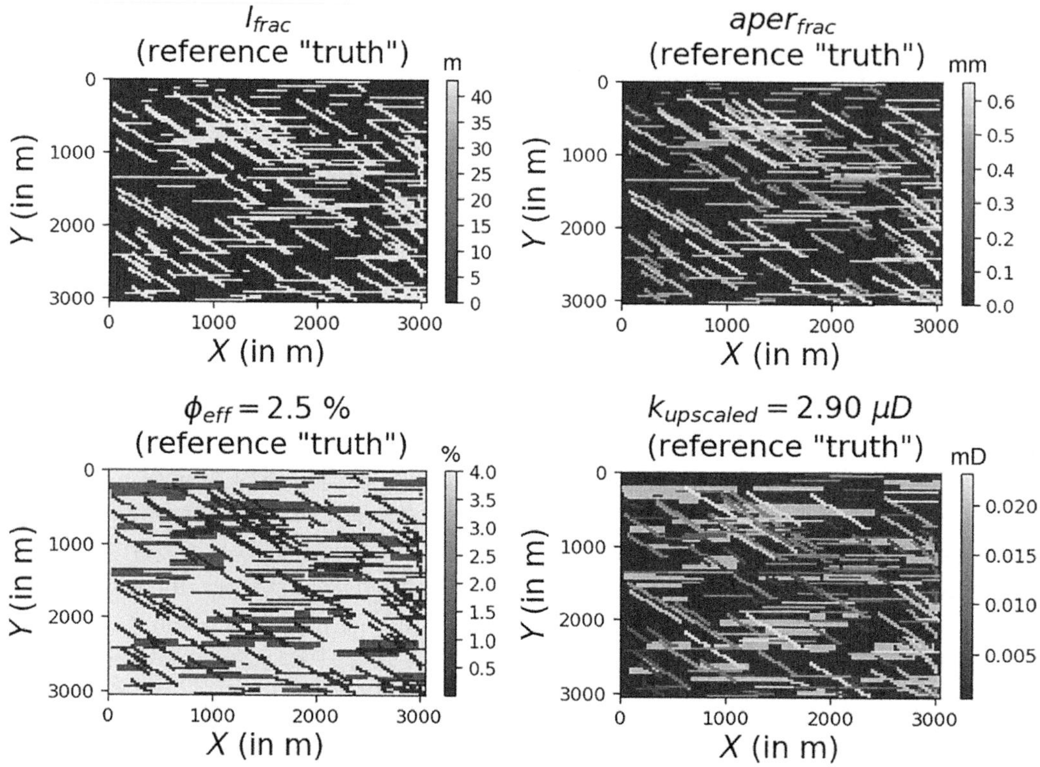

Fig. 14

Synthetic field (with fracture lengths, apertures, porosity and permeability) used as a reference "truth" (true solution) to simulate observed production history.

(or most-likely) permeability scenario by generating a production response of each realization of the geological model and comparing them against the observed field production history by computing the Euclidean norm metric between each realization and the observed production history. This "best-fitting" permeability scenario out of all realizations is the one that produces the closest matching production profile with the observed production history as shown by Fig. 15.

The best-fitting realization of the geological model that resulted in the best match for production profile with the field observation (as shown by the *dotted line* in Fig. 15) is presented in Fig. 16. The results of this realization include fracture lengths, apertures, porosity, and permeability.

Although the predicted permeability field and its upscaled value compare well against the reference truth, there will be some local differences between the reference truth and prediction because of the stochastic (random) nature of the permeability at field-scale, and also

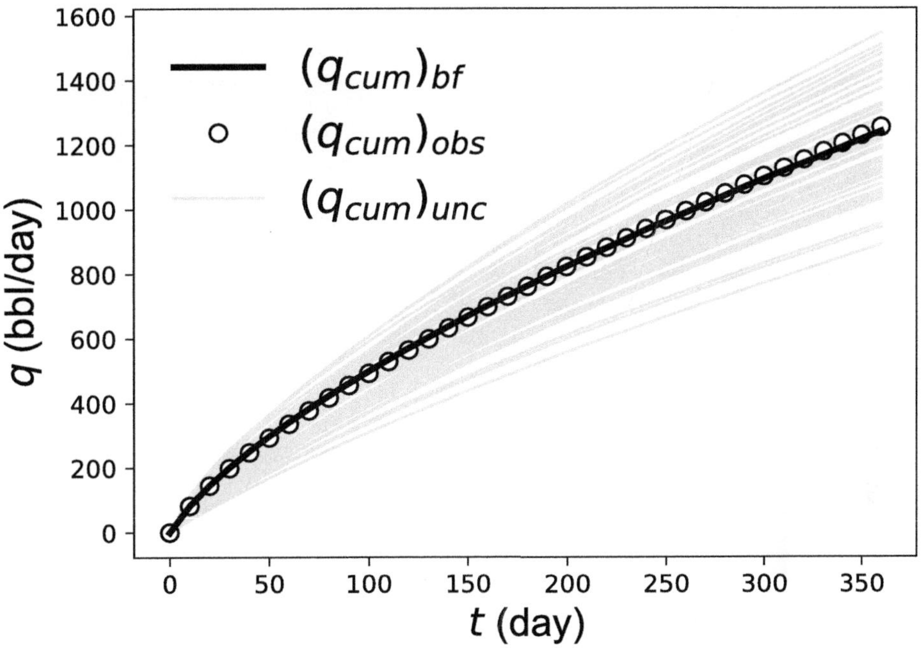

Fig. 15
Uncertainty depicted by 50 realizations of optimized fractured shale reservoir model. The "best-fitting" (most-likely) scenario of field permeability out of these 50 realizations is chosen as the one that produces the closest matching production profile with the field observation.

because the optimized fracture network is nonunique, i.e., there are multiple combinations of fracture parameters that result in identical field-scale permeability.

Spatial characterization of shale permeability in terms of its most common rock features (*iOM*, *OM*, fractures) allows for the exploitation of the shale reservoir in an economical way by screening out the areas that might otherwise prove to be an uneconomical prospect [3].

5.2 Permeability of Shale Using Well Logs

Permeability is generally not directly measured by well logs, but traditionally it has been estimated indirectly from well logs where a mathematical model acts as a bridge between the two [56]. The mathematical relationship that connects permeability with well logs is either obtained empirically from core samples or it is based on well-known models, such as Kozeny-Carmen (KC), that can work reasonably well in some conventional reservoirs. Advanced well logs such as nuclear magnetic resonance (NMR) have been used with empirical equations, such as Schlumber-Doll-Research (SDR) and Timur-Coates [19,57–59], to estimate permeability in reservoirs where the KC model does not work well. NMR measures the induced magnetic moment of hydrogen nuclei (protons) contained within the fluid-filled pore space, which is

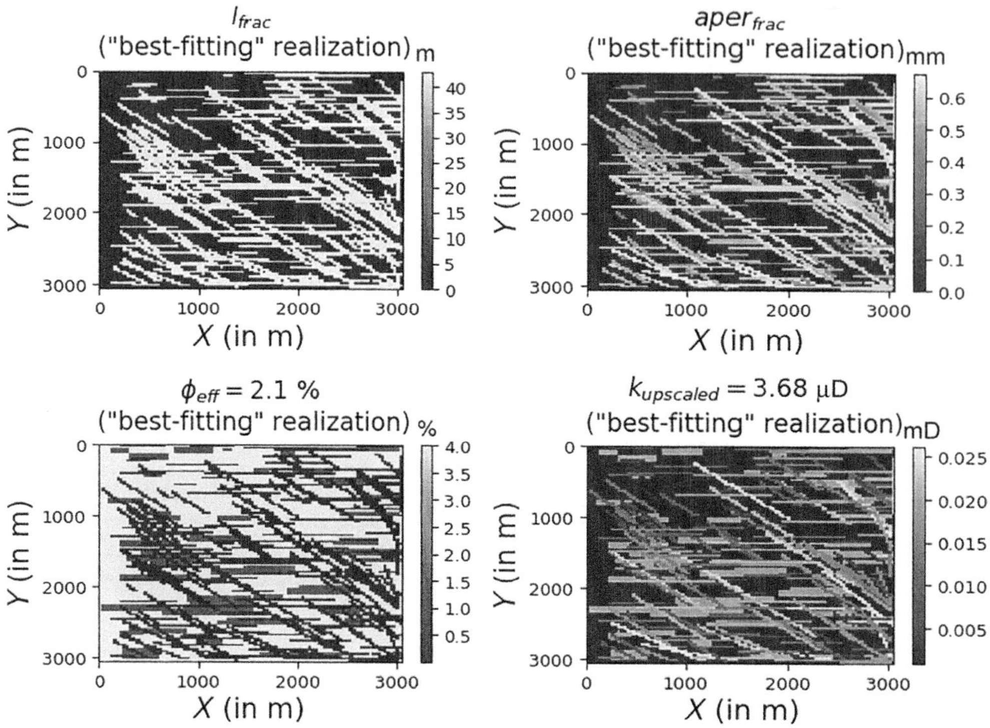

Fig. 16
"Best-fitting" realization of fracture lengths, apertures, porosity, and permeability from 50 realizations generated after optimization. The "best-fitting" realization is chosen as the one that produces the closest matching production profile with the field observation.

different from conventional well logs in that NMR only responds to the presence of hydrogen in fluid unlike conventional well logs (e.g., density, neutron, resistivity) that respond to both the matrix of the rock, hence its mineralogy, and fluid. NMR logs have been used to estimate permeability in tight sandstone reservoirs [60] with R10 and R35 models [61, 62] that represent pore throat radius at 10% and 35% mercury saturation [56], respectively. A high-resolution combinable magnetic resonance tool (CMR-Plus enhanced NMR) was used in Middle Bakken, southeastern Saskatchewan [63], to measure permeability with depth based on the SDR and Timur-Coates equation [19, 57–59]. It is found that permeability calculated by the SDR equation is usually higher than the permeability calculated by the Timur-Coates equation, where the difference between the two is significant for lower porosities (<9%).

5.2.1 Limitations of well log responses in shale

Unlike conventional reservoirs, permeability estimation using well logs for shale reservoirs is not a common practice, primarily because the reservoir properties of shale are difficult to characterize and often they do not depict a statistically significant correlation with permeability. Additionally, presence of kerogen, producible gas and high API gravity producible oils in shale reservoirs cause the logging tools to read unusually high (or low) values that present a challenge in interpreting logging tool responses [64]. This is because of the abundance of dispersed, insoluble sedimentary *OM* that is part of the solid matrix, which have log responses (e.g., density and hydrogen index) that are indistinguishable from pore fluids [1], which presents a major challenge for interpretation of well log responses in shale. For this reason, the empirical relationships that form a link between well logs and permeability in conventional reservoirs do not work well for shale. However, some recent studies [56, 65] have adapted the KC relationship used in conventional reservoirs by modifying the parameters that account for the unconventional behavior in reservoir properties of shale. Specifically, Shen et al. [56] modified the geometry term for the nonconductive medium in the factor of formation electrical resistivity and accounted for this modified parameter in the KC equation to obtain an equivalent relationship that can be used in unconventional glutenite reservoirs. Zhao et al. [65] used 38 samples from the Longmaxi formation in China to develop a correlation between specific surface area and rock composition based on the partial least square method, and the specific surface area was then used in Kozeny's equation to determine the permeability of shale. A similar approach had been used by Magara [66] in 1968 who treated shale as a medium with abnormally high pore pressure and modified the Archie's equation to estimate the permeability of shale using density porosity logs. Although these studies illustrate that their method works in the shale under study, their applicability to most shale reservoirs in general is questionable because of the two reasons explained earlier, which are (i) shale reservoir properties do not depict a statistically significant correlation with permeability, and (ii) the presence of kerogen, which is part of the solid matrix, exhibits log responses (e.g., density and hydrogen index) that are indistinguishable from pore fluid, therefore causing the logging tools to read unusually high (or low) values.

5.2.2 Prospective methods to estimate shale permeability using well logs

The first limitation in shale well logs regarding the lack of statistically significant correlation can be prospectively tackled by employing machine learning algorithms, e.g., using artificial neural networks (ANN) that can be trained with well log data, core permeability and other geological information [67, 68].

The second limitation related to shale well log (e.g., bulk densities, porosities, NMR) responses, which are unusually high (or low) due to the abundance of dispersed *OM*, can be prospectively tackled by employing image well logs that are not affected by the atypical response of *OM* to logs based on density or hydrogen index. Specifically, image well logs (e.g., FMI, FMS, EMI) can be used to extract fracture characteristics [69, 70] (e.g., fracture aperture, fracture intensity/abundance) that can be used to estimate shale permeability. Image well logs measure the fracture characteristics as a function of the button resistivity, tool current/voltage, mud resistivity and background rock resistivity. Fracture detection using image well logs can be verified with auxiliary methods such as mud loss data, which can be a reliable method to verify reservoir permeability, such that the depths exhibiting higher mud loss must correlate well with the fracture locations detected by the image well logs.

When image well logs are not available, a method proposed by Aghli et al. [71] can be used as a very rough proxy to detect fractures using conventional well logs, such as caliper, sonic and neutron density, by finding their second derivative. The second derivative of a variable helps to enhance the resolution of the data that can be used in conjunction with other auxiliary data (e.g., formation stratigraphy, cores, thin sections, mud loss versus depth) to verify the accuracy of fracture detection from conventional well logs. Although this method can provide some rough proxy to detect fractures, it cannot provide the aperture sizes required for permeability computation. It is recommended that a rough estimate of the field-scale shale permeability be obtained using the aperture sizes from the available thin sections or cores drilled from the formation along with the thicknesses of the depths where fractures are detected through conventional well logs.

6 Summary and Conclusions

This chapter critically discussed the permeability of shale from experimental and modeling perspectives, including the transport processes in shale that contribute to shale permeability at different time scales. This chapter also presented a physics-based model to predict two-phase relative permeability in shale fractures that does not require any experimental data or curve fitting. Further, this chapter discussed recent progress made in history-matching techniques to predict field-scale permeability of fractured shale in terms of its spatially varying distinct features (fractures, matrix, and organic matter) and the use of well logs in predicting field-scale shale permeability.

One of the major drawbacks in shale permeability modeling, with regards to the inability of various models to predict the upscaled permeability of a fractured shale with diverse permeable features, was removed by Singh and Cai [33] through a method that accounts for distinct permeable features through a finite difference grid containing fractures and matrix, and the permeability of the rock is estimated using an appropriate analytical expression of a feature described by each grid cell.

Two-phase relative permeability of scCO_2 and brine in the fracture was calculated using the model of Singh and Cai [33] for four different assumed values of residual brine saturation ($S_{wr}=0$, 0.1, 0.2, 0.4), which revealed that the relative permeability of gas is significantly reduced when the brine saturation increases beyond 0.2, and it is almost zero as brine saturation (S_w) approaches 0.7.

Field-scale permeability prediction was discussed using two methods: (i) history-match, and (ii) well logs. The history-match approach proposed by Singh and Cai [55] ensured consistency between the underlying fracture distribution and optimally-matched fracture lengths and their apertures, in addition to accounting for the random distribution of fractures and their abundance. Additionally, their method predicts the field-scale permeability in terms of different features of shale, including *iOM*, *OM*, and fractures. Two methods were recommended for estimating field-scale permeability of shale using well logs, which includes (i) machine learning approaches that can be trained with well log data, and (ii) detecting fractures using either image logs or conventional well logs in combination with aperture sizes obtained from thin sections or cores.

Acknowledgments

This research is supported in part by an appointment to the National Energy Technology Laboratory Research Participation Program, sponsored by the US Department of Energy and administered by the Oak Ridge Institute for Science and Education. This work is also supported by the National Natural Science Foundation of China (41722403; 41572116).

References

[1] P.R. Craddock, L. Mossé, C. Bernhardt, et al., Matrix-adjusted shale porosity measured in horizontal wells, Petrophysics (59) (2018) 288–307.
[2] H. Singh, A critical review of water uptake by shales, J. Nat. Gas Sci. Eng. 34 (2016) 751–766.
[3] H. Singh, J.C. Cai, Screening improved recovery methods in tight-oil formations by injecting and producing through fractures, Int. J. Heat Mass Transf. 116 (2018) 977–993.
[4] Y. Wang, D. Dasani, T.T. Tsotsis, et al., An experimental investigation of desorption kinetics and mass transfer in shale, in: SPE Low Perm Symposium, Denver, Colorado, USA 5-6 May, 2016.
[5] J. Wang, Z. Yang, M. Dong, et al., Experimental and numerical investigation of dynamic gas adsorption/desorption–diffusion process in shale, Energy Fuel 30 (2016) 10080–10091.
[6] F. Javadpour, D. Fisher, M. Unsworth, Nanoscale gas flow in shale gas sediments, J. Can. Pet. Technol. 46 (2007). https://doi.org/10.2118/07-10-06.
[7] H. Singh, F. Javadpour, A. Ettehadtavakkol, et al., Nonempirical apparent permeability of shale, SPE Reserv. Eval. Eng. 17 (2014) 414–424.
[8] R. Sander, Z. Pan, L.D. Connell, Laboratory measurement of low permeability unconventional gas reservoir rocks: a review of experimental methods, J. Nat. Gas Sci. Eng. 37 (2017) 248–279.
[9] Y. Gensterblum, A. Ghanizadeh, R.J. Cuss, et al., Gas transport and storage capacity in shale gas reservoirs—a review. Part A: transport processes, J. Unconv. Oil Gas Resour. 12 (2015) 87–122.
[10] S. Li, M. Dong, Z. Li, Measurement and revised interpretation of gas flow behavior in tight reservoir cores, J. Pet. Sci. Eng. 65 (2009) 81–88.
[11] W. Tanikawa, T. Shimamoto, Comparison of Klinkenberg-corrected gas permeability and water permeability in sedimentary rocks, Int. J. Rock Mech. Min. 46 (2009) 229–238.
[12] W.F. Brace, J.B. Walsh, W.T. Frangos, Permeability of granite under high pressure, J. Geophys. Res. 73 (1968) 2225–2236.
[13] R.L. Kranz, J.S. Saltzman, J.D. Blacic, Hydraulic diffusivity measurements on laboratory rock samples using an oscillating pore pressure method, Int. J. Rock Mech. Min. 27 (1990) 345–352.
[14] G.J. Fischer, The determination of permeability and storage capacity: pore pressure, in: Fault Mechanics and Transport Properties of Rocks, vol. 51, Academic Press, New York, 1992, pp. 187–211.
[15] D.L. Luffel, F.K. Guidry, Core-analysis results: comprehensive study wells, Devonian shale, Topical report, May, ResTech Houston, TX, https://www.osti.gov/biblio/7203749, 1989. Accessed 6 July 2018.

[16] D.L. Luffel, F.K. Guidry, New core analysis methods for measuring reservoir rock properties of Devonian shale, J. Pet. Technol. 44 (1992) 1184–1190.
[17] J. Kamath, Evaluation of accuracy of estimating air permeability from mercury-injection data, SPE Form. Eval. 7 (1992) 304–310.
[18] T.J. Marshall, A relation between permeability and size distribution of pores, J. Soil Sci. 9 (1958) 1–8.
[19] W.E. Kenyon, Petrophysical principles of applications of NMR logging, Log. Anal. 38 (1997) 21–43.
[20] X.J. Cui, A.M.M. Bustin, R.M. Bustin, Measurements of gas permeability and diffusivity of tight reservoir rocks: different approaches and their applications, Geofluids 9 (2009) 208–223.
[21] R. Eijpe, K.J. Weber, Mini-permeameters for consolidated rock and unconsolidated sand, AAPG Bull. 55 (1971) 307–309.
[22] R.M. Feng, J. Liu, S. Harpalani, Optimized pressure pulse-decay method for laboratory estimation of gas permeability of sorptive reservoirs: part 1—background and numerical analysis, Fuel 191 (2017) 555–564.
[23] R.M. Feng, S. Harpalani, J. Liu, Optimized pressure pulse-decay method for laboratory estimation of gas permeability of sorptive reservoirs: part 2—experimental study, Fuel 191 (2017) 565–573.
[24] F. Javadpour, Nanopores and apparent permeability of gas flow in mudrocks (shales and siltstone), J. Can. Pet. Technol. 48 (2009).
[25] F. Civan, Effective correlation of apparent gas permeability in tight porous media, Transp. Porous Media 82 (2010) 375–384.
[26] H. Darabi, A. Ettehad, F. Javadpour, et al., Gas flow in ultra-tight shale strata, J. Fluid Mech. 710 (2012) 641–658.
[27] A. Sakhaee-Pour, S. Bryant, Gas permeability of shale, SPE Reserv. Eval. Eng. 15 (2012) 401–409.
[28] K.L. Wu, X.F. Li, C.C. Wang, et al., A model for gas transport in microfractures of shale and tight gas reservoirs, AICHE J. 61 (2015) 2079–2088.
[29] M.E. Naraghi, F. Javadpour, A stochastic permeability model for the shale-gas systems, Int. J. Coal Geol. 140 (2015) 111–124.
[30] H. Singh, F. Javadpour, Langmuir slip-Langmuir sorption permeability model of shale, Fuel 164 (2016) 28–37.
[31] P. Cao, J.S. Liu, Y.K. Leong, A fully coupled multiscale shale deformation-gas transport model for the evaluation of shale gas extraction, Fuel 178 (2016) 103–117.
[32] Y. Wang, S.M. Liu, Y.X. Zhao, Modeling of permeability for ultra-tight coal and shale matrix: a multi-mechanistic flow approach, Fuel 232 (2018) 60–70.
[33] H. Singh, J.C. Cai, A feature-based stochastic permeability of shale: part 1—validation and two-phase permeability in a Utica shale sample, Transp. Porous Media (2018) 1–34.
[34] M. Sheng, G.S. Li, S.C. Tian, et al., A fractal permeability model for shale matrix with multi-scale porous structure, Fractals 24 (2016) 1650002.
[35] A. Beskok, G.E. Karniadakis, Report: a model for flows in channels, pipes, and ducts at micro and nano scales, Microsc. Thermophys. Eng. 3 (1999) 43–77.
[36] Y.D. Yuan, N.G. Doonechaly, S. Rahman, An analytical model of apparent gas permeability for tight porous media, Transp. Porous Media 111 (2016) 193–214.
[37] S.F. Wang, T. Wu, X.Y. Cao, et al., A fractal model for gas apparent permeability in microfractures of tight/shale reservoirs, Fractals 25 (2017) 1750036.
[38] L.D. Geng, G.S. Li, S.C. Tian, et al., A fractal model for real gas transport in porous shale, AICHE J. 63 (2017) 1430–1440.
[39] W.H. Song, J. Yao, Y. Li, et al., Fractal models for gas slippage factor in porous media considering second-order slip and surface adsorption, Int. J. Heat Mass Transf. 118 (2018) 948–960.
[40] J.J. Ma, Review of permeability evolution model for fractured porous media, J. Rock Mech. Geotech. Eng. 7 (2015) 351–357.
[41] L. Shi, Z.J. Zeng, B. Bai, et al., Effect of the intermediate principal stress on the evolution of mudstone permeability under true triaxial compression, Greenhouse Gases Sci. Technol. 8 (2018) 37–50.
[42] S. Jenni, L.Y. Hu, R. Basquet, et al., History matching of a stochastic model of field-scale fractures: methodology and case study, Oil Gas Sci. Technol. 62 (2007) 265–276.

[43] M.L. Porter, J. Jiménez-Martínez, R. Martinez, et al., Geo-material microfluidics at reservoir conditions for subsurface energy resource applications, Lab Chip 15 (2015) 4044–4053.
[44] C.-Y. Chen, R.N. Horne, Two-phase flow in rough-walled fractures: experiments and a flow structure model, Water Resour. Res. 42 (2006) W03430.
[45] D.S. Oliver, Y. Chen, Recent progress on reservoir history matching: a review, Comput. Geosci. 15 (2011) 185–221.
[46] R.W. Rwechungura, M. Dadashpour, J. Kleppe, Advanced history matching techniques reviewed, in: SPE Middle East Oil and Gas Show and Conference, Manama, Bahrain, 25-28 September, 2011.
[47] J. Caers, History matching under training-image-based geological model constraints, SPE J. 8 (2003) 218–226.
[48] B. Jafarpour, D.B. McLaughlin, History matching with an ensemble Kalman filter and discrete cosine parameterization, Comput. Geosci. 12 (2008) 227–244.
[49] S. Jung, K. Lee, C. Park, et al., Ensemble-based data assimilation in reservoir characterization: a review, Energies 11 (2018) 445.
[50] T. Gang, M.G. Kelkar, Efficient history matching in naturally fractured reservoirs, in: SPE/DOE Symposium on Improved Oil Recovery, Tulsa, Oklahoma, USA, 22-26 April, 2006.
[51] L. Lu, D. Zhang, Assisted history matching for fractured reservoirs by use of Hough-transform-based parameterization, SPE J. 20 (2015) 942–961.
[52] S. Nejadi, J.Y. Leung, J.J. Trivedi, et al., Integrated characterization of hydraulically fractured shale gas reservoirs production history matching, SPE Reserv. Eval. Eng. 18 (2015) 481–494.
[53] H. Chang, D. Zhang, History matching of stimulated reservoir volume of shale-gas reservoirs using an iterative ensemble smoother, SPE J. 23 (2018) 346–366.
[54] Z. Chai, B. Yan, J.E. Killough, et al., An efficient method for fractured shale reservoir history matching: the embedded discrete fracture multi-continuum approach, J. Pet. Sci. Eng. 160 (2018) 170–181.
[55] H. Singh, J.C. Cai, A feature-based stochastic permeability of shale: part 2—predicting field-scale permeability, Transp. Porous Media (2018) 1–18.
[56] B. Shen, D. Wu, Z.H. Wang, A new method for permeability estimation from conventional well logs in glutenite reservoirs, J. Geophys. Eng. 14 (2017) 1268.
[57] W.E. Kenyon, P.I. Day, C. Straley, et al., A three-part study of NMR longitudinal relaxation properties of water-saturated sandstones, SPE Form. Eval. 3 (1988) 622–636.
[58] G.R. Coates, L. Xiao, M.G. Prammer, NMR Logging: Principles and Applications, Halliburton Energy Services, Houston, TX, 1999.
[59] C. Smith, L. Hamilton, Carbonate reservoir permeability from nuclear magnetic resonance logs, in: International Petroleum Technology Conference, Kuala Lumpur, Malaysia, 10-12 December, 2014.
[60] L. Xiao, X.P. Liu, C.C. Zou, et al., Comparative study of models for predicting permeability from nuclear magnetic resonance (NMR) logs in two Chinese tight sandstone reservoirs, Acta Geophys. 62 (2014) 116–141.
[61] S.J. Kolodzie, Analysis of pore throat size and use of the Waxman-Smits equation to determine OOIP in Spindle Field, Colorado, in: SPE Annual Technical Conference and Exhibition, Dallas, Texas, 21-24 September, 1980.
[62] R. Rezaee, A. Saeedi, B. Clennell, Tight gas sands permeability estimation from mercury injection capillary pressure and nuclear magnetic resonance data, J. Pet. Sci. Eng. 88 (2012) 92–99.
[63] K. Hu, D. Kohlruss, C. Yang, et al., Porosity and Permeability Evaluation for the Devonian-Mississippian Lower Middle Bakken Member in the Viewfield Pool, Southeastern Saskatchewan, Geological Survey of Canada, Ottawa, Canada, 2017.
[64] R. Freedman, D. Rose, B. Sun, et al., Novel method for evaluating shale gas and shale tight oil reservoirs using well log data, in: SPE Annual Technical Conference and Exhibition, Dubai, UAE, 26-28 September, 2016.
[65] P.Q. Zhao, J.C. Cai, Z.H. Huang, et al., Estimating permeability of shale gas reservoirs from porosity and rock compositions, Geophysics 83 (2018) MR283–MR294.
[66] K. Magara, Porosity-permeability relationship of shale, J. Jpn. Assoc. Petrol. Technol. 34 (1) (1969) 1–7.
[67] E.-M. Shokir, A.A. Alsughayer, A. Al-Ateeq, Permeability estimation from well log responses, J. Can. Pet. Technol. 45 (2006) 41–46.

[68] X. Zhao, A. Popa, I. Ershaghi, et al., Permeability prediction in a heterogeneous reservoir using soft computing technologies, in: SPE Western Regional Meeting, Garden Grove, California, USA, 22-26 April, 2018.
[69] H. Xian, C. Tueckmantel, M. Giorgioni, et al., Fracture characterisation through image log conductive feature extraction, in: Offshore Technology Conference Asia, Kuala Lumpur, Malaysia, Offshore Technology Conference, 20-23 March, 2018.
[70] G. Aghli, B. Soleimani, S.S. Tabatabai, et al., Calculation of fracture parameters and their effect on porosity and permeability using image logs and petrophysical data in carbonate Asmari reservoir, SW Iran, Arab. J. Geosci. 10 (2017) 265.
[71] G. Aghli, B. Soleimani, R. Moussavi-Harami, et al., Fractured zones detection using conventional petrophysical logs by differentiation method and its correlation with image logs, J. Pet. Sci. Eng. 142 (2016) 152–162.

CHAPTER 7

Pore Structure, Wettability, and Their Coupled Effects on Tracer-Containing Fluid Migration in Organic-Rich Shale

Rui Yang*, Yongsheng Liu†, Sheng He*, Qinhong Hu‡, Li Zhang§

**Key Laboratory of Tectonics and Petroleum Resources, Ministry of Education, China University of Geosciences, Wuhan, P.R. China,* †*State Key Laboratory of Geological Processes and Mineral Resources, Faculty of Earth Sciences, China University of Geosciences, Wuhan, P.R. China,* ‡*Department of Earth and Environmental Sciences, The University of Texas at Arlington, Arlington, TX, United States,* §*Petroleum Exploration and Development, Jianghan Oilfield Branch Company, Sinopec, Wuhan, P.R. China*

Chapter Outline

Petrophysical Characterization and Fluids Transport in Unconventional Reservoirs. https://doi.org/10.1016/B978-0-12-816698-7.00007-3

1 Introduction

With the increasing demand in crude oil and natural gas, black shale formations are emerging as an important reservoir of energy across the world. As an extremely complex and heterogeneous porous medium, shale is often characterized by a high organic matter content, low porosity and permeability with pore sizes down to the nm scale, though some pores can reach up to the micrometer or sub-millimeter level [1–3]. A large amount of natural gas is usually found in typical shale nanoscale pore systems; however, understanding the pore structure (geometry and connectivity) of these shale reservoirs with an extremely low matrix permeability (commonly in the nanodarcy [nD] range) is still a challenging task because of the limitations of applicable characterization techniques [4, 5]. Recently, some experimental techniques have been applied to characterize pore systems in different gas-producing shale formations, including high-resolution imaging methods, such as field emission scanning electron microscopy (FE-SEM), CT-scan, focused ion beam-SEM (FIB-SEM), and transmission electron microscopy (TEM) and indirect methods, such as mercury intrusion capillary pressure (MICP), low-pressure gas physisorption, spontaneous imbibition (SI), and small-angle neutron scattering/ultrasmall angle neutron scattering (SANS/USANS) [2, 6–18].

Being a critical petrophysical parameter, wettability can affect rock-fluid properties (e.g., relative permeability and capillary pressure), which is significant for understanding the hydrocarbon production behavior [19–21]. Accurate evaluation of rock wettability is also important for production optimization, proper investigation of fracturing fluid loss, and selection of suitable fracturing fluids and additives [22]. For decades, several quantitative and qualitative methods have been used to evaluate rock wettability, such as contact angle measurement [23–25], SI [4, 26–28], displacement capillary pressure [29], Amott wettability index [30], and nuclear magnetic resonance (NMR) [31–33].

Though there are reports on the pore structure (geometry and connectivity) and wettability of shale in the United States (e.g., Gao and Hu [4]; Hu et al. [34]; Tinni et al. [35]), related studies on organic-rich gas-producing shale with thermal maturity levels over 2.0% within dry gas generation windows (referred to as over-mature shale [36]) are not available in the literature. In addition, it has been gradually accepted that a combination of the complex microscopic (e.g., pore size distribution and pore connectivity) and wettability characteristics control fluid flow and mass transport from the matrix storage area to the stimulated pore-fracture network [37]. However, experimental investigations on the coupled effects of pore structure and wettability are lacking in the literature.

Coupled with geochemical and mineralogical characterization, we conduct complementary tests to investigate the pore connectivity and wettability characteristics of the Longmaxi shale, including MICP, contact angle measurement, SI into initially dry shale, and tracer diffusion into fluid-saturated shale followed by the elemental mapping of laser ablation-inductively coupled plasma-mass spectrometry (LA-ICP-MS). Moreover, we also discuss the coupled effects of pore structure and wettability on tracer-containing fluid migration in organic-rich shale.

2 Samples and Methods

A total of 13 core samples with different depths from four shale gas wells (JY-A, PY-B, JY-C, and JY-D) drilled in the southeast Sichuan Basin are obtained for this study. All these samples are from Early Silurian Longmaxi shale and accurately represent this organic-rich gas-producing shale reservoir. Previous studies indicate that Longmaxi shale is rich in organic matter (0.46–6.79 wt%, with an average of 2.54 wt% [38]), over-mature [39, 40], and dominated by type I kerogen [38, 41]. More information about the geological setting can be found in previous works [42, 43]. According to the sample requirements of different tests, such as, cubic shape needed for MICP and SI tests, while powder for X-ray diffraction (XRD) and TOC measurement; each core sample is dry cut into several subsamples.

2.1 TOC and XRD Analyses

Using a CS-230 Analyzer with a precision of ±2%, the TOC content of the powder samples is determined by subtracting the total inorganic carbon (removed using 7% hydrochloric acid) from total carbon. According to the Chinese Oil and Gas Industry Standard (SY/T) 5163-2010, mineralogical composition is measured using an X-ray diffractometer (X'Pert PRO MPD from Panalytical Company) with a scan angle at a range of 3–65 degrees at a step of 0.02 degrees. The working voltage and current is 40 kV and 40 mA, respectively.

2.2 Mercury Intrusion Capillary Pressure

As a widely accepted and well established method for characterizing the pore structure of porous media, MICP is an effective method to investigate both the geometrical (e.g., porosity, pore-throat distribution, and pore volume) [44] and topological (e.g., permeability and tortuosity) characteristics [44–47]. Using an Automatic Mercury Porosimeter (Micromeritics AutoPore IV 9510 with the maximum pressure of approximately 60,000 psi; 1 psi = 0.006895 MPa), MICP tests are made following standard procedures with an equilibration time of 10 s in the low pressure analyses (5–30 psi) and 40 s in the high pressure (30–60,000 psi) analyses. The surface tension and contact angle of mercury are adopted as 485 dyne/cm and 130°, respectively [48]. Assuming that the pores are cylindrical in shape with a circular opening in cross-section, the pore-throat diameter at a certain applied pressure is calculated based on the Young-Laplace Equation [49, 50].

2.3 Spontaneous Imbibition and Contact Angle Measurement

SI is a capillary force controlled process, during which a wetting fluid displaces a nonwetting one in a porous media under the influence of capillary suction only [28, 51, 52]. Here, except for the top and bottom surfaces (transverse to the lamination), the other four sides of the cubic samples are coated with quick-cure epoxy to minimize vapor adsorption into the sample during the SI tests. A simple schematic diagram for the SI test is shown in Sun et al. [53]. Both deionized (DI) water and *n*-decane are used as the imbibing fluids to displace air in the pores. The weight change of the cubic samples over imbibition time is monitored and automatically recorded by a high-precision electronic balance. Repetitive measurements of DI water imbibition on the same sample (redrying between runs) are conducted to guarantee the reliability of the SI results, while *n*-decane imbibition is only performed once on a cube.

In addition, contact angles of the three types of fluid (DI water, API [*American Petroleum Institute*] brine, and *n*-decane) are also measured to quantitatively evaluate the relative adhesion of fluid and air to the Longmaxi shale surfaces, as well as the fractional wettability behavior [32, 54]. The physical properties of these fluids involved in imbibition and contact angle measurement analyses are shown in Table 1. Fluid contact angle tests are conducted on an optical contact angle and interface tension meter (SL200KB, KINO Industry) under ambient laboratory conditions. To minimize the effect of heterogeneity of a rock surface, fluid contact angles for each sample are measured on three different random spots on the flat surface. Thus the measured contact angles generally reflect the averaged wettability characteristics.

2.4 Saturated Diffusion With Tracer-Containing API Brine

Saturated diffusion with tracer-containing API brine (an aqueous phase) in the Longmaxi shale is conducted to investigate the spatial continuity of water-wet pores [55]. With the concentration gradient as a driving force, saturated diffusion tests are also used to quantify different tracer diffusion behaviors in the connected hydrophilic pores. The side-epoxied cubic samples are oven-dried, evacuated and vacuum saturated with tracer-free API brine under atmospheric pressure [53]; the bottom face of the samples are exposed to a tracer-containing API brine reservoir to maintain a constant-concentration boundary condition.

Table 1 Physical properties of fluids involved in imbibition tests

Fluid	Viscosity (mPa s)	Interfacial Tension With Air (mN/m)	Density (g/cm^3)
Air (STP)	0.018	/	0.0018
Deionized (DI) water	1.002	72.80	0.998
API brine	1.193	~65.00	1.071
n-decane	0.840	23.83	0.730

API = American Petroleum Institute; STP = Standard temperature and pressure.

After a diffusion time of 26 h, the cubic samples are removed from the fluid reservoir, flash-frozen by liquid nitrogen, freeze-dried at vacuum (−53°C and 9 Pa) for over 48 h, and stored dry in a desiccator before LA-ICP-MS tracer mapping analysis.

The LA-ICP-MS (LA: UP-213, New Wave; ICP-MS: PerkinElmer/SCIEX ELAN DRC II) tracer mapping analysis is used to investigate the tracer migration in connected pores that are not visible from SEM images. LA-ICP-MS tracer mapping analysis is also conducted to map the tracer concentration at a spatial distribution to generate 2D/3D maps (including top, bottom, and interior surfaces). The detection limit of the LA-ICP-MS technique can reach as low as mg/kg level and data reduction follows the work of Peng et al. [56].

3 Composition Characteristics in the Longmaxi Shale

Overall, the Longmaxi shale, rich in organic matter, is mainly dominated by quartz and clay minerals. According to the mineral composition characteristics, the lithology of these shale samples can be classified into three types: argillaceous siliceous shale, silica-rich argillaceous shale, and siliceous shale. The TOC contents for these argillaceous siliceous shales vary from 1.02 to 3.22 wt% with an average of 2.17 wt%, and from 0.93 to 2.63 wt% (average 1.94 wt%) for silica-rich argillaceous shales. In contrast, siliceous shales have much higher TOC contents, varying from 3.2 to 4.06 wt%, with an average of 3.52 wt%.

4 Pore Structure Characteristics

The results of typical pore structure parameters from the MICP tests are compiled in Table 2. Overall, the Longmaxi shales have extremely low porosities and the pore-throat diameters for the connected pore networks are predominantly in the nm size range (Table 2). The porosities of the argillaceous siliceous shales (average 1.71%) are comparable with the measured values in siliceous shales (average 1.80%). In contrast, the silica-rich argillaceous shales have slightly larger porosities and smaller average pore-throat diameters. The connected pore spaces with pore-throat sizes of 3–10 nm generally contribute to more than half of the total pore volume in Longmaxi shale (Table 2).

Fig. 1 shows the MICP results for the three types of lithology in the Longmaxi shale with obvious hysteresis observed in the plots of cumulative intrusion and extrusion volume vs. the applied pressure. Approximately 50% of the intruded mercury remain trapped inside the pore spaces after depressurization (Fig. 1A), indicating the relatively complex pore structure in the Longmaxi shale. During the extrusion process, as for ink-bottle shaped pores (large pores connected by a narrow pore throat), the wider pore body cannot empty until a smaller pressure is reached, which can result in mercury entrapment in the wider and inner portion of the pore space [57]. Thus, the observed hysteresis is most likely related to the

Table 2 Pore structure characteristics from MICP analyses

Well Name	Sample ID	Lithology	TOC (wt%)	Porosity (%)	Total Pore Volume (cm^3/100 g)	Total Pore Area (cm^2/g)	Average Pore-Throat Diameter (nm)[a]	Pore Network of 3–10 nm Pore-Throat Sizes			
								Volume (%)	Permeability (nD)	Effective Tortuosity τ	Geometrical Tortuosity L_e/L[b]
JY-A	LM-A2371	Argillaceous siliceous shale	1.53	1.87	0.70	4.19	6.7	62.0	1.71	1702	5.63
JY-A	LM-A2379	Silica-rich argillaceous shale	2.63	2.36	0.91	5.50	6.6	58.6	1.93	3941	9.64
JY-A	LM-A2383	Argillaceous siliceous shale	3.22	1.65	0.63	3.93	6.4	57.6	1.18	4439	8.55
JY-A	LM-A2402	Siliceous shale	4.06	2.35	0.94	7.26	5.2	73.0	1.67	2879	8.23
PY-B	LM-B2076	Silica-rich argillaceous shale	0.93	1.31	0.50	2.14	9.3	59.4	17.5	914	3.46
PY-B	LM-B2138	Argillaceous siliceous shale	2.34	1.66	0.64	2.93	8.7	52.6	17.3	689	3.38
PY-B	LM-B2142	Argillaceous siliceous shale	2.89	1.29	0.50	1.73	11.5	57.8	10.2	1114	3.80
JY-C	LM-C2303	Argillaceous siliceous shale	1.02	2.05	0.79	5.68	5.6	68.4	1.75	2615	7.32
JY-C	LM-C2329	Siliceous shale	3.20	1.64	0.64	4.05	6.4	56.9	1.14	3504	7.57
JY-C	LM-C2347	Siliceous shale	3.35	2.23	0.88	6.25	5.6	65.5	1.66	2511	7.49
JY-D	LM-D2579	Silica-rich argillaceous shale	2.27	2.08	0.80	3.88	8.3	44.7	1.57	6880	11.95
JY-D	LM-D2590	Argillaceous siliceous shale	2.03	1.73	0.66	3.10	8.5	43.3	1.21	4242	8.57
JY-D	LM-D2601	Siliceous shale	3.45	0.99	0.39	1.92	8.0	48.2	7.18	3612	5.97

TOC = wt% total organic carbon.

[a]Average pore-throat diameter is calculated as 4V/A (V: total pore volume; A: total pore area) based on the assumption of cylindrical pores.

[b]L: the straight-line distance between the two points; L_e: the actual diffusion distance between the same two points and is usually larger than L ($L_e > L$). Geometrical tortuosity is defined as $L_e/L = \sqrt{\emptyset D_o / D_e}$ (Hu et al., 2015); Ø: Porosity.

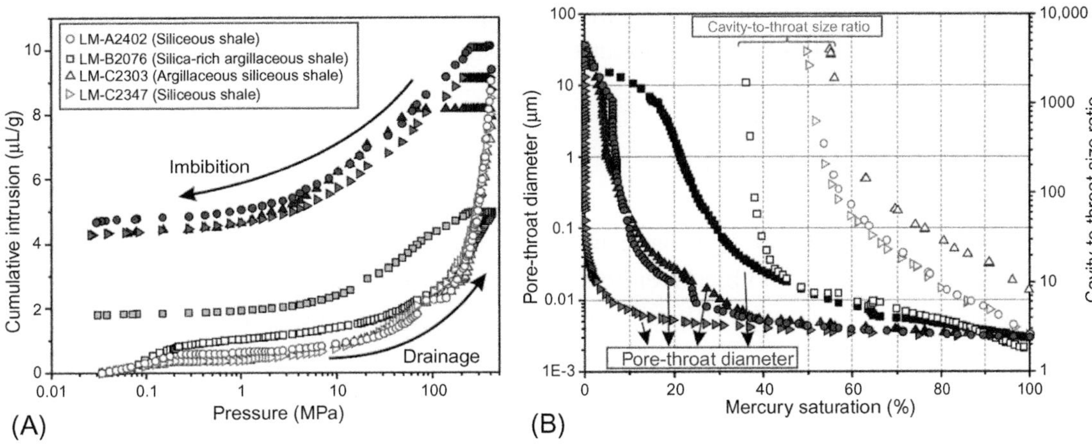

Fig. 1

MICP results for three types of lithology in the Longmaxi shale. (A) Curves of drainage (*mercury intrusion, open symbols*) and imbibition (*mercury extrusion, solid symbols*) from MICP data with obvious hysteresis; (B) Relationships between mercury saturation vs. pore-throat diameter (*solid symbols*) and cavity-to-throat size ratio (*open symbols*).

ink-bottle effect caused by ink-bottle shaped pores, which is confirmed by the high resolution FE-SEM images on the Longmaxi shale [52, 58].

The plots between mercury saturation vs. pore-throat diameter also suggest that most of the pores in the Longmaxi shale are in the meso-macropore size range (Fig. 1B). As for the siliceous and argillaceous siliceous shales, with an increase of applied pressure, approximately 80%–90% of the mercury intrudes the connected pores with pore-throat sizes smaller than 20 nm. In contrast, silica-rich argillaceous shales develop some macropores with pore-throat diameters from about 10 μm to 100 nm, accounting for about 40% of the intruded mercury. The micrometer-sized macropores in the Longmaxi shales are visible in previous FE-SEM images [13, 59], which are probably related to: (1) the microfractures formed from abnormal high pressure in situ [60]; (2) stress relaxation caused by a decrease in the over burden load [10, 61]; (3) ex-situ factors (sample preparation or drying), or a combination of these factors [60]. The cavity-to-throat size ratios drop rapidly with the decrease of pore-throat diameter, which is especially obvious in silica-rich argillaceous shales.

The cavity-to-throat size ratio values in Longmaxi shale decrease rapidly from several thousands to several tens with the increase of mercury saturation (Fig. 1B). Based on percolation theory, the estimated matrix permeability for the pore spaces with a pore-throat diameter of 3–10 nm from MICP data for all the Longmaxi shales are in the nD range (1.14–17.5 nD), which is comparable to Barnett shale (2–5 nD) [34]. The nD-scale permeability values indicate that it is extremely difficult for fluids (e.g., mercury) to flow through the connected pore spaces in shale. In addition, the effective tortuosity and geometrical tortuosity obtained from MICP data for the pore network with pore-throat size of 3–10 nm vary from 689 to 6880 and from 3.38 to 11.95, respectively, also suggesting the presence of highly tortuous pathways.

5 Connectivity from SI Tests

Pore connectivity, an important topology parameter, can affect fluid migration and movement (advection and diffusion) in both conventional and unconventional reservoirs. Many approaches (e.g., gas adsorption, FIB-SEM images, SI, nano-CT, and pore-scale network modeling) have been attempted to evaluate the pore connectivity of a porous media [35, 62–65]. Among these methods, the imbibition slope obtained from the SI test is a quick, economic, and effective method to evaluate pore connectivity [52, 64].

As a capillary-force-dominated process, the SI rate is often collectively controlled by both the properties and interactions of porous media and fluid (e.g., pore connectivity, wettability, fluid viscosity, and interfacial tension) [4, 5, 51, 66–68]. Pore-scale modeling studies reveal that the progress of diffusion-type fronts in a porous medium can be used to investigate the pore connectivity levels [52, 69]. Especially, when the gravitational effects are negligible, SI characteristics can be mathematically analogous to the process of molecular diffusion, which exhibits macroscopic behaviors from microscale pore connections [34]. Namely, the pore connectivity of a porous medium can be effectively evaluated by fluid imbibition behaviors, as denoted by the resulting slope of the log scale cumulative imbibition height vs. the log scale imbibition time [64].

Considering the predominant nanopores in the Longmaxi shale, capillary forces act as the driving forces behind SI tests while gravitational forces are negligible [70, 71]. In a water-air system with negligible gravitational forces, Handy model [70, 71] is widely used to describe the cocurrent imbibition process:

$$V_{imb} = \left[\frac{2P_C k_w \emptyset S_w A_C^2}{\mu_w} t\right]^{0.5} \quad (1)$$

where V_{imb} is the total volume of imbibed fluid into the shale sample; P_c is the capillary pressure; k_w is the effective permeability of the porous medium towards a wetting fluid; ϕ is the sample porosity (fraction); S_w is the fractional water content of the pores; parameter A_c is the cross-sectional area; t is the imbibition time; and μ_w is the fluid viscosity.

Philip [72] developed a simplified equation to describe fluid imbibition/infiltration into soils:

$$H = St^{0.5} + At \quad (2)$$

where H is the average cumulative imbibition height of the wetting front (equals V_{imb}/A_C); S is a constant referred to as sorptivity; and A is an empirical parameter that is closely related to the sample properties (e.g., lithology, porosity, and permeability). When the gravitational force is much smaller than the capillary pressure gradient, the intercept (At) in Eq. 2 can be neglected. Thus, a linear relationship can be theoretically obtained between the cumulative imbibition height and the square root of the imbibition time with a slope of 0.5. Here, SI data are plotted as cumulative imbibition height vs. imbibition time in a log-log plot to compare the pore connectivity characteristics between different samples.

The DI water imbibition curves in log-log scales for the three types of Longmaxi shale are shown in Fig. 2. Fig. 2A presents the water imbibition curves of parallel and vertical silica-rich argillaceous shale to compare the effect of bedding on the DI water imbibition behaviors. A much higher slope value is obtained in the parallel sample, suggesting that the DI water imbibition behavior is closely related to the bedding direction, and the parallel sample has better pore connectivity than the vertical one. As for DI water imbibition, a repetitive imbibition test is conducted on each shale sample (oven-dried between runs) to examine the reproducibility of the imbibition slopes. Table 3 shows that the first DI water imbibition slopes varied from 0.316 to 0.671 (average 0.53), while the second imbibition slopes are between 0.191 and 0.605 (average 0.40). The Longmaxi shale has a relatively high clay content, in which illite is the dominant clay mineral [38]. Therefore, the strong interaction between the water molecular and clay minerals could change the pore structure (e.g., pore blocking caused by clay mineral swelling) and finally affects the imbibition behaviors and slope values. Thus the movement of DI water within the connected pore network of the Longmaxi shale could be inhibited to some extent, resulting in the decrease of DI water imbibition slopes during the second imbibition experiment. In other words, the strong water-rock interaction probably accounts for the decrease of the water-wet pore spaces in the Longmaxi shale. The different imbibition slopes between runs also suggest that the water-rock interaction is quite complex and can exert an important effect on the pore structure in the Longmaxi shale.

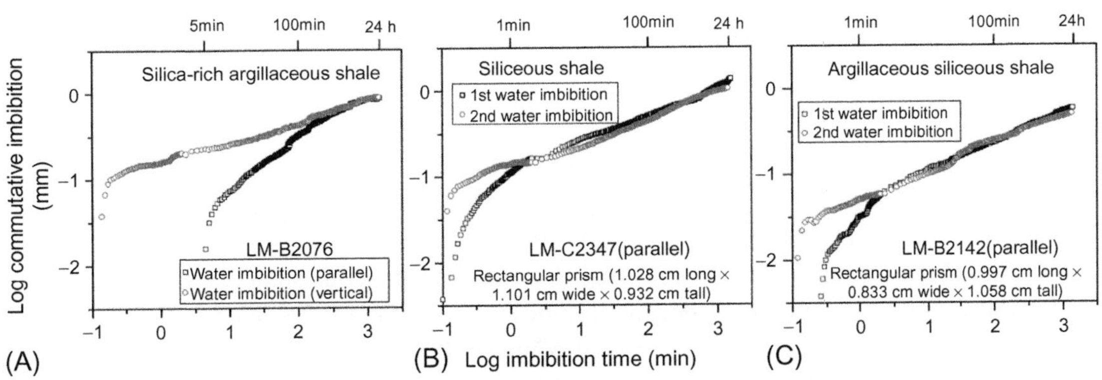

Fig. 2

Cumulative DI water imbibition (mm) vs time (min) in log scales for the typical Longmaxi shales: (A) silica-rich argillaceous shale; (B) siliceous shale; (C) argillaceous siliceous shale.

Table 3 Equilibrium contact angles and imbibition results

Sample ID	DI Water Imbibition Slope		*n*-Decane Imbibition Slope	Contact Angle (degree)		
	1st	2nd		DI Water	API Brine	*n*-Decane
LM-A2371	0.589	0.354	0.529	2.8	23.1	~0
LM-A2379	0.598	0.477	0.484	3.4	16.3	~0
LM-A2383	0.469	0.417	0.496	20.7	41.2	~0
LM-A2402	0.493	0.407	0.632	10.4	25.8	~0
LM-B2076	0.605	0.483	0.529	9.5	16.7	~0
LM-B2138	0.649	0.605	0.551	4.9	11.2	~0
LM-B2142	0.649	0.334	0.499	10.4	11.6	~0
LM-C2303	0.437	0.340	0.557	7.1	15.8	~0
LM-C2329	0.436	0.376	0.448	16.5	23.7	~0
LM-C2347	0.316	0.340	0.716	9.7	46.1	~0
LM-D2579	0.444	0.392	0.490	12.8	20.1	~0
LM-D2590	0.671	0.469	0.917	9.4	16.8	~0
LM-D2601	0.524	0.191	0.792	12.3	31.1	~0

Contact angle values after fluid-sample contact time of 30 s are presented.

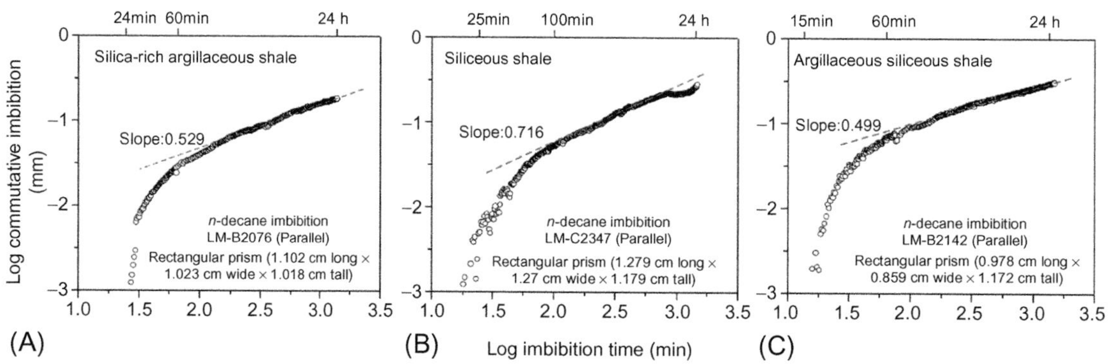

Fig. 3

Cumulative *n*-decane imbibition (mm) vs time (min) in log scales for the typical Longmaxi shales: (A) Silica-rich argillaceous shale; (B) Siliceous shale; (C) Argillaceous siliceous shale.

To minimize the effect of water-rock interaction on the pore structure (pore connectivity) of the Longmaxi shale, in this work, we use the DI water imbibition slopes in the first run to evaluate the water-wet pore connectivity and *n*-decane imbibition slopes for oil-wet pore connectivity (Figs. 2 and 3; Table 3). As for silica-rich argillaceous shales, the slopes for the first DI water imbibition and the *n*-decane imbibition are around 0.5, indicating that both the water-wet and oil-wet pore networks are well-connected. Similar fluid (DI water and the *n*-decane) imbibition slopes are obtained in the argillaceous siliceous shales. In contrast, the siliceous shales have slightly smaller DI water imbibition slopes and slightly larger *n*-decane imbibition slopes, indicating the good pore connectivity for oil-wet pore spaces and moderately good pore connectivity for water-wet pores. It is noted that both the DI water imbibition and *n*-decane show extremely high imbibition slopes, with some of the values even larger than the theoretical value of 0.5, which is probably related to the pore spaces between bedding planes to generate a good pathway for fluid movement in the Longmaxi shale.

According to the percolation theory [73, 74], both the water-wet and oil-wet pores in the Longmaxi shale generally exhibit a well-connected pore network. Compared to water-wet pores, oil-wet pore spaces in siliceous shale have a much better pore connectivity, while water-wet pores in silica-rich argillaceous shale have slightly higher DI water imbibition slopes than *n*-decane, suggesting a better water-wet pore connectivity.

A weak negative correlation occurs between the DI water imbibition slopes and MICP porosity values (Fig. 4), which probably indicates that the effects of the water-rock interaction on Longmaxi shale samples with higher porosity values are extremely obvious, resulting in the decrease of water-wet pore connectivity. In contrast, the *n*-decane imbibition slopes for other Longmaxi shales generally exhibit a weakly positive relationship with the MICP porosity, indicating that the Longmaxi shale with high porosities have better connected oil-wet pathways for *n*-decane migration. Even though two shale samples do not fall in the same areas with other samples, they also show a positive tendency with an increase of porosity. Considering the abundant organic matter-related nanopores and the strongly positive correlation between porosity and the TOC contents in the Longmaxi shale [13, 52], it is reasonable to speculate that those developed organic matter-related nanopores have significantly influences on the oil-wet pore connectivity by increasing the total porosity.

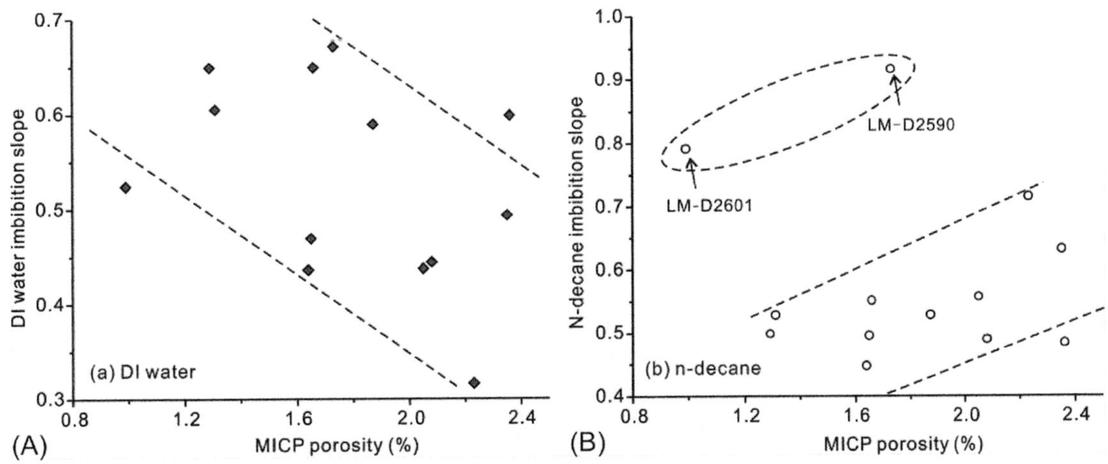

Fig. 4
Relationships between fluid imbibition slope and porosity obtained from MICP tests: (A) DI water imbibition; (B) *n*-decane imbibition.

6 Wettability

6.1 Wettability From SI Tests

Both DI water (an aqueous phase) and *n*-decane (a surrogate oil phase) are used in the SI tests, and air is regarded as occupying the displaced nonwetting phase in capillary imbibition. Thus, the wettability of shale matrices can be evaluated based on the imbibition behaviors towards different imbibing fluids. To compensate for differences in porosity, sample volumes, fluid densities between the samples, Gao and Hu [4] proposed a dimensionless weight term W_D to compare the wettability characteristics of Barnett shales for different imbibing fluids, which is defined as:

$$W_D = \frac{W}{\rho \varnothing V} \tag{3}$$

where W is the imbibed fluid weight (g); ρ is the fluid density (g/cm^3); and V is the sample volume (cm^3). The dimensionless time is defined as [4]:

$$t_D = t\sqrt{\frac{k}{\varnothing}}\frac{\sigma}{L_c^2\sqrt{\mu_w \mu_{nw}}} \tag{4}$$

where t_D and t are dimensionless time and imbibition time (s), respectively; k is the absolute permeability (m^2); σ is the interfacial tension (N/m); and μ_w and μ_{nw} are the viscosity of wetting and nonwetting phases (Pa s), respectively. L_c is the characteristic length (m).

Fig. 5 shows the plots of dimensionless weight (W_D) vs. dimensionless time (t_D) in the log scales for both DI water and *n*-decane imbibition tests in the three types of Longmaxi shale. As for silica-rich argillaceous shale, the DI water imbibition slope is slightly larger than *n*-decane in the log dimensionless W_D vs. log dimensionless t_D plot (Fig. 5A), suggesting that the

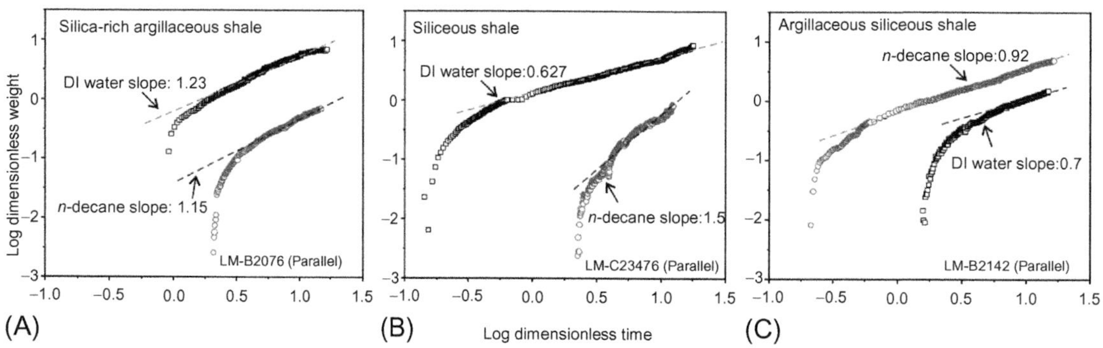

Fig. 5
Plots of log dimensionless weight (W_D) vs log dimensionless time (t_D) for both DI water and *n*-decane imbibition tests in the three types of Longmaxi shale: (A) silica-rich argillaceous shale; (B) siliceous shale; (C) argillaceous siliceous shale.

water-wet pores are well developed in the silica-rich argillaceous shale. Due to the relatively low TOC and high clay contents in the silica-rich argillaceous shales (average 1.94 wt%), organic matter-related pores are poorly developed and more water-wet inorganic pores are preserved, which is also confirmed by the high resolution SEM images [75]. Siliceous shales have the highest TOC contents (average of 3.54 wt%), and the nanoscale organic matter-related pores are well developed, with most of them irregular, bubble-like, elliptical and faveolate in shape [52]. Considering the high thermal maturities of the Longmaxi shale [76], it is appropriate to postulate that residual kerogen particles in the Longmaxi shale are oil-wetting. Therefore, the high contents of residual kerogen particles and abundant oil-wet organic matter-related pores account for the obviously higher *n*-decane imbibition slopes than DI water in siliceous shales (Fig. 5B), resulting in the higher oil-wet property. In contrast, argillaceous siliceous shales have moderate TOC contents (average 2.17 wt%), and both the water-wet inorganic and oil-wet organic matter-related pores are well developed. Fig. 5C exhibits comparable imbibition slopes between DI water and *n*-decane, suggesting the pronounced mixed-wet characteristics but also strongly oil-wet.

6.2 Wettability From Contact Angle Measurement

The results of contact angles for three types of fluids (DI water, API brine, and *n*-decane) on the Longmaxi shale are given in Table 3. As for silica-rich argillaceous shales, droplets of DI water and API brine on air-dry flat surfaces of the Longmaxi shale spread quickly and the measured contact angles for API brine are slightly larger than the DI water, and similar wettability characteristics are observed in the argillaceous siliceous shales. In contrast, DI water and API brine remain as a bead in the presence of air on the air-dry flat surfaces of siliceous shales and their measurable contact angles are shown in Table 3. According to the wettability classifications from contact angles [54, 77, 78], silica-rich argillaceous shales and argillaceous siliceous shales exhibit strong water-wetting characteristics, while siliceous shales are moderately strong water-wetting, which is generally consistent with the fluid imbibition slopes shown in Fig. 5.

In contrast, the rapid and complete spreading behaviors of *n*-decane on all the sample surfaces result in the development of a zero contact angle, revealing that the Longmaxi shale has a strong affinity towards *n*-decane and can be classified as strongly oil-wet. The strongly oil-wetting for all the Longmaxi shale is also confirmed by the *n*-decane imbibition slope values being much larger than the theoretical value of 0.5.

Overall, based on the different spreading behaviors of fluids, the Longmaxi shale are mixed-wet but are also strongly oil-wet. In addition, it is noted that even though the accuracy of the contact angles is difficult, which may have been affected by the surface roughness, complex geometry and heterogeneity of the shale [66, 79], the averaged contact angles reported here reflect the samples' wettability characteristics recorded at mm observational scales.

7 Coupled Effects of Pore Structure and Wettability on Tracer-Containing API Brine Migration

Characterized by low porosity and extremely low permeability (on a nD scale), diffusive flux is one of the major mass transport mechanisms within shale matrixes [80, 81]. Figs. 6–8 show the LA-ICP-MS mapping results for the tracer concentration profiles after a diffusion time of 26h for the silica-rich argillaceous shale, siliceous shale and argillaceous siliceous shale, respectively. Color variations observed at different locations (especially for the interior face) directly reflect the migration behaviors of different tracers within the shale samples. Thus, tracer distributions, including nonsorbing (ReO_4^- on the left), weakly sorbing (Cs^+ in the middle) and strongly sorbing (Eu^{3+} on the right), on the lower, interior and upper faces in the API brine saturated shale samples are easy to identify. Due to the different reactivities for these tracers, the nonsorbing ReO_4^- only occupies connected pore spaces and does not interact

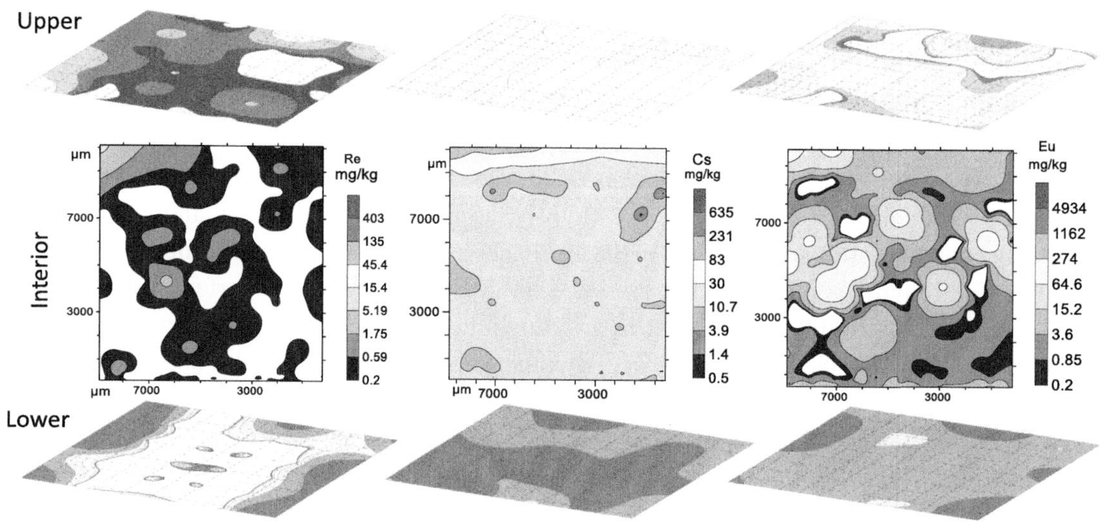

Fig. 6
Tracer concentration distributions after 26h of diffusion into saturated argillaceous siliceous shale (LM-C2303) from the Longmaxi Formation. Upper and lower surfaces are shown as horizontal cross-sections of the sample, and the interior is shown as a vertical cross-section of the sample. Panels for each tracer include three sample faces (upper and lower surfaces denote horizontal cross-sections of the sample, and the interior is shown as a vertical cross-section of the sample; each dashed line for the upper and lower faces is positioned 1000 μm apart). The LA-ICP-MS detection limits for Re, Cs, and Eu tracer elements are 0.1, 0.5, and 0.2 mg/kg, respectively (shown as the lowest values in the scale bars and the change in color directly denotes the extent of tracer penetration); white areas represent regions with concentrations falling below detection limits. The depth of a crater formed from laser ablation on a shale sample is 100 μm in spot size and the number of laser pluses is 20 pulses.

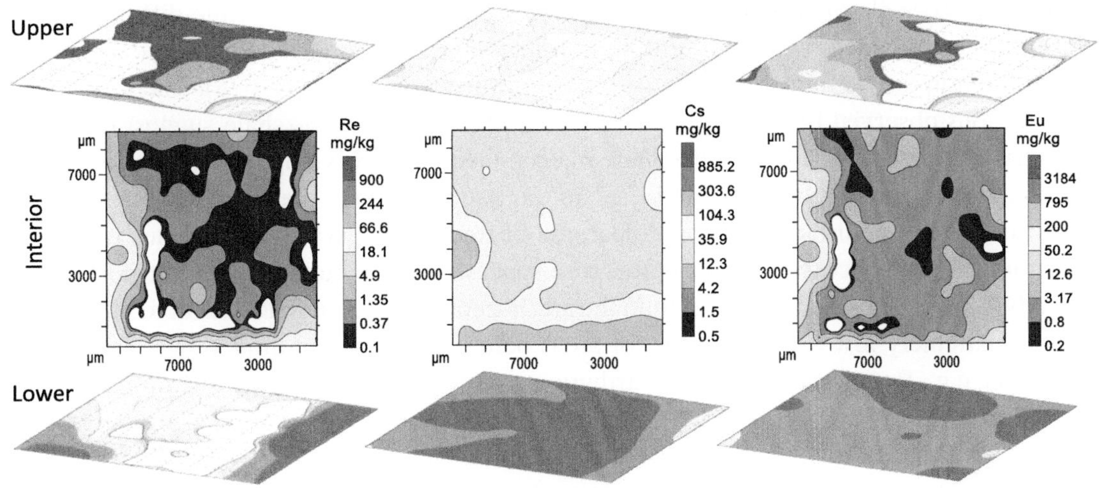

Fig. 7

Tracer concentration distributions after 26 h of diffusion into saturated silica-rich argillaceous shale (LM-A2379) from the Longmaxi Formation. Upper and lower surfaces are shown as horizontal cross-sections of the sample, and the interior is shown as a vertical cross-section of the sample.

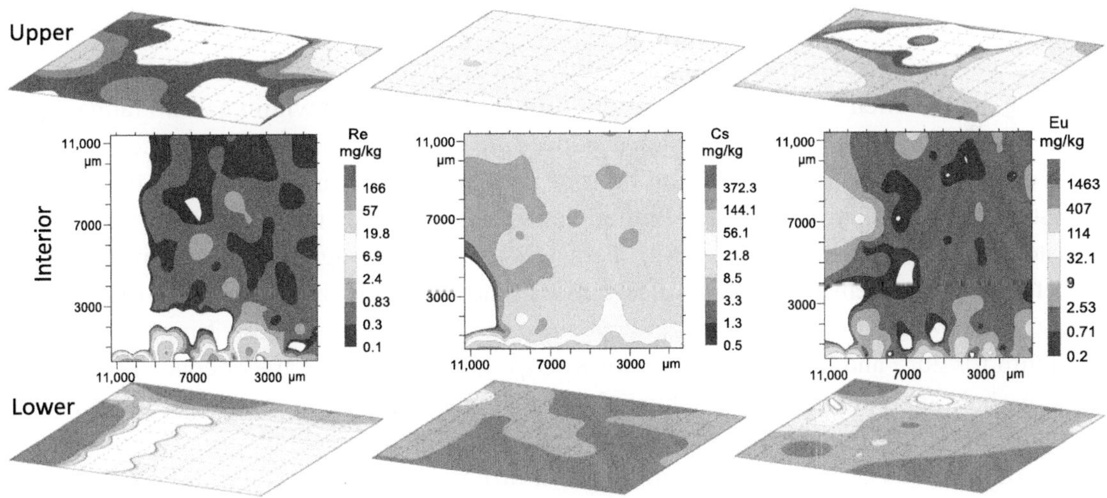

Fig. 8

Tracer concentration distributions after 26 h of diffusion into saturated siliceous shale (LM-A2402) from the Longmaxi Formation. Upper and lower surfaces are shown as horizontal cross-sections of the sample, and the interior is shown as a vertical cross-section of the sample.

with the shale matrix, while the cations (the weakly sorbing Cs^+ and strongly sorbing Eu^{3+}) could interact with shale minerals for retarded transport. Thus, the different tracer distribution can also demonstrate the effect between the cations and the clay minerals on the movement of tracers.

Overall, the tracer distribution profiles exhibit sporadic characteristics for the diffusion front in the interior faces after 26 h of diffusion into saturated samples, with the highest concentration in the bottom faces and the lowest in the top faces (Figs. 6–8). The low tracer concentrations observed in the interior faces of argillaceous siliceous shale suggest that few tracers have diffused through into the shale matrix. Overall, the average concentrations of nonsorbing ReO_4^- and sorbing Cs^+ and Eu^{3+} in the bottom surface are much larger than the values in the interior surface. The rapid decrease of tracer concentrations indicates the distribution of edge-accessible hydrophilic pore spaces and very limited levels of pore connectivity for API brine diffusion within the argillaceous siliceous shale. It is noted that there are some extremely low concentrations (or areas below the detection limit shown as white areas, e.g., ReO_4^- concentration distribution in the Fig. 6) also observable in the interior and top profiles, which is probably caused by the complex pore structure. Evidence from MICP, SEM and gas physisorption show that the Longmaxi shale are dominated by nanoscale organic matter-related pores and most of them are irregular, bubble-like, elliptical, and faveolate in shape [13, 15, 52], resulting in the high levels of geometrical tortuosity and small pore-throat size. In addition, the SANS, MICP and imbibition results show that Longmaxi shale are very impermeable and the percentage of dead pores increases with the TOC content, which can be associated with the presence of numerous dead-end pores that form "backbone pores" for fluid flow and chemical transport [15, 27, 52].

A similar phenomenon is observed for the silica-rich argillaceous shale (Fig. 7) and siliceous shale (Fig. 8), in which relative concentrations drop roughly 10-fold within a 2 mm range from the lower faces. When the tracer diffusion distances are larger than about 2 mm, concentrations barely exceed the background detection limit of both the sorbing and nonsorbing tracers, suggesting that hydrophilic pores are limited-connected for molecular movement; this is consistent with the moderately water-wet characteristics of the Longmaxi shale found from the contact angle measurements. As for silica-rich argillaceous shale, high concentrations are mainly limited within the sub-mm range above the base of the sample (within 2 mm for Cs^+, less than 1 mm for Eu^{3+} and less than 1.5 mm for ReO_4^-). In contrast, high tracer concentrations in siliceous shale are mainly limited within <3 mm from the bottom edge (within 3 mm for Cs^+, less than 1 mm for Eu^{3+}, and less than 2 mm for ReO_4^-). The different diffusion distances for tracer (sorbing and nonsorbing) concentrations in different Longmaxi shales reflect the variation of edge-connected hydrophilic pores for tracer movement. As cationic Cs^+ has a smaller molecular size than ReO_4^-, its migration distances are slightly larger than those of ReO_4^- in silica-rich argillaceous shale and siliceous shale. The comparable distances for nonsorbing and sorbing tracers may also suggest that Cs^+-shale matrix interactions are not that strong, though Cs^+ can exhibit retardation from cation exchange.

Fig. 9 also shows that relative concentrations (the detected concentration divided by the averaged base concentration) of sorbing and nonsorbing tracers drop rapidly in the Longmaxi shales from the bottom edge upward, a factor of roughly 50 with a diffusion distance of 2–3 mm upward from the interior face. It is noted that a few locations (or sites) in both the

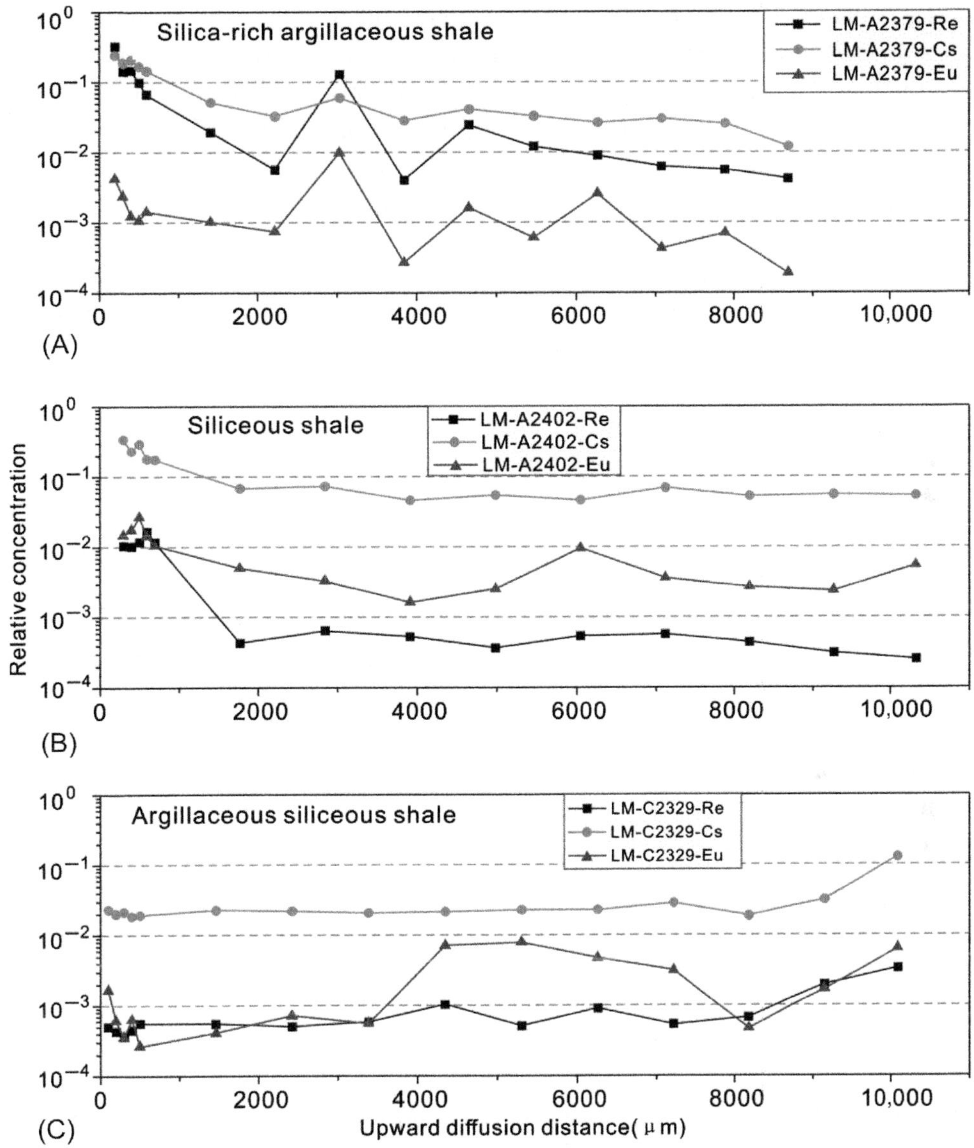

Fig. 9
The relative concentration (ratio of the average bottom concentration) of the nonsorbing (ReO_4^-) and sorbing tracers (Cs^+ and Eu^{3+}) profiles after 26 h of diffusion across an upward imbibition distance in the interior surface into three typical shale types (silica-rich argillaceous shale, siliceous shale and argillaceous siliceous shale).

interior and the top faces show relatively high concentrations, which is most likely to be related to the presence of preferential pathways along the wall. Overall, the small pore size, low porosity, and limited edge-pore connectivity reported here may partially account for the extremely slow movement of the fluid within the shale matrix.

Although tracer movement observed from the saturated-diffusion tests in Longmaxi shale is a poor proxy for running experiments under in situ conditions, our results can provide a reference for fluids (oil, gas, and water) flow and mass transport processes (from either imbibition or diffusion) occurring during hydraulic fracturing and shut-in periods. The practical implication is that due to the limited edge-connected pore spaces in the Longmaxi shale, the natural gas will migrate very slowly from the shale matrix's interior to the hydraulically created fractures, and initially produced natural gas is mainly stored within or close to the connected pore spaces near the fracture-matrix interface. Thus, the coupled effect of complex pore structure (nanoscale pore-throat size, high tortuosity, and limited pore connectivity) and mixed wettability may explain the steep decrease of gas production after its initial years with a relatively high recovery rate of gas extraction and production. However, the natural gas molecules need to overcome a high capillary force from water-wetting inorganic minerals before gas molecules transport into a pore-fracture network.

8 Conclusions

We have investigated the pore structure (connectivity) and wettability of several Longmaxi shale samples from the southeast of the Sichuan Basin, as well as their coupled effects on tracer-containing fluid migration. Consistent with literature reports on many shale types, pore spaces in Longmaxi shale are predominantly at the nanometer scale with the permeability values in the nD range; pores with pore-throat sizes of 3–10 nm contribute to more than half of the total pore volume.

Longmaxi shale is strongly oil-wetting and moderately strong water-wetting. Concentrations of nonsorbing and sorbing tracers decrease rapidly a few mm from the edge of the sample, which is associated with the predominant nanopore space and associated high tortuosity (highly restricted pathways), as well as the mixed wettability in Longmaxi shale. Thus, due to the limited edge-connected pore spaces and low diffusion coefficients, the migration of gas could be very slow from the shale matrix to hydraulically created fractures, with the initially high production of hydrocarbons from connected pore spaces near the sample edge. To our best knowledge, this work is beneficial to understand the migration and storage of fluids (oil, gas, and water) in matrix-fracture systems, which also provide scientific basis for sustainable shale oil/gas development in other shale gas basins.

Acknowledgments

We thank SINOPEC Exploration Company and SINOPEC Jianghan Oilfield Branch Company for providing the shale samples used in this study. We are particularly grateful for the financial support from the National Postdoctoral Program for Innovative Talents (No. BX20180279), Project Funded by China Postdoctoral Science Foundation (No. 2018M640740), China Geological Survey Project Grants (No. DD20160185), and China National Science and Technology Major Projects (No. 2016ZX05034-002-003).

References

[1] K.L. Milliken, M. Rudnicki, D.N. Awwiller, et al., Organic matter-hosted pore system, Marcellus formation (Devonian), Pennsylvania, AAPG Bull. 97 (2013) 177–200.

[2] R.G. Loucks, R.M. Reed, S.C. Ruppel, et al., Morphology, genesis, and distribution of nanometer-scale pores in siliceous mudstones of the Mississippian Barnett shale, J. Sediment. Res. 79 (2009) 848–861.

[3] L.T. Ko, R.G. Loucks, S.C. Ruppel, et al., Origin and characterization of Eagle Ford pore networks in the south Texas Upper Cretaceous shelf, AAPG Bull. 101 (2017) 387–418.

[4] Z. Gao, Q. Hu, Wettability of Mississippian Barnett shale samples at different depths: investigations from directional spontaneous imbibition, AAPG Bull. 100 (2016) 101–114.

[5] J. Cai, B. Yu, A discussion of the effect of tortuosity on the capillary imbibition in porous media, Transp. Porous Media 89 (2011) 251–263.

[6] T. Dong, N.B. Harris, K. Ayranci, et al., Porosity characteristics of the Devonian Horn River shale, Canada: insights from lithofacies classification and shale composition, Int. J. Coal Geol. 141 (2015) 74–90.

[7] R.G. Loucks, R.M. Reed, S.C. Ruppel, et al., Spectrum of pore types and networks in mudrocks and a descriptive classification for matrix-related mudrock pores, AAPG Bull. 96 (2012) 1071–1098.

[8] M. Pommer, K. Milliken, Pore types and pore-size distributions across thermal maturity, Eagle Ford Formation, southern Texas, AAPG Bull. 99 (2015) 1713–1744.

[9] S. Bernard, B. Horsfield, H.-M. Schulz, et al., Geochemical evolution of organic-rich shales with increasing maturity: a STXM and TEM study of the Posidonia shale (lower Toarcian, northern Germany), Mar. Petrol. Geol. 31 (2012) 70–89.

[10] J. Klaver, G. Desbois, J.L. Urai, et al., BIB-SEM study of the pore space morphology in early mature Posidonia shale from the Hils area, Germany, Int. J. Coal Geol. 103 (2012) 12–25.

[11] S. Hemes, G. Desbois, J. Urai, et al., Variations in the morphology of porosity in the Boom Clay Formation: insights from 2D high resolution BIB-SEM imaging and Mercury injection Porosimetry, Neth. J. Geosci. 92 (2013) 275–300.

[12] H.E. King, A.P.R. Eberle, C.C. Walters, et al., Pore architecture and connectivity in gas shale, Energ. Fuel. 29 (2015) 1375–1390.

[13] R. Yang, S. He, J. Yi, et al., Nano-scale pore structure and fractal dimension of organic-rich Wufeng-Longmaxi shale from Jiaoshiba area, Sichuan Basin: investigations using FE-SEM, gas adsorption and helium pycnometry, Mar. Petrol. Geol. 70 (2016) 27–45.

[14] M. Mastalerz, A. Schimmelmann, A. Drobniak, et al., Porosity of Devonian and Mississippian New Albany shale across a maturation gradient: insights from organic petrology, gas adsorption, and mercury intrusion, AAPG Bull. 97 (2013) 1621–1643.

[15] R. Yang, S. He, Q. Hu, et al., Applying SANS technique to characterize nano-scale pore structure of Longmaxi shale, Sichuan Basin (China), Fuel 197 (2017) 91–99.

[16] J. Cai, B. Yu, M. Zou, et al., Fractal characterization of spontaneous co-current imbibition in porous media, Energ. Fuel. 24 (2010) 1860–1867.

[17] J. Cai, E. Perfect, C.-L. Cheng, et al., Generalized modeling of spontaneous imbibition based on Hagen-Poiseuille flow in tortuous capillaries with variably shaped apertures, Langmuir 30 (2014) 5142–5151.

[18] M. Sun, B. Yu, Q. Hu, et al., Pore structure characterization of organic-rich Niutitang shale from China: small angle neutron scattering (SANS) study, Int. J. Coal Geol. 186 (2018) 115–125.

[19] K. Wu, Z. Chen, J. Li, et al., Wettability effect on nanoconfined water flow, P. Natl. Acad. Sci. U. S. A. (2017). 201612608.

[20] Q. Lan, H. Dehghanpour, J. Wood, et al., Wettability of the Montney tight gas formation, SPE Reserv. Eval. Eng. 18 (2015) 417–431.

[21] W.G. Anderson, Wettability literature survey part 5: The effects of wettability on relative permeability, J. Petrol. Technol. 39 (1987) 1453–1468.

[22] A. Javaheri, H. Dehghanpour, J.M. Wood, Tight rock wettability and its relationship to other petrophysical properties: a Montney case study, J. Earth Sci. 28 (2017) 381–390.

[23] R. Johnson, R. Dettre, Wettability and contact angles, Surf. Colloid Sci. 2 (1969) 85–153.
[24] M.R. Yassin, M. Begum, H. Dehghanpour, Organic shale wettability and its relationship to other petrophysical properties: a Duvernay case study, Int. J. Coal Geol. 169 (2017) 74–91.
[25] L. Liang, D. Luo, X. Liu, et al., Experimental study on the wettability and adsorption characteristics of Longmaxi formation shale in the Sichuan Basin, China, J. Nat. Gas Sci. Eng. 33 (2016) 1107–1118.
[26] T. Chaturvedi, J.M. Schembre, A.R. Kovscek, Spontaneous imbibition and wettability characteristics of Powder River Basin coal, Int. J. Coal Geol. 77 (2009) 34–42.
[27] R. Yang, Q. Hu, S. He, et al., Pore structure, wettability and tracer migration in four leading shale formations in the Middle Yangtze Platform, China, Mar. Petrol. Geol. 89 (2018) 415–427.
[28] S. Ma, X. Zhang, N. Morrow, et al., Characterization of wettability from spontaneous imbibition measurements, J. Can. Petrol. Technol. 38 (1999). 94–47.
[29] S.C. Jones, W.O. Roszelle, Graphical techniques for determining relative permeability from displacement experiments, J. Petrol. Technol. 30 (1978) 807–817.
[30] E. Amott, Observations relating to the wettability of porous rock, AIME 216 (1960) 156–162.
[31] R. Brown, I. Fatt, Measurements of fractional wettability of oil fields rocks by the nuclear magnetic relaxation method, in: Fall Meeting of the Petroleumr Branch of AIME, Los Angeles, California, 14–17 October, 1956.
[32] A. Tinni, E. Odusina, I. Sulucarnain, et al., Nuclear-magnetic-resonance response of brine, oil, and methane in organic-rich shales, SPE Reserv. Eval. Eng. 18 (2015) 400–406.
[33] S.H. Al-Mahrooqi, C.A. Grattoni, A.H. Muggeridge, et al., Pore-scale modelling of NMR relaxation for the characterization of wettability, J. Pet. Sci. Eng. 52 (2006) 172–186.
[34] Q. Hu, R.P. Ewing, H.D. Rowe, Low nanopore connectivity limits gas production in Barnett formation, J. Geophys. Res-Solid Earth 120 (2015) 8073–8087.
[35] A. Tinni, C. Sondergeld, C. Rai, Pore connectivity between different wettability systems in organic-rich shales, SPE Reserv. Eval. Eng. 20 (2017) 1020–1027.
[36] B.P. Tissot, D.H. Welte, Petroleum Formation and Occurrence, second ed., Springer, Berlin, 1984.
[37] F. Javadpour, Nanopores and apparent permeability of gas flow in mudrocks (shales and siltstone), J. Can. Petrol. Technol. 48 (2009) 16–21.
[38] T. Guo, H. Zhang, Formation and enrichment mode of Jiaoshiba shale gas field, Sichuan Basin, Petrol. Explor. Dev. 41 (2014) 31–40.
[39] T. Guo, Evaluation of highly thermally mature shale-gas reservoirs in complex structural parts of the Sichuan Basin, J. Earth Sci. 24 (2013) 863–873.
[40] M. Wang, X. Xiao, Q. Wei, et al., Thermal maturation of solid bitumen in shale as revealed by Raman spectroscopy, Nat. Gas Geosci. 26 (2015) 1712–1718.
[41] T. Guo, Discovery and characteristics of the Fuling shale gas field and its enlightenment and thinking, Front. Earth Sci. 23 (2016) 29–43.
[42] D. Yan, D. Chen, Q. Wang, et al., Predominance of stratified anoxic Yangtze Sea interrupted by short-term oxygenation during the Ordo-Silurian transition, Chem. Geol. 291 (2012) 69–78.
[43] X. Chen, J. Rong, Y. Li, et al., Facies patterns and geography of the Yangtze region, South China, through the Ordovician and Silurian transition, Palaeogeogr. Palaeoclimatol. Palaeoecol. 204 (2004) 353–372.
[44] P.A. Webb, An Introduction to the Physical Characterization of Materials by Mercury Intrusion Porosimetry with Emphasis on Reduction and Presentation of Experimental Data, Micromeritics Instrument Corp, Norcross, Georgia, 2001.
[45] A. Abell, K. Willis, D. Lange, Mercury intrusion porosimetry and image analysis of cement-based materials, J. Colloid Interf. Sci. 211 (1999) 39–44.
[46] Z. Gao, Q. Hu, Estimating permeability using median pore-throat radius obtained from mercury intrusion porosimetry, J. Geophys. Eng. 10 (2013) 025014.
[47] Q. Hu, X. Liu, Z. Gao, et al., Pore structure and tracer migration behavior of typical American and Chinese shales, Pet. Sci. 12 (2015) 651–663.
[48] H. Gan, S.P. Nandi, P.L. Walker, Nature of the porosity in American coals, Fuel 51 (1972) 272–277.

[49] G.R. Chalmers, R.M. Bustin, I.M. Power, Characterization of gas shale pore systems by porosimetry, pycnometry, surface area, and field emission scanning electron microscopy/transmission electron microscopy image analyses: examples from the Barnett, Woodford, Haynesville, Marcellus, and Doig units, AAPG Bull. 96 (2012) 1099–1119.

[50] K. Bui, I. Akkutlu, B. Li, Capillary pressure in nanopores: deviation from Young-Laplace Equation, in: SPE Europec featured at 79th EAGE Conference and Exhibition, Paris, France, 12–15 June, 2017.

[51] J. Cai, S. Guo, L. You, et al., Fractal analysis of spontaneous imbibition mechanism in fractured-porous dual media reservoir, Acta Phys. Sin. 62 (2013) 014701.

[52] R. Yang, F. Hao, S. He, et al., Experimental investigations on the geometry and connectivity of pore space in organic-rich Wufeng and Longmaxi shales, Mar. Petrol. Geol. 84 (2017) 225–242.

[53] M. Sun, B. Yu, Q. Hu, et al., Pore connectivity and tracer migration of typical shales in South China, Fuel 203 (2017) 32–46.

[54] A.Y. Dandekar, Petroleum Reservoir Rock and Fluid Properties, second ed., Taylor & Francis Group, LLC, 2013.

[55] Q. Hu, X. Mao, Application of laser ablation-inductively coupled plasma-mass spectrometry to studies of chemical diffusion, sorption, and transport in natural rock, Geochem. J. 46 (2012) 459–475.

[56] S. Peng, Q. Hu, R.P. Ewing, et al., Quantitative 3-D elemental mapping by LA-ICP-MS of a basaltic clast from the Hanford 300 area, Washington, USA, Environ. Sci. Technol. 46 (2012) 2025–2032.

[57] F. Moro, H. Böhni, Ink-bottle effect in mercury intrusion porosimetry of cement-based materials, J. Colloid Interf. Sci. 246 (2002) 135–149.

[58] P. Wang, Z. Jiang, W. Ji, et al., Heterogeneity of intergranular, intraparticle and organic pores in Longmaxi shale in Sichuan Basin, South China: evidence from SEM digital images and fractal and multifractal geometries, Mar. Petrol. Geol. 72 (2016) 122–138.

[59] Y. Ma, Z. Pan, N. Zhong, et al., Experimental study of anisotropic gas permeability and its relationship with fracture structure of Longmaxi shales, Sichuan Basin, China, Fuel 180 (2016) 106–115.

[60] A. Ougier-Simonin, F. Renard, C. Boehm, et al., Microfracturing and microporosity in shales, Earth-Sci. Rev. 162 (2016) 198–226.

[61] M.E. Houben, A. Barnhoorn, J. Lie-A-Fat, et al., Microstructural characteristics of the Whitby mudstone formation (UK), Mar. Petrol. Geol. 70 (2016) 185–200.

[62] H. Liu, L. Zhang, N. Seaton, Determination of the connectivity of porous solids from nitrogen sorption measurements—II. Generalisation, Chem. Eng. Sci. 47 (1992) 4393–4404.

[63] M.V. López-Ramón, J. Jagiełło, T.J. Bandosz, et al., Determination of the pore size distribution and network connectivity in microporous solids by adsorption measurements and Monte Carlo simulation, Langmuir 13 (1997) 4435–4445.

[64] Q. Hu, R.P. Ewing, S. Dultz, Low pore connectivity in natural rock, J. Contam. Hydrol. 133 (2012) 76–83.

[65] J. Klaver, S. Hemes, M. Houben, et al., The connectivity of pore space in mudstones: insights from high-pressure Wood's metal injection, BIB-SEM imaging, and mercury intrusion porosimetry, Geofluids 15 (2015) 577–591.

[66] H. Dehghanpour, H. Zubair, A. Chhabra, et al., Liquid intake of organic shales, Energ. Fuel. 26 (2016) 5750–5758.

[67] C.C. Mattax, J. Kyte, Imbibition oil recovery from fractured, water-drive reservoir, SPE J. 2 (1962) 177–184.

[68] J. Cai, X. Hu, D.C. Standnes, et al., An analytical model for spontaneous imbibition in fractal porous media including gravity, Colloids Surf. A Physicochem. Eng. Asp. 414 (2012) 228–233.

[69] R.P. Ewing, R. Horton, Diffusion in sparsely connected pore spaces: temporal and spatial scaling, Water Resour. Res. 38 (2002) 1285.

[70] L. Handy, Determination of effective capillary pressures for porous media from imbibition data, Trans. AIME 219 (1960) 75–80.

[71] K. Li, R.N. Horne, An analytical scaling method for spontaneous imbibition in gas/water/rock systems, SPE J. 9 (2004) 322–329.

[72] J.R. Philip, The theory of infiltration: 4. Sorptivity and algebraic infiltration equations, Soil Sci. 84 (1957) 257–264.
[73] R. Larson, L. Scriven, H. Davis, Percolation theory of two phase flow in porous media, Chem. Eng. Sci. 36 (1981) 57–73.
[74] A. Hunt, R. Ewing, B. Ghanbarian, Percolation Theory for Flow in Porous Media, Springer, 2014.
[75] S. Zhang, Z. Meng, Z. Guo, et al., Characteristics and major controlling factors of shale reservoirs in the long maxi Fm, Fuling area, Sichuan Basin, Nat. Gas Ind. 34 (2014) 16–24.
[76] R. Yang, S. He, Q. Hu, et al., Geochemical characteristics and origin of natural gas from Wufeng-Longmaxi shales of the Fuling gas field, Sichuan Basin (China), Int. J. Coal Geol. 171 (2017) 1–11.
[77] W.G. Anderson, Wettability literature survey-part 3: the effects of wettability on the electrical properties of porous media, J. Pet. Technol. 38 (1986) 1371–1378.
[78] I.D. Sulucarnain, C.H. Sondergeld, C.S. Rai, An NMR study of shale wettability and effective surface relaxivity, in: SPE Canadian Unconventional Resources Conference, Calgary, Alberta, Canada, 30 October-1 November, 2012.
[79] M. Xu, H. Dehghanpour, Advances in understanding wettability of gas shales, Energ. Fuel. 28 (2014) 4362–4375.
[80] F. Javadpour, D. Fisher, M. Unsworth, Nanoscale gas flow in shale gas sediments, J. Can. Petrol. Technol. 46 (2007) 55–61.
[81] J. Wang, B. Wang, Y. Li, et al., Measurement of dynamic adsorption–diffusion process of methane in shale, Fuel 172 (2016) 37–48.

CHAPTER 8

Tight Rock Wettability and Its Relationship With Petrophysical Properties

Ali Habibi, Ali Javaheri, Hassan Dehghanpour

Department of Civil and Environmental Engineering, School of Mining and Petroleum Engineering, University of Alberta, Edmonton, Alberta, Canada

Chapter Outline

1 Introduction

The soaring global energy demand and decreasing oil production from conventional resources during the last decade have motivated the industry to produce hydrocarbon from unconventional resources hosting huge hydrocarbon reserves [1, 2]. Hydrocarbon production from unconventional resources has been extended by the advent of horizontal drilling and multistage hydraulic fracturing operations. Although the estimated hydrocarbon reserves in these resources are huge, the primary oil recovery factor is very low (less than 10% of original

Petrophysical Characterization and Fluids Transport in Unconventional Reservoirs. https://doi.org/10.1016/B978-0-12-816698-7.00008-5

oil in place) [3–5]. Prior to developing any enhanced oil recovery (EOR) techniques, it is imperative to understand the rock-fluid interactions affected by the natural wettability of tight rocks.

Wettability is the tendency of a fluid to preferentially wet a solid surface in the presence of other fluid(s) [6, 7]. Understanding the wettability of natural rocks has drawn great attention from different disciplines such as petroleum reservoir engineering [8–10], cleanup of underground reservoirs contaminated by nonaqueous phase liquids [11, 12], and CO_2 sequestration underground [13, 14]. Wettability affects (a) pore-scale fluids distribution, (b) the irreducible water saturation (S_{wirr}), (c) efficiency of an immiscible displacement such as water flooding operations, (d) rock-fluid properties such as capillary pressure and relative permeability, and (e) electrical properties of porous media [15]. There is a wide spectrum in wettability states changing from strongly water-wet to strongly oil-wet states. The four general wettability states are water-wet, fractional-wet, mixed-wet, and oil-wet [7, 16, 17].

There are different methods for wettability characterization of a rock-fluid system including (a) equilibrium contact angle [18], (b) Amott wettability index [19], (c) USBM (United States Bureau of Mines) index [20], (d) spontaneous imbibition [21], (e) hysteresis of the relative permeability curves [22], and (f) nuclear magnetic relaxation (NMR) [16, 23]. It is challenging to apply the conventional methods, involving forced displacements, for wettability evaluation of tight rocks since it may lead to invalid results. This is mainly due to the complex pore structure, complex mineralogy, and low porosity and permeability of tight rocks. Equilibrium contact angle measurement is the simplest technique for wettability evaluation but may need complementary analysis, as it does not necessarily represent the wettability of the effective pore network [24, 25]. Recent studies show that comparative imbibition tests can provide useful information regarding the wettability of tight rocks [8, 26, 27].

Extensive experimental and modeling studies have been conducted to evaluate the main parameters influencing oil displacement during the spontaneous imbibition of aqueous phase into tight rocks [28–32]. The Bond number is generally used to compare the effect of capillary and gravity forces acting as two main driving forces for fluid displacement during the spontaneous imbibition process [33, 34]. However, the rock-fluid interactions in tight rocks are more complicated than those in conventional resources. The existence of organic matter (OM) and clay can strongly affect the wetting affinity of tight rocks. Therefore, investigation into the possibility of correlations between the natural wettability of tight rocks and their petrophysical properties is needed.

In this chapter, we review recent studies that focused on wettability evaluation of tight rocks collected from the Montney Formation located in the WCSB. First, we explain the experimental procedure performed for wettability evaluation. Then, we present the experimental results and possible correlations between wetting affinities and petrophysical properties.

2 Materials and Methods

2.1 Core Properties

Extensive studies have recently been conducted to evaluate the natural wettability of downhole core plugs drilled in different areas located in the Montney Formation [10, 35–38]. The world-class Montney Formation (Fig. 1) is a stratigraphical unit of Lower Triassic age in the WCSB. This formation has abundant hydrocarbon reserves and is one of the largest economically feasible resources in North America [39]. the Montney tight oil play (with a size of more than 90,000 km^2) has an estimated in-place hydrocarbon resource of almost 2133 tcf of natural gas, 28.9 billion bbl of natural gas liquids, and 136.3 billion bbl of oil in the province of Alberta [40]. The Montney Formation is composed of siltstone, minor intervals of shaly siltstone and very fine-grained sandstone [41, 42].

Javaheri et al. [36, 37] evaluated the wettability of five twin core plugs drilled from a well located in the Montney tight gas play (Gordondale area). Table 1 lists the mineralogy and total organic carbon (TOC) of the samples evaluated by X-ray diffraction analysis and Rock-Eval 6

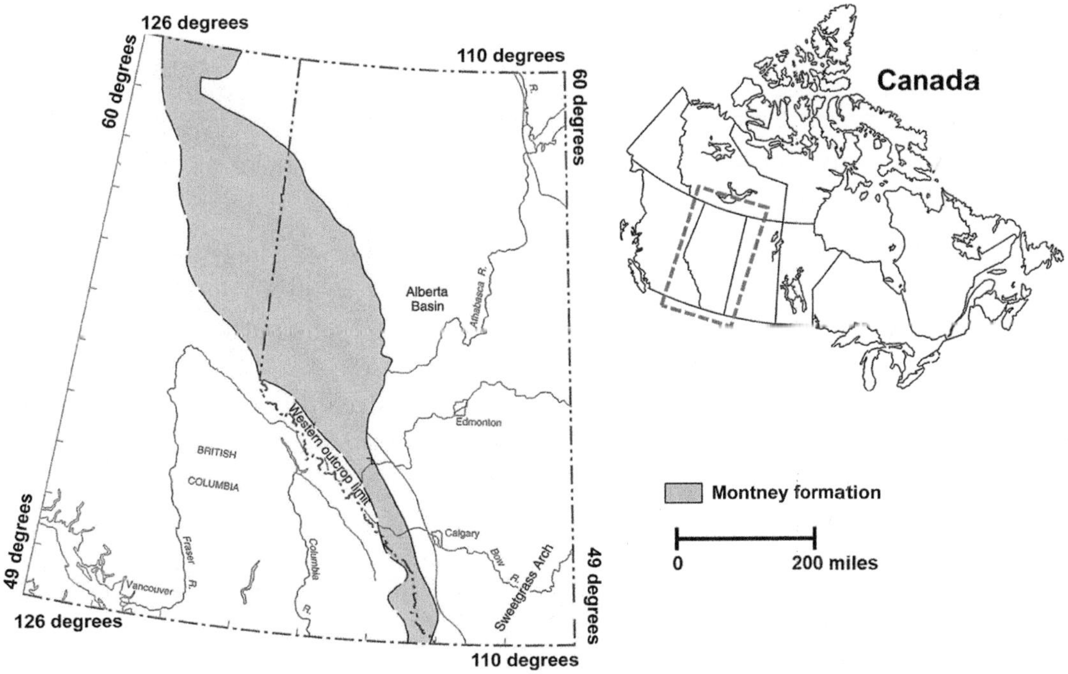

Fig. 1

The Montney Formation located in the Western Canadian Sedimentary Basin. *(Modified after D.E. Edwards, J.E. Barclay, D.W. Gibson, et al., Triassic Strata of the Western Canadian Sedimentary Basin. in: G.D. Mossop, I. Shetsen (Eds.), Geological Atlas of the Western Canada Sedimentary Basin, Canadian Society of Petroleum Geologists and Alberta Research Council, 1994.)*

Table 1 TOC content and mineral composition of the core plugs

Mineral (wt%)	Depth(m)				
	2138 MT-1	2144 MT-2	2149 MT-3	2174 MT-4	2192 MT-5
Quartz	40.4	40.4	44.8	41.6	43.1
K-Feldspar	9.0	12.1	10.0	14.9	12.8
Plagioclase	13.6	15.1	10.9	14.0	11.6
Calcite	3.9	3.0	5.7	0.0	2.0
Dolomite & Fe-Dolomite	22.5	14.0	14.2	14.5	16.5
Pyrite	1.2	1.5	1.5	1.4	1.7
Anhydrite	0.0	0.8	0.5	0.4	0.0
Illite/Mica	4.86	7.37	6.43	8.53	7.13
Chlorite	4.63	5.72	5.86	4.67	5.16
TOC	0.22	0.2	0.24	0.32	0.44

Table 2 Petrophysical properties of the MT core plugs

Sample No.	Permeability, $K_{air} \times 10^{-4}$ mD	Porosity, Helium (fraction)
MT-1	4.90	0.045
MT-2	0.40	0.031
MT-3	5.60	0.046
MT-4	0.30	0.030
MT-5	4.30	0.040

technique, respectively. Cores are mainly composed of quartz, dolomite, and plagioclase. The concentration of organic matter is very low (less than 0.5 wt%).

Porosity and permeability of the core plugs were measured with the Helium porosimeter and air Swanson method, respectively. Table 2 lists the petrophysical properties of the core plugs. The Mercury Injection Capillary Pressure (MICP) technique was used to characterize the pore-throat size distributions. The MICP plots were used to measure the pore-throat size (r_{P-T}) distribution of the core plugs. Fig. 2 shows the pore-throat size distribution of the core plugs. Most of the pores (more than 70%) were categorized as mesopores (2 nm $< r_{P-T} <$ 50 nm).

2.2 Fluid Properties

Reservoir brine (TDS = 126,000–130,000 ppm) and ultra-light reservoir oil (API = 45.07) were used in the spontaneous imbibition experiments.

2.3 Comparative Behavior of Oil and Water Imbibition in Tight Rocks

First, comparative oil and water spontaneous imbibition experiments were performed using dry twin core plugs. Next, the water (or oil) spontaneous imbibition into core plugs partly saturated with oil (or water) was conducted. Finally, the contact angle measurement as a complementary test was performed.

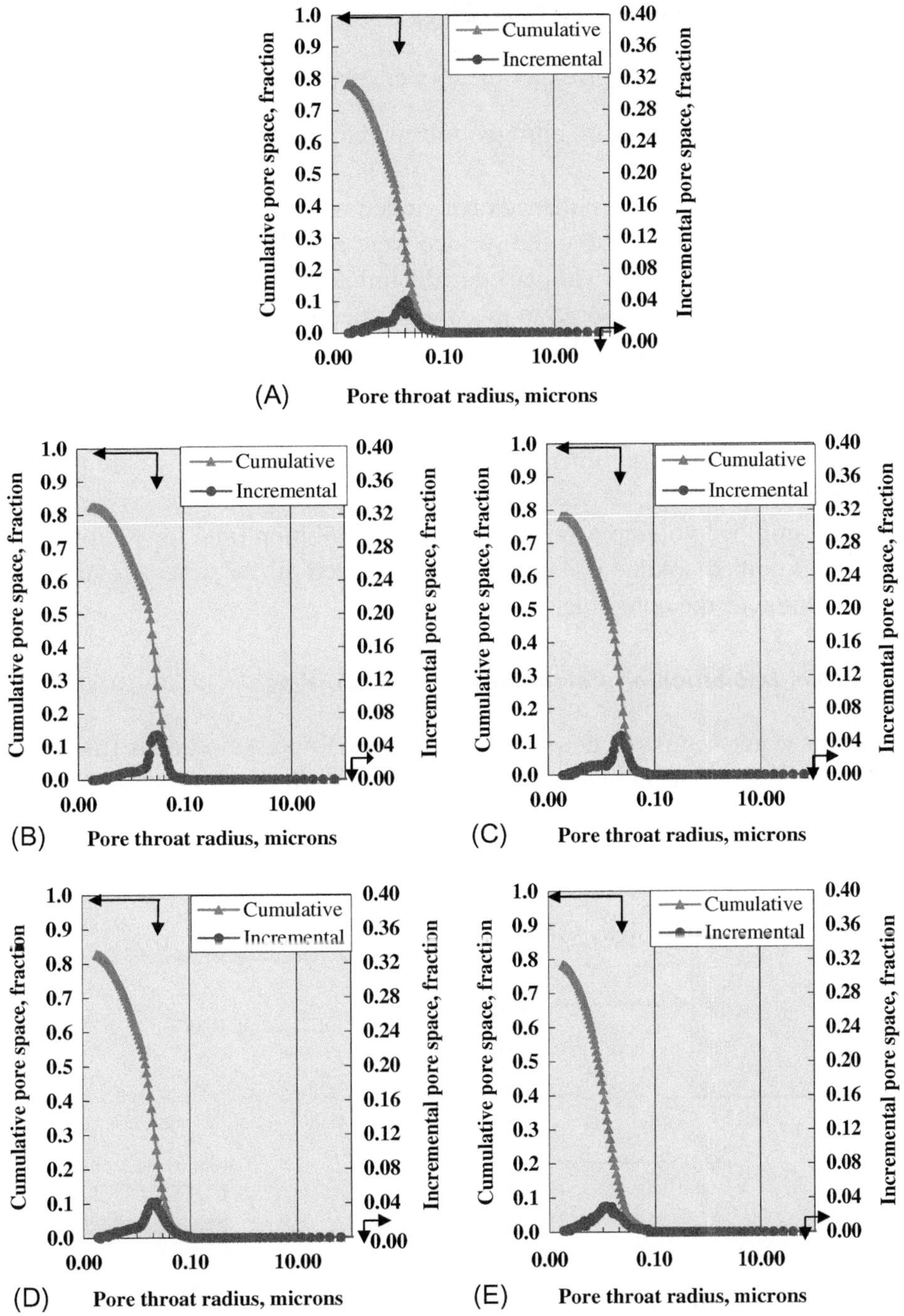

Fig. 2

Pore-throat size distribution of the five core plugs (A) MT1, (B) MT2, (C) MT3, (D) MT4, and (E) MT5. Both Cumulative (*triangle*) and incremental (*circle*) distributions are plotted for each rock sample [36].

2.4 Spontaneous Imbibition on Dry Core Plugs

Comparative spontaneous imbibition experiments were performed on dry core plugs as follows:

(1) Twin core plugs were cut using nitrogen for conducting comparative spontaneous imbibition experiments.
(2) Surface of the core plugs was polished on a graded wheel plate using 240 grit sandpapers to make sure the pores located at the surface were not clogged.
(3) One plug of each twin sample was placed in the imbibition cell filled with water. Only one face of the core plug was exposed to the water. The other plug of each twin sample was placed in the imbibition cell filled with the oil.
(4) The imbibition cell was sealed to avoid evaporation of liquid. The weight change of the core plugs was measured periodically with an accurate balance (±0.001 g). The spontaneous imbibition experiment stopped when the weight of the core plugs did not change over time.
(5) Normalized imbibed volume was plotted versus imbibition time for each twin plug. Normalized imbibed volume is defined as the imbibed oil (or water) volume divided by the pore volume of the core plug.

2.5 Spontaneous Imbibition on Partly Saturated Core Plugs

The partly oil (or water)-saturated core plugs were placed into Amott cells filled with water (or oil) as shown in Fig. 3. The produced volume of the oil during water spontaneous imbibition

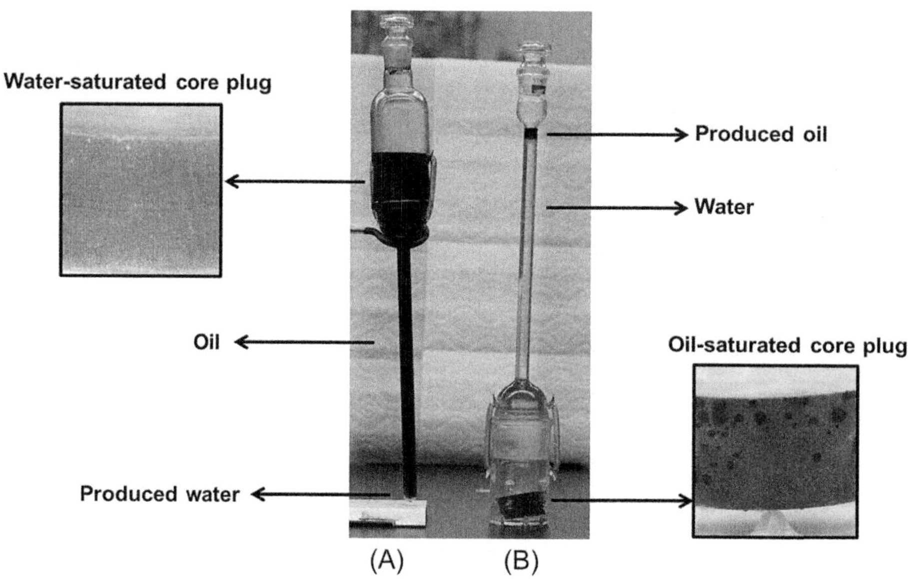

Fig. 3
Imbibition cells used for soaking of (A) water-saturated and (B) oil-saturated core plugs in oil and water, respectively [10].

and that of water during oil imbibition were collected at the top and bottom of the cell, respectively. The produced volume of oil/water was periodically measured during the experiment and plotted versus time. The oil recovery factor is defined as the produced oil volume divided by the final oil volume imbibed during the spontaneous imbibition process conducted on dry core plugs.

2.6 Contact Angle Measurement

Two different sets of contact angle measurements were conducted on polished end-pieces of the core plugs. First, the wetting affinities of oil and water toward the dry rock surface were investigated by analyzing the profile of a water/oil droplet equilibrated on the rock surface using a high-resolution camera. Second, the liquid/liquid contact angle measurements were performed using an Attention Theta (Biolin Scientific) instrument. The oil-saturated end-pieces of the core plug were immersed into a cell filled with water. Then a fixed volume of oil droplet was injected using a J-shape needle from the bottom of the cell. The OneAttention software analyzed the shape of the oil droplet over the course of time. The final contact angle equilibrated on the surface was reported.

3 Results and Discussions

3.1 Tight Rocks Wettability Evaluated by Spontaneous Imbibition Experiments

Two sets of spontaneous imbibition experiments were performed on dry and partially oil (or water) saturated core plugs. The results of the oil and water spontaneous imbibition in dry core plugs show that the total imbibed oil volume was higher than that of water. Water imbibes faster than oil and reaches equilibrium after 200–300h, earlier than oil, while oil imbibition continues for more than 1000h (Fig. 4). Continuing the oil imbibition for a longer period of time compared with water suggests that the oil imbibes into small-scale pores, which have a high affinity to oil. The existence of TOC content can explain the higher oil affinity of pores compared with water. Although the TOC content is very low in these cores (<0.5wt%) as listed in Table 1, organic matter with high specific surface area can cover the surface of pores. As a result, pores covered with organic matter show the oil-wet behavior [43] and can imbibe oil more than water. The results of contact angle measurements conducted on dry end-pieces of rock were in agreement with the comparative spontaneous imbibition experiments conducted on dry core plugs. Oil droplet completely spread on the rock surface while the water droplets partially wet the rock surface, indicating higher rock affinity toward oil compared with water at dry conditions (Fig. 5).

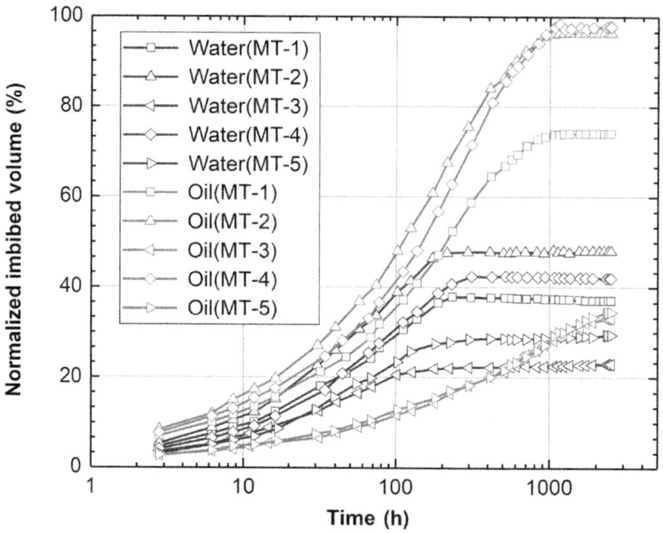

Fig. 4

Oil and water spontaneous imbibition-time profiles for (A) MT1, (B) MT2, (C) MT3, (D) MT4, and (E) MT5. Although water reaches equilibrium faster than oil, final imbibed water volume is lower than the oil. *(Modified after A. Javaheri, A. Habibi, H. Dehghanpour, Imbibition oil recovery from tight rocks with dual-wet pore-network, J. Pet. Sci. Eng. 167 (2018) 180–191.)*

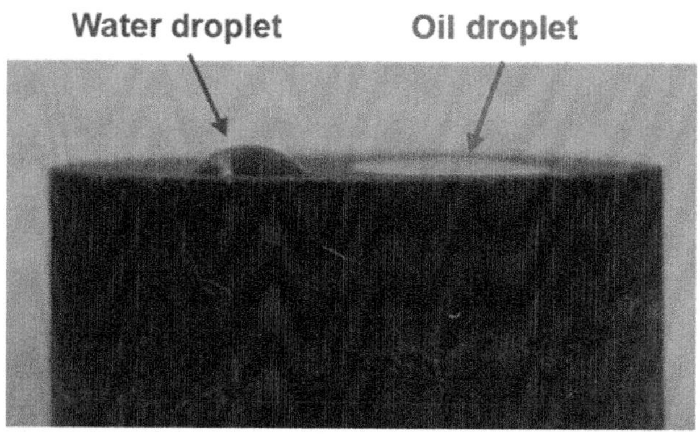

Fig. 5

Oil and brine droplets equilibrated on the surface of a core plug in air at ambient conditions. Oil completely spread on the rock surface while water partially wet the rock surface [36].

3.2 Spontaneous Imbibition on Partly Saturated Core Plugs

In this step, the spontaneous imbibition experiments into the oil- (or water-) saturated core plugs were performed to investigate the possibility of oil (or water) displacement by water (or oil) during the spontaneous imbibition experiment. Fig. 6 shows the oil droplets produced during the spontaneous imbibition process and attached to the outer surface of a core plug.

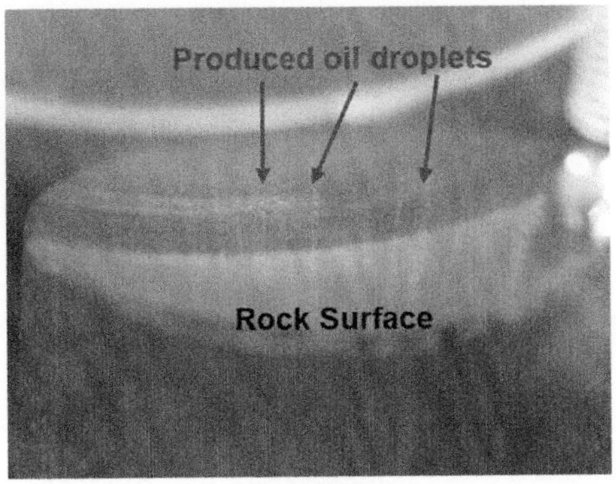

Fig. 6
Produced oil droplets, shown by arrows, when an oil-saturated core plug is immersed in the water [36].

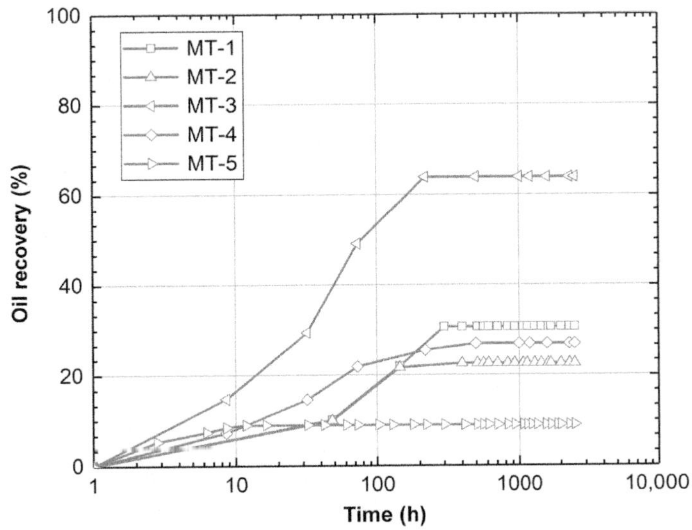

Fig. 7
The oil recovery plots for the oil-saturated core plugs immersed in water. The oil recovery factor changes from 9% to 64% of the oil volume imbibed into the dry core plugs during the spontaneous imbibition experiment *(Modified after A. Javaheri, A. Habibi, H. Dehghanpour, Imbibition oil recovery from tight rocks with dual-wet pore-network, J. Pet. Sci. Eng. 167 (2018) 180–191.)*

This indicates that the water can spontaneously imbibe into the oil-saturated core plugs and expel the oil out. However, oil cannot imbibe into the water-saturated core plugs and displace the water out. Fig. 7 shows the oil recovery-time profiles for all core plugs used in this study.

Moreover, the results of liquid/liquid contact angle measurements suggest the water-wet behavior of the rock samples (Fig. 8). The water droplet partially wet the rock surface when an end-piece of the rock is immersed in the oil. However, the oil contact angle is higher than

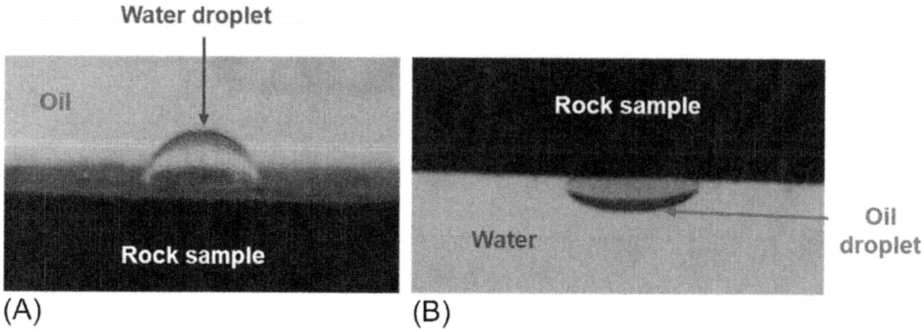

Fig. 8
Liquid/liquid contact angle of (A) a water droplet and (B) an oil droplet on an end-piece of the rock sample.

90 degrees when the end-piece of the rock is immersed in the water, indicating that the rock surface is oil repellent.

3.3 Relationship Between the Spontaneous Imbibition Results and Other Petrophysical Properties

Multivariate analysis was used to evaluate the relationship between the results of oil/water spontaneous imbibition and petrophysical properties. I_w and I_o are defined as the total imbibed volume of water and oil during the spontaneous imbibition process performed on dry core plugs, respectively. To find the possible correlations between the imbibition results and petrophysical properties of the core plugs, gamma ray (GR) and neutron porosity (NPOR) logs were used. In addition, porosity values evaluated by the Helium porosimeter were used for the multivariate analysis. Fig. 9 shows the multivariate scatterplots between five parameters: I_w, I_o, NPOR, GR, and porosity. Each parameter is positively or negatively correlated to the other parameters. Diagonal histograms represent the variation of each parameter for five core plugs; while off-diagonal scatter plots represent the binary correlations between the five parameters. The results show that I_o vs porosity exhibits the strongest correlation, suggesting higher oil uptake occurs in more porous core plugs.

To compare the oil and water wetting affinities, the equilibrium wetting affinity indices [25] were defined.

$$\text{Equilibrium wetting affinity index of water}\,(WI_w) = \frac{I_w}{I_w + I_o} \tag{1}$$

$$\text{Equilibrium wetting affinity index of oil}\,(WI_o) = \frac{I_o}{I_w + I_o} \tag{2}$$

where I_w and I_o are normalized imbibed volume of water and oil, respectively. Fig. 10 shows the calculated wetting indices versus depth for each set of twin core plugs.

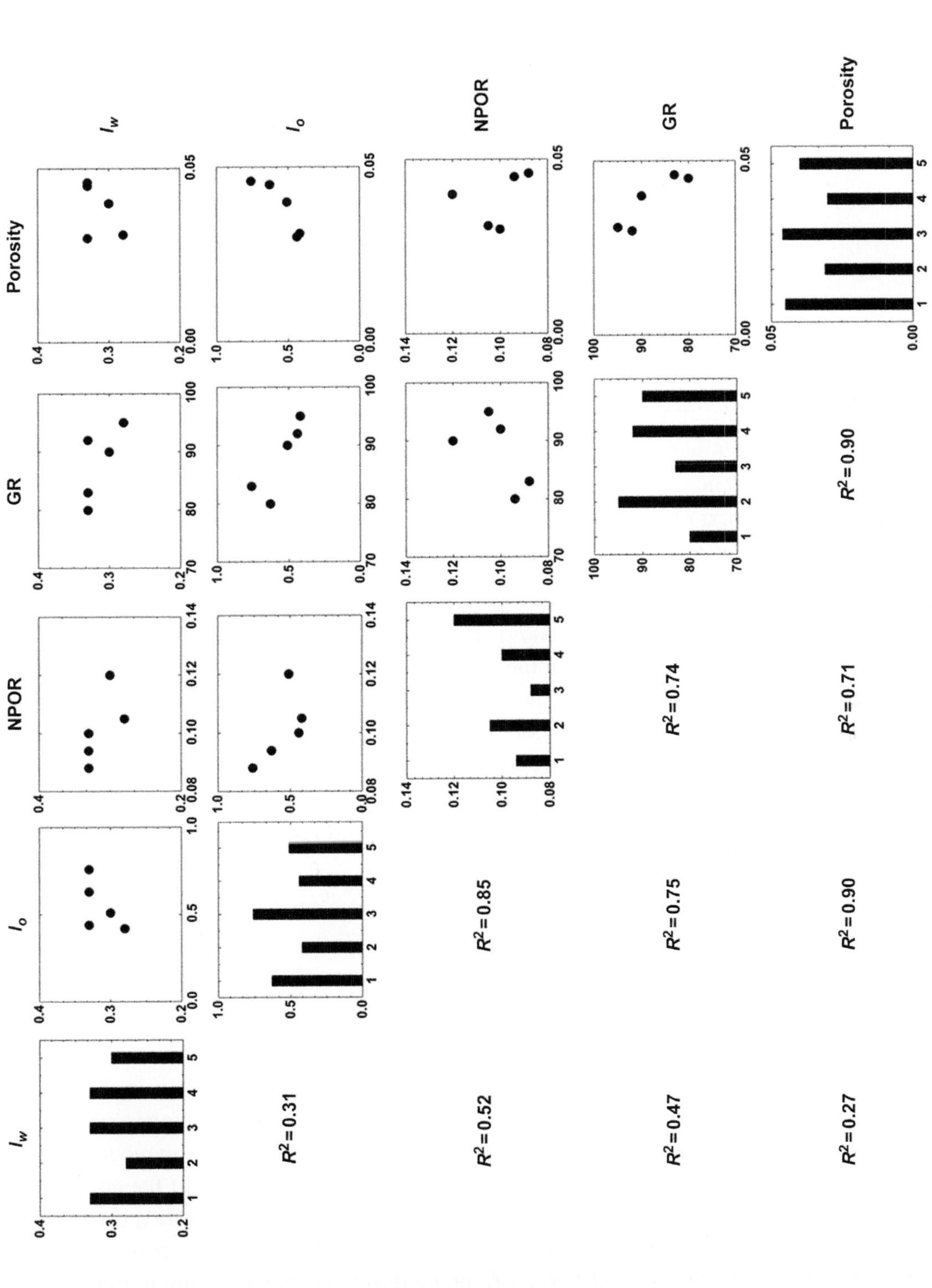

Fig. 9
Multivariate scatterplots for five variables: I_w, I_c, NPOR, GR, and porosity. I_w and I_o represent the final normalized imbibed volume of water and oil into the dry core plugs, respectively [36].

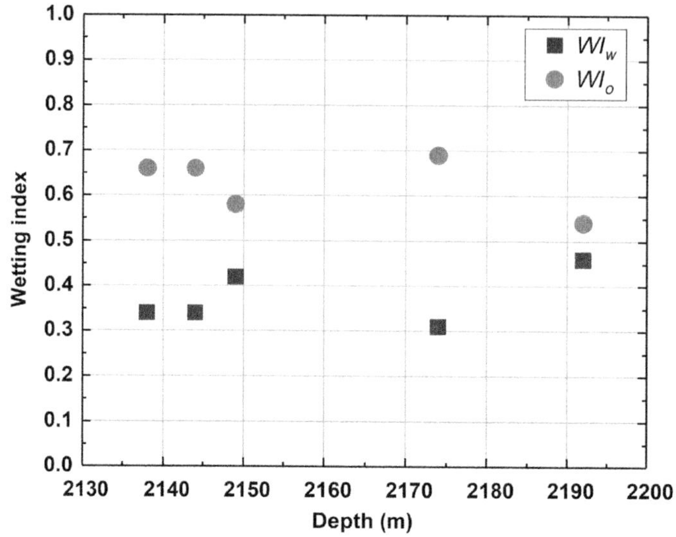

Fig. 10

Oil and water wetting indices calculated for the core plugs. Higher oil wetting indices indicate higher oil-wet behavior of the dry core plugs.

To understand how higher oil wetting affinity is correlated with other petrophysical properties, the cross plots between oil wetting index (WI_o) and permeability, NPOR, GR, and porosity are presented in Fig. 11. The results show that the oil-wetting index (WI_o) increases by increasing the porosity and permeability, indicating more available pore volume for oil imbibition. However, the oil-wetting index (WI_o) decreases to some extent by increasing the NPOR and GR values. This means that when the clay concentration increases, the oil-wetting index decreases. The water affinity index of the core plugs increases by increasing the clay concentration because clay minerals are mainly composed of hydrophilic clays; i.e., illite/mica and chlorite.

Although the dry core plugs exhibit the oil-wet behavior (Fig. 4), water can spontaneously imbibe into the oil-saturated core plugs, suggesting the water-wet behavior of the oil-saturated core plugs. To understand how petrophysical properties can affect the oil recovery factor during the spontaneous imbibition experiment, the cross plots between the final oil recovery factor (presented in Fig. 7) and permeability, NPOR, GR, and porosity are plotted Fig. 12. The results show that the oil recovery factor is not strongly correlated with porosity and permeability.

Moreover, the oil recovery factor decreases by increasing the GR and NPOR values. It is expected that the oil recovery increases by increasing the clay content. To explain why increasing the clay content (representing high values in GR and NPOR logs) can reduce the oil recovery factor, it is required to investigate the type of clay. Clay minerals of the core plugs used in this study are mainly composed of illite/mica (pore bridging) and chlorite (pore lining). The existence of these types of clay in narrow pores can significantly reduce the permeability of the core plugs. This can lead to oil and water trapping during spontaneous imbibition.

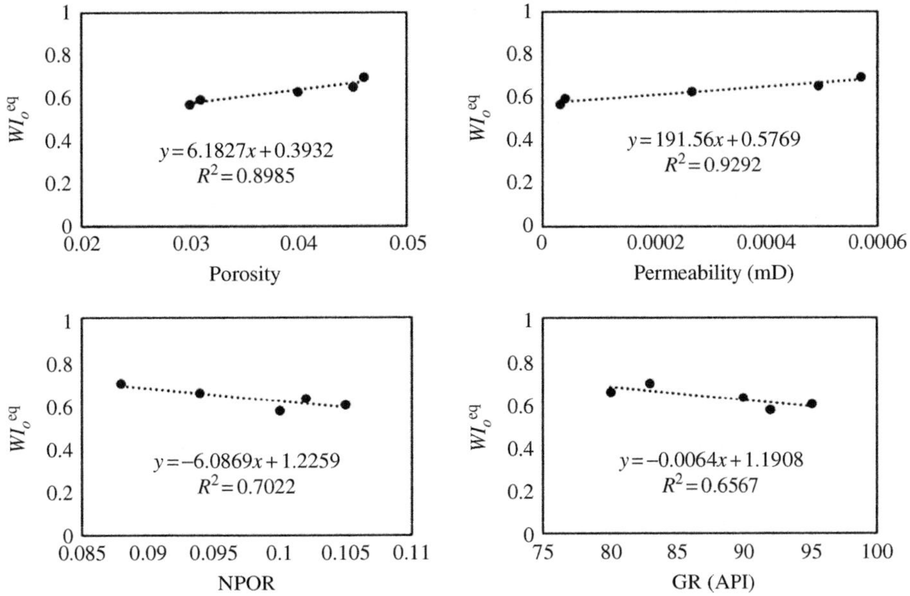

Fig. 11

Equilibrium oil wetting indices vs petrophysical properties [37].

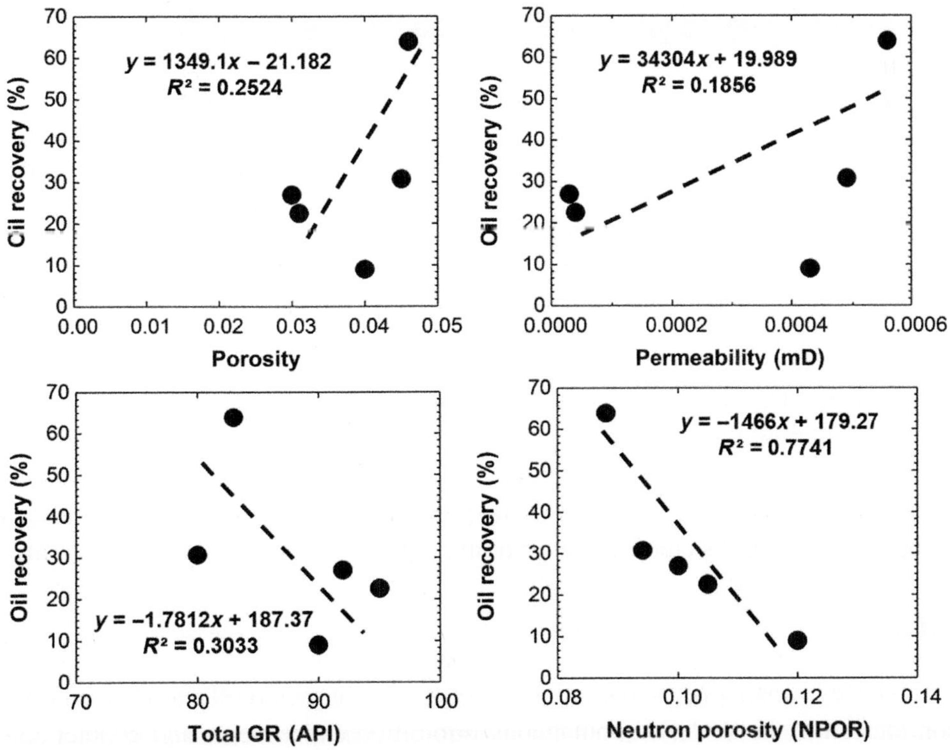

Fig. 12

Oil recovery factor versus porosity, permeability, total GR, and neutron porosity.

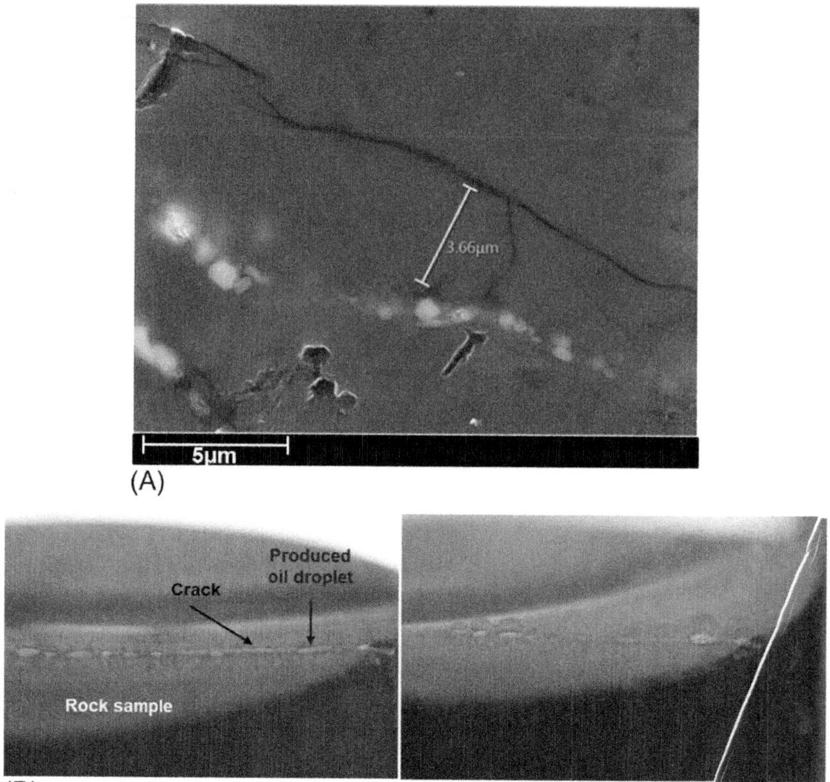

(A)

(B)

(C)

Fig. 13

(A) A microcrack shown on the surface of core plug 3. (B) Produced oil droplets along a microcrack of core plugs 3 after $t=10$ days. (C) Produced oil droplets along the microcrack after $t=20$ days [36].

Therefore, a higher concentration of clay minerals results in more oil trapping and less oil recovery factor.

In addition, the presence of micro cracks/fractures may explain the highest oil recovery factor observed for the core MT-3 even though these cracks might only have limited influence on the total porosity. During spontaneous imbibition of wetting fluid, induction of such cracks or enlargement of the existing microfractures can enhance total rock permeability [44]. Subsequently, such permeability enhancement of the rock can enhance hydrocarbon recovery from low-permeability rocks. Fig. 13 shows the produced oil droplets along the cracks at two different times. These cracks create a preferential path for oil production by water imbibition.

4 Conclusion

In this chapter, we investigated the natural wettability of tight rocks collected from the Montney Formation located in the WCSB. Spontaneous imbibition experiments and contact angle measurements were performed to evaluate the wettability of the core plugs. To understand the effect of petrophysical properties on oil/water spontaneous imbibition results, we present the correlations between the oil/water imbibition and NPOR, GR, porosity, and permeability properties. The key observations are summarized as follows:

(1) The results of oil/water spontaneous imbibition into dry core plugs show that the oil imbibition is higher than water imbibition, showing higher oil wetting indices than water. The results of air/liquid contact angle measurements also suggest that the dry core plugs are oil-wet.
(2) The results of water imbibition into the oil-saturated core plugs show that water can imbibe into the oil-saturated core plugs and expel the oil out. The results of liquid/liquid contact angle measurements suggest that core plugs are water-wet.
(3) The oil-wetting index (WI_o) evaluated for the oil imbibition into the dry core plugs is well correlated with porosity and permeability. WI_o increases by increasing the porosity and permeability, indicating more available pore volume for oil imbibition.
(4) The produced oil volume during water imbibition into the oil-saturated core plugs decreases by increasing the clay contents. Illite/mica (pore bridging) and chlorite (pore lining) as the main clay components of the cores used in this study can significantly reduce the permeability of the core plugs, leading to oil trapping during spontaneous imbibition process.

Acknowledgments

We thank Natural Sciences and Engineering Research Council (NSERC) of Canada and Encana Corporation for supporting this study. We also thank James Wood for his constructive comments.

References

[1] K. He, L. Xu, Y. Gao, et al., Evaluation of surfactant performance in fracturing fluids for enhanced well productivity in unconventional reservoirs using rock-on-a-Chip approach, J. Pet. Sci. Eng. 135 (2015) 531–541.
[2] Drilling productivity report, U.S. Energy Information Administration, Washington, DC, 2018, p. 11.
[3] P. Kathel, K.K. Mohanty, Wettability alteration in a tight oil reservoir, Energ. Fuel. 27 (2013) 6460–6468.
[4] B.T. Hoffman, Comparison of various gases for enhanced recovery from shale oil reservoirs, in: SPE-154329, SPE Improved Oil Recovery Symposium, Tulsa, Oklahoma, USA, 14–18 April, 2012.
[5] L. Jin, S. Hawthorne, J. Sorensen, et al., Advancing CO2 enhanced oil recovery and storage in unconventional oil play—experimental studies on Bakken shales, Appl. Energ. 208 (2017) 171–183.
[6] W.G. Anderson, Wettability literature survey-part 1: rock/oil/brine interactions and the effects of core handling on wettability, J. Petrol. Technol. 38 (1986) 1125–1144.
[7] E.C. Donaldson, W. Alam, Wettability, Gulf Publishing Company, 2008, pp. 1–55.
[8] M.R. Yassin, M. Begum, H. Dehghanpour, Organic shale wettability and its relationship to other petrophysical properties: a Duvernay case study, Int. J. Coal Geol. 169 (2017) 74–91.
[9] H. Dehghanpour, M. Xu, A. Habibi, Wettability of gas shale reservoirs, in: Fundamentals of Gas Shale Reservoirs, 2015, pp. 341–359.
[10] A. Habibi, H. Dehghanpour, M. Binazadeh, et al., Advances in understanding wettability of tight oil formations: a Montney case study, SPE Reserv. Eval. Eng. 19 (2016) 583–603.

[11] H.I. Essaid, B.A. Bekins, I.M. Cozzarelli, Organic contaminant transport and fate in the subsurface: evolution of knowledge and understanding, Water Resour. Res. 51 (2015) 4861–4902.

[12] A. Al-Futaisi, T.W. Patzek, Secondary imbibition in NAPL-invaded mixed-wet sediments, J. Contam. Hydrol. 74 (2004) 61–81.

[13] S. Iglauer, C.H. Pentland, A. Busch, CO2 wettability of seal and reservoir rocks and the implications for carbon geo-sequestration, Water Resour. Res. 51 (2015) 729–774.

[14] T.K. Tokunaga, DLVO-based estimates of adsorbed water film thicknesses in geologic CO2 reservoirs, Langmuir 28 (2012) 8001–8009.

[15] E.J. Peters, Advanced Petrophysics, Live Oak Book Company, 2012.

[16] R.J.S. Brown, I. Fatt, Measurements of fractional wettability of oil fields' rocks by the nuclear magnetic relaxation method, in: SPE-743,Fall Meeting of the Petroleum Branch of AIME, Los Angeles, California, 14–17 October, 1956.

[17] R.A. Salathiel, Oil recovery by surface film drainage in mixed-wettability rocks, J. Petrol. Technol. 25 (1973) 1216–1224.

[18] W.G. Anderson, Wettability literature survey- part 2: Wettability measurement, J. Pet. Technol. 38 (1986) 1246–1262.

[19] E. Amott, Observations relating to the wettability of porous rock, Petrol. Trans. AIME 216 (1959) 156–162.

[20] E.C. Donaldson, R.D. Thomas, P.B. Lorenz, Wettability determination and its effect on recovery efficiency, Soc. Petrol. Eng. J. 9 (1969) 13–20.

[21] N.R. Morrow, Wettability and its effect on oil recovery, J. Petrol. Eng. 42 (1990) 1476–1484.

[22] S.C. Jones, W.O. Roszelle, Graphical techniques for determining relative permeability from displacement experiments, J. Pet. Technol. 30 (1978) 807–817.

[23] R. Freedman, N. Heaton, M. Flaum, et al., Wettability, saturation, and viscosity from NMR measurements, SPE J. 8 (2003) 317–327.

[24] M. Xu, H. Dehghanpour, Advances in understanding wettability of gas shales, Energ. Fuel. 28 (2014) 4362–4375.

[25] Q. Lan, H. Dehghanpour, J. Wood, et al., Wettability of the Montney tight gas formation, SPE Reserv. Eval. Eng. 18 (2015) 417–431.

[26] K. Makhanov, A. Habibi, H. Dehghanpour, et al., Liquid uptake of gas shales: a workflow to estimate water loss during shut-in periods after fracturing operations, J. Unconvent. Oil Gas Resour. 7 (2014) 22–32.

[27] E. Ghanbari, H. Dehghanpour, The fate of fracturing water: a field and simulation study, Fuel 163 (2016) 282–294.

[28] C.R. Clarkson, N. Solano, R.M. Bustin, et al., Pore structure characterization of north American shale gas reservoirs using USANS/SANS, gas adsorption, and mercury intrusion, Fuel 103 (2013) 606–616.

[29] Z. Song, L. Liu, M. Wei, et al., Effect of polymer on disproportionate permeability reduction to gas and water for fractured shales, Fuel 143 (2015) 28–37.

[30] F.P. Wang, R.M. Reed, Pore networks and fluid flow in gas shales, in: SPE-124253, SPE Annual Technical Conference and Exhibition , New Orleans, Louisiana, 4–7 October, 2009.

[31] M. Binazadeh, M. Xu, A. Zolfaghari, et al., Effect of electrostatic interactions on water uptake of gas shales: The interplay of solution ionic strength and electrostatic double layer, Energ. Fuel. 30 (2016) 992–1001.

[32] M. Xu, M. Binazadeh, A. Zolfaghari, et al., Effects of dissolved oxygen on water imbibition in gas shales, Energ. Fuel. 32 (2018) 4695–4704.

[33] D.S. Schechter, D. Zhou, F.M. Orr, Low IFT drainage and imbibition, J. Pet. Sci. Eng. 11 (1994) 283–300.

[34] D.C. Standnes, L.A.D. Nogaret, H.-L. Chen, et al., An evaluation of spontaneous imbibition of water into oil-wet carbonate reservoir cores using a nonionic and a cationic surfactant, Energ. Fuel. 16 (2002) 1557–1564.

[35] A. Habibi, M. Xu, H. Dehghanpour, et al., Understanding rock-fluid interactions in the Montney tight oil play, in: SPE-175924, SPE/CSUR Unconventional Resources Conference, Calgary, Alberta, Canada, 20–22 October, 2015.

[36] A. Javaheri, A. Habibi, H. Dehghanpour, Imbibition oil recovery from tight rocks with dual-wet pore-network, J. Pet. Sci. Eng. 167 (2018) 180–191.

[37] A. Javaheri, H. Dehghanpour, J.M. Wood, Tight rock wettability and its relationship to other petrophysical properties: a Montney case study, J. Earth Sci. 28 (2017) 381–390.

[38] H.J. Deglint, C.R. Clarkson, C. DeBuhr, et al., Live imaging of micro-wettability experiments performed for low-permeability oil reservoirs, Sci. Rep. 7 (2017) 4347.

[39] A. Keneti, R.C.K. Wong, Investigation of anisotropic behavior of Montney shale under indirect tensile strength test, in: SPE-138104, Canadian Unconventional Resources and International Petroleum Conference, Calgary, Alberta, Canada, 19–21 October, 2010.

[40] M. Reynolds, R. Bachman, J. Buendia, et al., The full montney—a critical review of well performance by production analysis of over 2,000 montney multi-stage fractured horizontal gas wells, in: SPE-175948, SPE/CSUR Unconventional Resources Conference, Calgary, Alberta, Canada, 20–22 October, 2015.

[41] J. Wood, Water distribution in the Montney tight gas play of the western Canadian sedimentary basin: significance for resource evaluation, SPE Reserv. Eval. Eng. 16 (2013) 290–302.

[42] H. Sanei, J.M. Wood, O.H. Ardakani, et al., Characterization of organic matter fractions in an unconventional tight gas siltstone reservoir, Int. J. Coal Geol. 150 (2015) 296–305.

[43] A. Mitchell, L. Hazell, K. Webb, Wettability determination: pore surface analysis, in: SPE Annual Technical Conference and Exhibition, Society of Petroleum Engineers, New Orleans, LA, 1990.

[44] H. Dehghanpour, Q. Lan, Y. Saeed, et al., Spontaneous imbibition of brine and oil in gas shales: effect of water adsorption and resulting microfractures, Energ. Fuel. 27 (2013) 3039–3049.

PART 2

Porous Flow Dynamics

CHAPTER 9

Flow Mechanism of Fractured Low-Permeability Reservoirs

Wanjing Luo*, Changfu Tang†, Cheng Lu‡, Bo Ning§

**School of Energy Resources, China University of Geosciences (Beijing), Beijing, P.R. China, †Exploration Research Institute, Anhui Provincial Bureau of Coal Geology, Hefei, P.R. China, ‡Oil & Gas Survey, China Geological Survey, Beijing, P.R. China, §Research Institute of Petroleum Exploration and Development, PetroChina, Beijing, P.R. China*

Chapter Outline

Petrophysical Characterization and Fluids Transport in Unconventional Reservoirs. https://doi.org/10.1016/B978-0-12-816698-7.00009-7

1 Introduction

In recent years, hydraulic fractures have become key to enhancing oil and gas recovery for low permeability reservoirs. Fractures can create more surface area and improve well productivity effectively. Pressure transient analysis (PTA) is one of the diagnostic techniques used to improve the understanding of hydraulic fracture behaviors and evaluate fracture characteristic parameters such as effective fracture length and fracture conductivity. Many analytical (semi-analytical) and numerical solutions have been reported for transient flow analysis of fractured wells, which will be briefly introduced in the next section.

1.1 Simple Fracture Model

For the bi-wing symmetry fracture, source functions for transient pressure analysis of uniform-flux fracture and infinite-conductivity fracture were first introduced by Gringarten and Ramey [1]. Taking account of more realistic cases, Cinco-Ley et al. [2] extended the Green's functions and presented the transient pressure solutions for a well with a finite-conductivity vertical fracture in an infinite-acting reservoir using a discretized fracture approach. Cinco-Ley [3] also proposed a bilinear flow model for analyzing early-time pressure data and type curves for identifying all the flow regimes for wells intersected by a finite-conductivity fracture. Furthermore, more general solutions accounting for the effects of wellbore storage and formation damage around the fracture were presented [4].

For the multi-wing planar fractures, it has been proven that the horizontal well intercepted by multiple fractures is key for the successful and efficient development of low-permeability reservoir. The fluid flow mechanism of a horizontal well with multiple fractures has been discussed in detail by many researchers [5–9]. Some significant results have been presented, such as the flow-regime analysis and production analysis of infinite/finite-conductivity vertical/inclined fractures. Note that all the fractures presented in those papers are symmetrically distributed from the horizontal wellbore.

1.2 Complex Fracture Model

It has been observed that the fractures can have more complex patterns in the field and experiments, such as asymmetrical fracture, nonplanar fracture, and multiple fractures connected to a horizontal wellbore or a vertical wellbore.

1.2.1 Bi-wing asymmetrical fracture

The effects of fracture asymmetry, which could be an important consideration for fracture characterization and design, have been reported in some papers (Fig. 1).

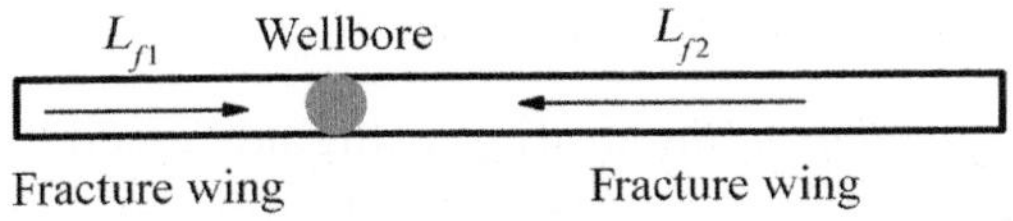

Fig. 1
Schematic diagram of an asymmetrical fracture with two wings.

Crawford and Landrum [10] first noticed and discussed the effect of fracture asymmetry. Later, the numerical approach was used to analyze the problem of asymmetrical fractures [11, 12]. However, it was not until the early 1990s, that the first comprehensive study of the PTA of asymmetrically fractured wells was achieved [13]. The study presented a semi-analytical solution (a graphical technique) to evaluate asymmetry of hydraulically fractured wells in an infinite reservoir. Resurreicão and Fernando [14] examined the effects of the asymmetry factor on the reciprocal of the rate of a fractured well producing under constant wellbore pressure condition. Berumen et al. [15] also investigated the pressure behavior of a well intercepting an asymmetrical fracture with both infinite and finite conductivity under constant rates by numerical methods. Based on Rodriguez's work, Tiab [16] applied the Tiab's Direct Synthesis technique to evaluate fracture asymmetry of a finite-conductivity fractured well producing at a constant-rate.

1.2.2 Multibranch fracture

The models proposed by the authors mentioned above were based on the assumption that the fractures were vertical and coplanar fractures. In fact, a number of microseismic and laboratory studies have shown that the complex fractures can be observed and most of these fractures are neither symmetrical nor planar (Fig. 2).

Some microseismic fracture maps show that hydraulic fracture treatments create fracture networks [17–19]. By direct observation on fractured cores, geological evidence, and laboratory experiments, Germanovich et al. [20, 21] indicated that hydraulic fractures are not single, symmetrical or planar features in rocks, instead they have irregular branches.

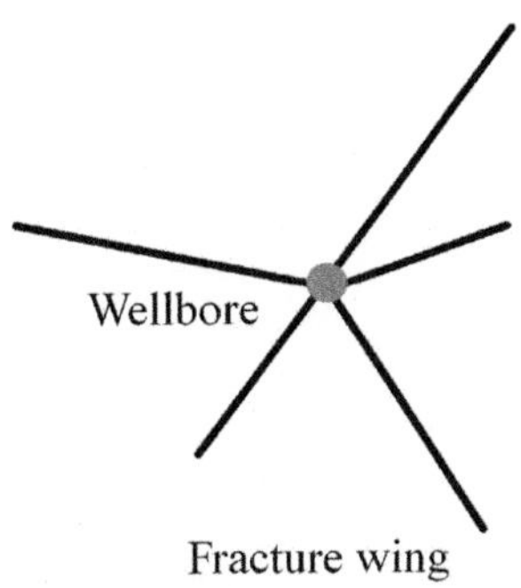

Fig. 2
Schematic diagram of multiple nonplanar fractures connected to a vertical well.

1.2.3 Horizontal wells with multiple curved fractures

Moreover, recent fracture stimulation monitoring using advanced technologies such as microseismic monitoring, fiber optics temperature, and strain sensing have shown that multiple fractures in a horizontal well do not grow and develop in the same way [22, 23].

Complex in situ stress field and formation heterogeneities affected the placement and growth of multiple fractures. The mechanism of the fracture growth and interaction with each other has been discussed in detail [24, 25]. These studies pointed out that the action of multiple fractures may result in significant shearing displacements along the fracture surface causing the fracture growth to follow a curved path. Note that less horizontal stress difference, stiffer formations, and shorter fracture spacing (FS) lead to more curve fracture growth. A numerical model of 3D nonplanar hydraulic fracture growth in multiple layers with fluid and proppant transport description was developed for optimal fracture design [25]. Another new comprehensive numerical simulator for nonplanar fracture designs has been devised to solve the almost intractable problem of multiple fractures [24].

2 Mathematical Model of a Fracture Wing

In this section, we focus on the fluid flow of a fracture wing shown in Fig. 3.

The following assumptions are made for flow models in the reservoir and fracture wing.

1. The reservoir is isotropic, homogeneous with impermeable upper and lower boundaries.
2. The reservoir is an infinite slab of constant thickness h, constant porosity φ, and constant permeability k. The reservoir contains a slightly compressible single-phase fluid of constant compressibility c_t and constant viscosity μ. The flow in the formation is assumed to obey Darcy's law.

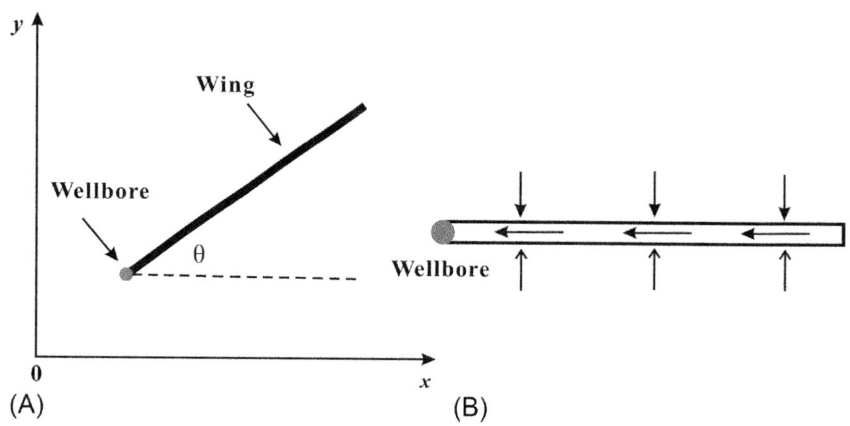

Fig. 3
(A) Location of a fracture wing in reservoir and (B) flow inside a fracture wing.

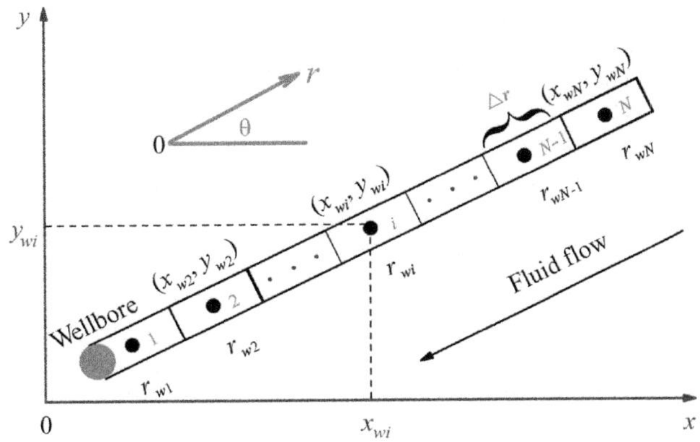

Fig. 4
Schematic diagram of the discretized fracture wing.

3. The reservoir is fully penetrated by finite-conductivity vertical fractures. As shown in Fig. 3, the wing originates from the wellbore at angle θ with x axis. The wing is of length L_f, constant width w_f, and constant permeability k_f.
4. No fluid is assumed to flow into the fracture wing at the tip. Moreover, flow in fracture wing is assumed to be incompressible. The total flow contribution of the fracture wing to the wellbore is q_{fw}.

In the following section, the semi-analytical solution of single fracture wing will be presented. For the sake of simplicity, the dimensionless variables will be used (Appendix A). The fracture wing is equally divided into N segments and the mid-point of the ith segment is located at (x_{wi}, y_{wi}) with length of L_{fi} ($L_{fi} = L_f/N$) (Fig. 4). The pressure of the ith segment is p_{fi} and the flow from the reservoir into the ith segment is q_{fi}.

2.1 Reservoir Flow Model

Each segment can be considered as a uniform-flux fracture. Therefore, according to the superposition principle, the dimensionless pressure of the ith segment in the reservoir can be written in Laplace domain as:

$$\overline{p}_{Di} = \sum_{j=1}^{N} \overline{\overline{q}}_{Dj} \cdot s\overline{p}_{uDij}\left(x_{Di}, y_{Di}, x_{wDj}, y_{wDj}, L_{fDj}\right) \tag{1}$$

where $\overline{p}_{uDij}$ denotes the dimensionless pressure change in the ith segment caused by the production of the jth segment in Laplace domain.

For a fully penetrating vertical fracture with uniform flux, Ozkan and Raghavan [26] presented the basic source function of pressure response in Laplace domain under constant flow rate q_D:

$$\overline{p}_{uD} = \left(\frac{\overline{q}_D}{s}\right)\left(\frac{1}{L_{fD}}\right)\int_{x_{wD}-L_{fD}/2}^{x_{wD}+L_{fD}/2} K_0\left(\sqrt{s}\sqrt{(x_D-u)^2+(y_D-y_{wD})^2}\right)du \tag{2}$$

where overline '–' indicates variables in Laplace space and s is the Laplace variable. Eq. (2) is based on the assumption that the fracture is parallel to the x axis ($\theta = 0$). For the case that there is an angle θ between the fracture and x axis, we can obtain the following equation through coordinate rotation:

$$\overline{p}_{uD} = \left(\frac{\overline{q}_D}{s}\right)\left(\frac{1}{L_{fD}}\right)\int_{x'_{wD}-L_{fD}/2}^{x'_{wD}+L_{fD}/2} K_0\left(\sqrt{s}\sqrt{(x'_D-u)^2+(y'_D-y'_{wD})^2}\right)du \tag{3}$$

with

$$x'_D = x_D\cos\theta + y_D\sin\theta, y'_D = y_D\cos\theta - x_D\sin\theta \tag{4}$$

$$x'_{wD} = x_{wD}\cos\theta + y_{wD}\sin\theta, y'_{wD} = y_{wD}\cos\theta - x_{wD}\sin\theta \tag{5}$$

The matrix form of Eq. (1) is expressed as

$$\vec{\overline{p}}_D = AR \cdot \vec{\overline{q}}_D \tag{6}$$

where $\vec{\overline{p}}_D$ is the pressure vector in the reservoir

$$\vec{\overline{p}}_D = (\overline{p}_{D1} \cdots \overline{p}_{DN})^T \tag{7}$$

and $\vec{\overline{q}}_D$ is the flow rate vector in the reservoir

$$\vec{\overline{q}}_D = (\overline{q}_{D1} \cdots \overline{q}_{DN})^T \tag{8}$$

The coefficient matrix in the reservoir can be expressed as

$$AR = \begin{pmatrix} s\overline{p}_{uD11} & s\overline{p}_{uD12} & \cdots & s\overline{p}_{uD1N} \\ s\overline{p}_{uD21} & s\overline{p}_{uD22} & \cdots & s\overline{p}_{uD2N} \\ \vdots & \vdots & \ddots & \vdots \\ s\overline{p}_{uDN1} & s\overline{p}_{uDN2} & \cdots & s\overline{p}_{uDNN} \end{pmatrix} \tag{9}$$

2.2 Fracture Wing Flow Model

Flow inside the fracture is one-dimensional and can be described in the one-dimensional coordinate originating from the wellbore. In Appendix B, we present the derivation of the flow equation in the fracture wing and the following equation in the Laplace domain yields

$$\overline{p}_{wD} - \overline{p}_{fD}(r_D) = \frac{2\pi}{C_{fD}}\left(\overline{q}_{fwD} r_D - \int_0^{r_D}\int_0^{v} \tilde{\overline{q}}_{fD}(u)dudv\right) \tag{10}$$

Eq. (10) can be further transformed into the discretized form and the pressure of the ith segment can be expressed as [2]

$$\overline{p}_{wD}-\overline{p}_{fDi}=\left(\frac{2\pi}{C_{fD}}\right)\cdot\left[\begin{array}{l} r_{Di}\cdot\sum_{k=1}^{N}\overline{q}_{fDk}-\left(\frac{\Delta r_{Di}}{8}\right)\cdot\overline{q}_{fDi} \\ -\sum_{k=1}^{i-1}\overline{q}_{fDk}\cdot\left[\frac{\Delta r_{Di}}{2}+\left(r_{Di}-k\cdot\Delta r_{Di}\right)\right] \end{array}\right],\quad i=1,2,\ldots,N \tag{11}$$

where

$$r_{Di}=\sum_{k=1}^{i-1}L_{fDk}+L_{fDi}/2,\quad \Delta r_{Di}=L_{fDi} \tag{12}$$

Also, the fracture flow equation (11) can be written in matrix form as

$$\begin{pmatrix}\overline{p}_{wD}\\ \overline{p}_{wD}\\ \vdots\\ \overline{p}_{wD}\end{pmatrix}-\vec{\overline{p}}_{fD}=AF\cdot\vec{\overline{q}}_{fD} \tag{13}$$

where $\vec{\overline{p}}_{fD}$ is the pressure vector in the fracture wing

$$\vec{\overline{p}}_{fD}=\left(\overline{p}_{fD1}\ \cdots\ \overline{p}_{fDN}\right)^{T} \tag{14}$$

and $\vec{\overline{q}}_{fD}$ is the flow rate vector in the fracture wing

$$\vec{\overline{q}}_{fD}=\left(\overline{q}_{fD1}\ \cdots\ \overline{q}_{fDN}\right)^{T} \tag{15}$$

and the coefficient matrix AF is

$$AF=\frac{2\pi}{C_{fD}}\begin{pmatrix} \frac{1}{2N}-\frac{1}{8N} & \frac{1}{2N} & \cdots & \frac{1}{2N} \\ \frac{3}{2N}-\left(\frac{1}{2N}+\frac{1}{2N}\right) & \frac{3}{2N}-\frac{1}{8N} & \cdots & \frac{3}{2N} \\ \vdots & \vdots & \ddots & \vdots \\ \frac{2N-1}{2N}-\left(\frac{1}{2N}+\frac{N-1-0.5}{N}\right) & \frac{2N-1}{2N}-\left(\frac{1}{2N}+\frac{N-2-0.5}{N}\right) & \cdots & \frac{2N-1}{2N}-\frac{1}{8N} \end{pmatrix} \tag{16}$$

It is noted that our discretized model is based on the flow rate of each segment instead of the flow rate strength. The coefficient matrix AF is only related to the number of the discretized segments of fracture wing and can be tabulated.

2.3 Coupling

According to the continuity condition that the pressure and the flux must be continuous along the fracture surface, the following conditions must hold along the fracture plane:

$$\vec{\overline{p}}_{fD} = \vec{\overline{p}}_D, \quad \vec{\overline{q}}_{fD} = \vec{\overline{q}}_D \tag{17}$$

and the flow rate in the wellbore is

$$\sum_{i=1}^{N} \overline{q}_{fDi} = \overline{q}_{fwD} \tag{18}$$

Combining Eqs. (6), (13), (17), and (18), we can obtain

$$\begin{pmatrix} A & -I \\ I^T & 0 \end{pmatrix} \cdot \begin{pmatrix} \vec{\overline{q}}_{fD} \\ \overline{p}_{wD} \end{pmatrix} = \begin{pmatrix} 0 \\ \overline{q}_{fwD} \end{pmatrix} \tag{19}$$

where

$$A = AR + AF \tag{20}$$

and unit vector I

$$I = (\ 1, \quad 1, \quad \cdots, \quad 1\ )^T \tag{21}$$

While the flow q_{fwD} is specified, we can obtain wellbore pressure p_{wD} and flow rate in Laplace domain by solving Eq. (19).

3 Mathematical Model of Multi-Wing Fractures

We have assumed that a complex fracture system consists of M wings. The total flow rate is equal to Q in the wellbore. The mth wing is equally divided into N_m segments. The transient pressure solution for the complex fracture system can be obtained from the basic fracture wing solution according to the superposition principle.

3.1 Semi-Analytical Solution

According to the superposition principle, the dimensionless pressure in the reservoir can be written in the Laplace domain as follows:

$$\overline{p}_{Dij} = \sum_{k=1}^{M} \sum_{m=1}^{N_k} \overline{q}_{Dkm} \cdot s\overline{p}_{uD(\,ij,\,km)} = \sum_{k=1}^{M} \sum_{m=1}^{N_k} \overline{q}_{Dkm} \cdot s\overline{p}_{uD(\,ij,\,km)} \tag{22}$$

where subscript "ij" indicates the jth segment of the ith fracture wing:

$$s\overline{p}_{uD(ij,km)} = s\overline{p}_{uD}\left(x_{wDij}, y_{wDij}, x_{wDkm}, y_{wDkm}, L_{fDkm}\right) \tag{23}$$

$\overline{p}_{uD(ij,km)}$ is the dimensionless pressure change at the jth segment of the ith fracture wing caused by the production of the mth segment of the kth fracture wing in the Laplace domain.

Eq. (22) can be further written in the following matrix form:

$$\vec{\overline{P}}_{MD} = AMR \cdot \vec{\overline{q}}_{MD} \tag{24}$$

where $\vec{\overline{P}}_{MD}$ is the pressure vector and $\vec{\overline{q}}_{MD}$ is the flow rate vector

$$\vec{\overline{p}}_{MD} = \left(\vec{\overline{p}}_{D1} \cdots \vec{\overline{p}}_{DM}\right)^T \tag{25}$$

$$\vec{\overline{q}}_{MD} = \left(\vec{\overline{q}}_{D1} \cdots \vec{\overline{q}}_{DM}\right)^T \tag{26}$$

with subvector $\vec{\overline{p}}_{Di}$ denoting pressure vector of the ith fracture wing

$$\vec{\overline{P}}_{Di} = \left(\overline{p}_{Di1} \quad \cdots \quad \overline{p}_{DiN_i}\right)^T \tag{27}$$

and subvector $\vec{\overline{q}}_{Di}$ denoting flow rate vector of the ith fracture wing

$$\vec{\overline{q}}_{Di} = \left(\overline{q}_{Di1} \quad \cdots \quad \overline{q}_{DiN_i}\right)^T \tag{28}$$

The coefficient matrix AMR in Eq. (24) can be written as

$$AMR = \begin{pmatrix} AR^{11} & AR^{12} & \cdots & AR^{1M} \\ AR^{21} & AR^{22} & \cdots & AR^{2M} \\ \vdots & \vdots & \ddots & \vdots \\ AR^{M1} & AR^{M2} & \cdots & AR^{MM} \end{pmatrix} \tag{29}$$

where AR^{ik} denotes the effect of the production from the kth fracture wing on the ith fracture wing. It can be expressed as

$$AR^{ik} = \begin{pmatrix} s\overline{p}_{uD(i1,k1)} & s\overline{p}_{uD(i1,k2)} & \cdots & s\overline{p}_{uD(i1,kN_k)} \\ s\overline{p}_{uD(i2,k1)} & s\overline{p}_{uD(i2,k2)} & \cdots & s\overline{p}_{uD(i2,kN_k)} \\ \vdots & \vdots & \ddots & \vdots \\ s\overline{p}_{uD(iN_i,k1)} & s\overline{p}_{uD(iN_i,k2)} & \cdots & s\overline{p}_{uD(iN_i,kN_k)} \end{pmatrix} \tag{30}$$

Unlike the flow in the reservoir, there is no influence between flow inside each fracture wing, and according to Eq. (11), the pressure equation for the mth wing can be written as

$$\begin{pmatrix} \overline{p}_{wD} \\ \overline{p}_{wD} \\ \vdots \\ \overline{p}_{wD} \end{pmatrix} - \vec{\overline{p}}_{fDm} = AF^m \cdot \vec{\overline{q}}_{fDm}, \quad m = 1,2,\ldots,M \tag{31}$$

where $\vec{\overline{p}}_{fDm}$ is the pressure subvector representing the pressure response of the mth fracture wing

$$\vec{\overline{p}}_{fDm} = \left(\overline{p}_{fDm1} \quad \cdots \quad \overline{p}_{fDmN_m}\right)^T \tag{32}$$

and $\vec{\overline{q}}_{fDm}$ is the flow rate subvector, which denotes the flow rate of the mth fracture wing.

$$\vec{\overline{q}}_{fDm} = \left(\overline{q}_{fDm1} \quad \cdots \quad \overline{q}_{fDmN_m}\right)^T \tag{33}$$

According to the flow Eq. (11), the coefficient matrix of the mth wing, AF^{m}, can be written as

$$AF^m = \left(\frac{2\pi}{C_{fDm}}\right)\begin{pmatrix} \frac{1}{2N_m}-\frac{1}{8N_m} & \frac{1}{2N_m} & \cdots & \frac{1}{2N_m} \\ \frac{3}{2N_m}-\left(\frac{1}{2N_m}+\frac{1}{2N_m}\right) & \frac{3}{2N_m}-\frac{1}{8N_m} & \cdots & \frac{3}{2N_m} \\ \vdots & \vdots & \ddots & \vdots \\ \frac{2N_m-1}{2N_m}-\left(\frac{1}{2N_m}+\frac{N_m-1-0.5}{N_m}\right) & \frac{2N_m-1}{2N_m}-\left(\frac{1}{2N_m}+\frac{N_m-2-0.5}{N_m}\right) & \cdots & \frac{2N_m-1}{2N_m}-\frac{1}{8N_m} \end{pmatrix} \tag{34}$$

Combing all the flow equations in the fracture, we can obtain

$$\begin{pmatrix} \overline{p}_{wD} \\ \overline{p}_{wD} \\ \vdots \\ \overline{p}_{wD} \end{pmatrix} - \vec{\overline{p}}_{MfD} = AMF \cdot \vec{\overline{q}}_{MfD} \tag{35}$$

where $\vec{\overline{P}}_{MfD}$ is the pressure vector and $\vec{\overline{q}}_{MfD}$ is the flow rate vector. The two vectors can be expressed as

$$\vec{\overline{p}}_{MfD} = \left(\vec{\overline{p}}_{fD1} \cdots \vec{\overline{p}}_{fDM}\right)^T \tag{36}$$

$$\vec{\overline{q}}_{MfD} = \left(\vec{\overline{q}}_{fD1} \cdots \vec{\overline{q}}_{fDM}\right)^T \tag{37}$$

and the coefficient matrix can be written into

$$AMF = \begin{pmatrix} AF^1 & 0 & \cdots & 0 \\ 0 & AF^2 & \cdots & 0 \\ \vdots & \vdots & \ddots & \vdots \\ 0 & 0 & \cdots & AF^M \end{pmatrix} \tag{38}$$

According to the continuity condition that the pressure and the flux must be continuous along the fracture surface, the following conditions must hold along the fracture plane

$$\vec{\overline{p}}_{MfD} = \vec{\overline{p}}_{MD}, \quad \vec{\overline{q}}_{MfD} = \vec{\overline{q}}_{MD} \tag{39}$$

In addition, the total flow rate must satisfy the unity condition and in the Laplace domain yields

$$\sum_{m=1}^{M}\sum_{n=1}^{N_m}\overline{q}_{fDmn}=1/s \tag{40}$$

Thus, we can obtain the following equation for multi-wing system by combining Eqs. (24), (35), (39), and (40):

$$\begin{pmatrix} A & -I \\ I^T & 0 \end{pmatrix}\cdot\begin{pmatrix} \vec{\overline{q}}_{MfD} \\ \overline{p}_{wD} \end{pmatrix}=\begin{pmatrix} 0 \\ 1/s \end{pmatrix} \tag{41}$$

where

$$A=AMR+AMF=\begin{pmatrix} AR^{11}+AF^{1} & AR^{21} & \cdots & AR^{M1} \\ AR^{12} & AR^{22}+AF^{2} & \cdots & AR^{M2} \\ \vdots & \vdots & \ddots & \vdots \\ AR^{1M} & AR^{2M} & \cdots & AR^{MM}+AF^{M} \end{pmatrix} \tag{42}$$

and unit vector I

$$I=(\ 1,\quad 1,\quad \cdots,\quad 1\ )^T \tag{43}$$

The pressure solution can be obtained by solving Eq. (41) using the Gaussian elimination method and further be inverted to the time domain using the Stehfest numerical algorithm [27].

4 Fluid Flow in Low Permeability Reservoir

The best advantage of the fracture wing model is that it can handle complex fractures using a combination of different wing lengths and azimuths. In this section, we present some applications of the new model.

4.1 PTA of an Asymmetrical Planar Fracture

Fig. 1 presents a vertical well stimulated by a fracture consisting of two wings with different lengths L_{f1} and L_{f2}. As proposed in the literature, an asymmetric factor α is introduced to measure the offsetting of the well about the center of the fracture:

$$a=\frac{|L_{f1}-L_{f2}|}{L_{f1}+L_{f2}}=\frac{|L_{fD1}-L_{fD2}|}{L_{fD1}+L_{fD2}} \tag{44}$$

In addition, apparent fracture conductivity is also introduced according to the conventional definition of fracture conductivity [2]:

$$C_{fDa} = \frac{k_f w_f}{k x_f} \tag{45}$$

$$x_f = \frac{L_{f1} + L_{f2}}{2} \tag{46}$$

For a bi-wing asymmetrical fracture (Fig. 1), we assume that each wing has the same fracture parameters, k_f and w_f. According to the different definitions of the dimensionless fracture conductivity (Eqs. (A.3) and 45), the relationship between C_{fDa} and C_{fD} can be expressed as

$$C_{fDa} = \frac{2C_{fD1}C_{fD2}}{C_{fD1} + C_{fD2}} \tag{47}$$

We present the pressure response of an asymmetrical fracture [15]. Fig. 5 illustrates the effect of the fracture asymmetry factor on the dimensionless wellbore pressure under constant flow conditions when $C_{fDa} = 50$. It was found that the asymmetric factor had significant influence on the pressure response in the bilinear flow period. In addition, analysis shows that the end of the bilinear flow period occurs earlier as the asymmetric factor increases except $\alpha = 1$. For the case of $C_{fDa} = 50$, the formation of linear flow and radial flow can be identified.

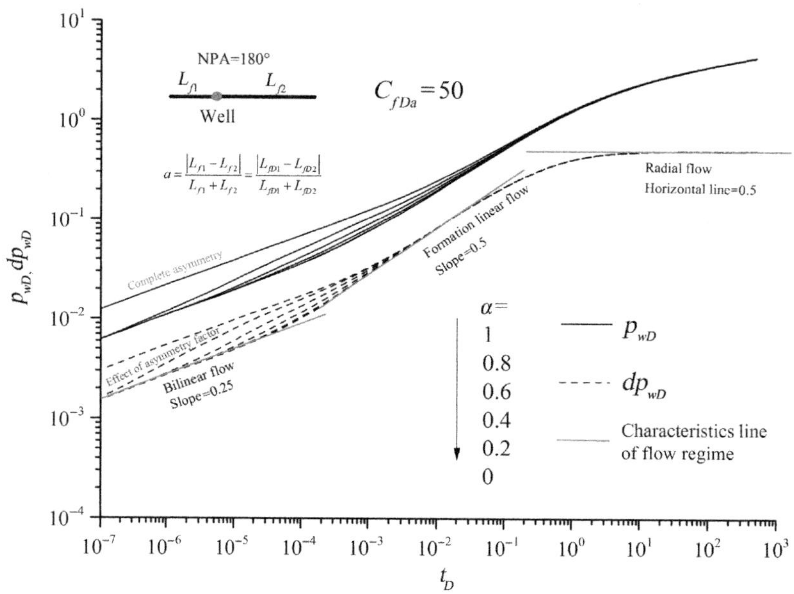

Fig. 5
Effect of the fracture asymmetry factor on the dimensionless wellbore pressure under constant flow conditions, $C_{fDa} = 50$.

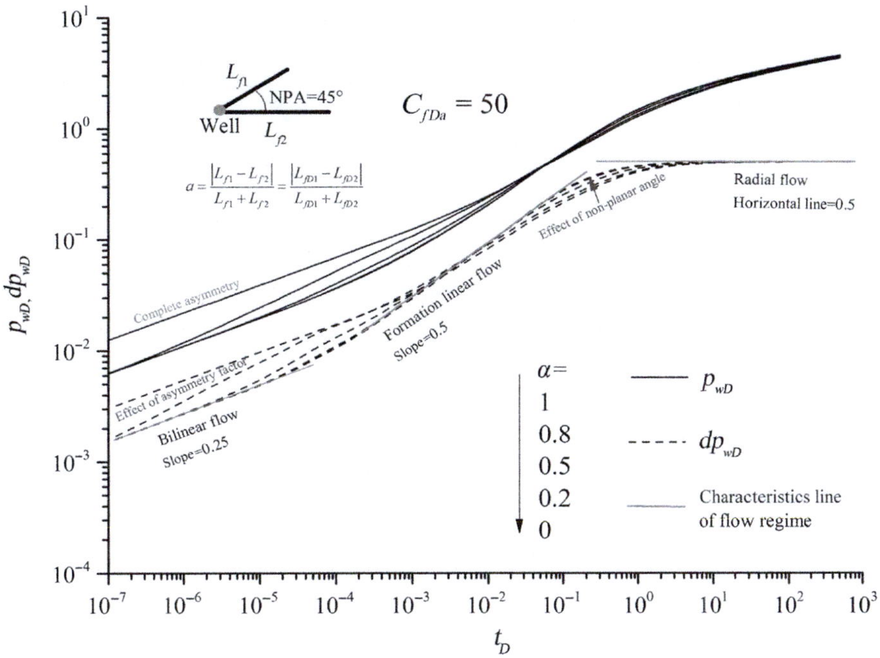

Fig. 6

Effect of the fracture asymmetry factor and NPA on the dimensionless wellbore pressure under constant flow conditions, $C_{fDa} = 50$, NPA $= \pi/4$.

4.2 PTA of an Asymmetrical Nonplanar Fracture

Fig. 6 presents the effect of the asymmetry factor and nonplanar angle (NPA) on the dimensionless wellbore pressure under constant flow conditions for the case of $C_{fDa} = 50$ and NPA $= \pi/4$. It is shown that the value of the NPA mainly affects the curve shape of the transition flow period from the formation linear flow to radial flow. The existence of the NPA strengthens the interaction of the fracture wings and may lead to the absence of the linear flow. In this case, only the radial flow period can be identified.

4.3 PTA of Multiple Fracture Wings Connected to a Vertical Well

In this part, we discuss the pressure transient behavior of multiple fracture wings connected to a vertical well. Fig. 7 presents the pressure derivative curves for a vertical well intercepted by multi-wing fractures with fracture conductivity equal to 50. For simplicity, we assume that the angle between each fracture wing is equal, for example, 60 degrees for six wings, 72 degrees for five wings. As can be seen from Fig. 7, fracture number (NF), the bilinear flow, formation linear flow, and pseudo-radial flow can be observed for different fracture numbers. As the wing numbers increase, the pressure derivative decreases, which indicates that the well productivity can be enhanced by multi-wing fractures under the same production pressure drop.

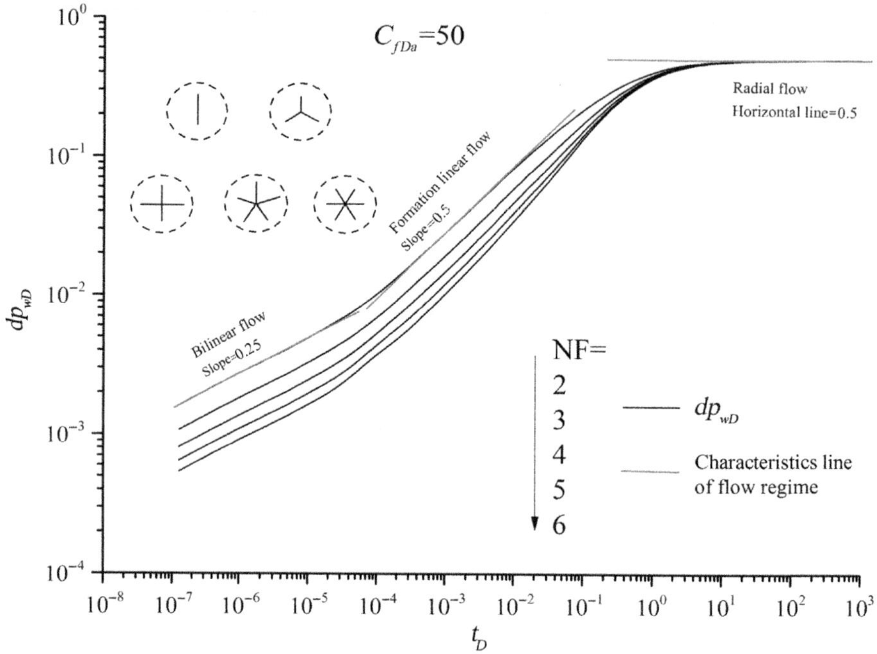

Fig. 7
The effect of fracture numbers on the pressure derivative at $C_{fDa} = 50$.

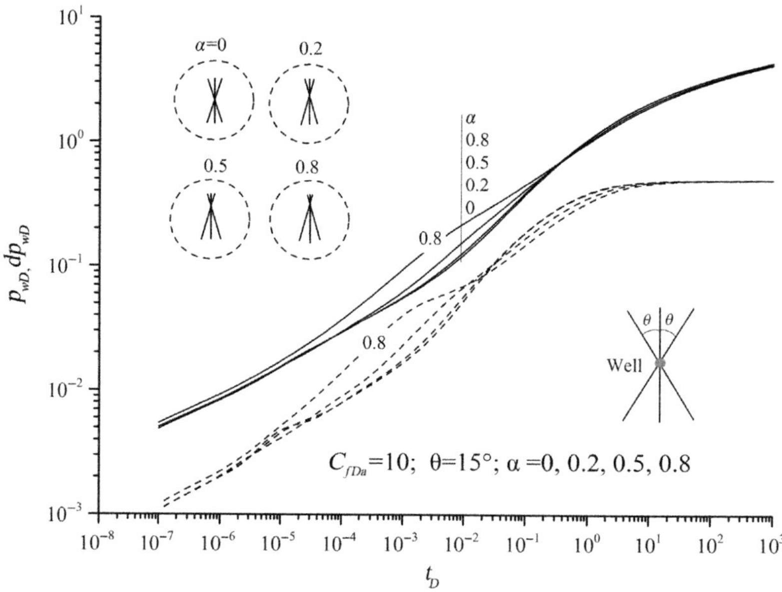

Fig. 8
The effect of fracture asymmetry factor on the pressure and pressure derivative for 6-wing fractures.

Fig. 8 shows the impact of the asymmetry factor of fracture cluster on the pressure behaviors. For comparison, four values of the asymmetry factor have been used, 0, 0.2, 0.5, and 0.8, respectively. It can be seen that there is a transition point on the pressure curves at the dimensionless time $t_D = 2 \times 10^{-2}$. For time less than the transition point, the bigger the asymmetry factor is, the larger the dimensionless pressure and pressure derivative are; while the dimensionless time is larger than the transition point, the dimensionless pressure increases as the asymmetry factor reduces.

4.4 PTA of Multi-Wing Nonplanar Fractures in a Horizontal Well

Recent fracture stimulation monitoring using advanced technologies such as microseismic monitoring, fiber optics temperature, and strain sensing have shown that multiple fractures do not grow and develop in the same way. Complex in situ stress field and formation heterogeneities affected the placement and growth of multiple fractures. The mechanism of the fracture growth and interaction with each other has been discussed in detail [24, 25]. These studies pointed out that the action of multiple fractures may result in significant shearing displacements along the fracture surface causing the fracture growth to follow a curved path.

In order to eliminate the effect of fracture interaction caused by short fracture spacing (FS) on the pressure response, we set the FS to be 50. Fig. 9 shows the pressure derivative curve for a horizontal well intercepted by three nonplanar fractures (six wings, $C_{fDa}=5$, FS$=50$) with varied NPA. As can be seen from Fig. 9, the NPA mainly affects the pressure at the early flow

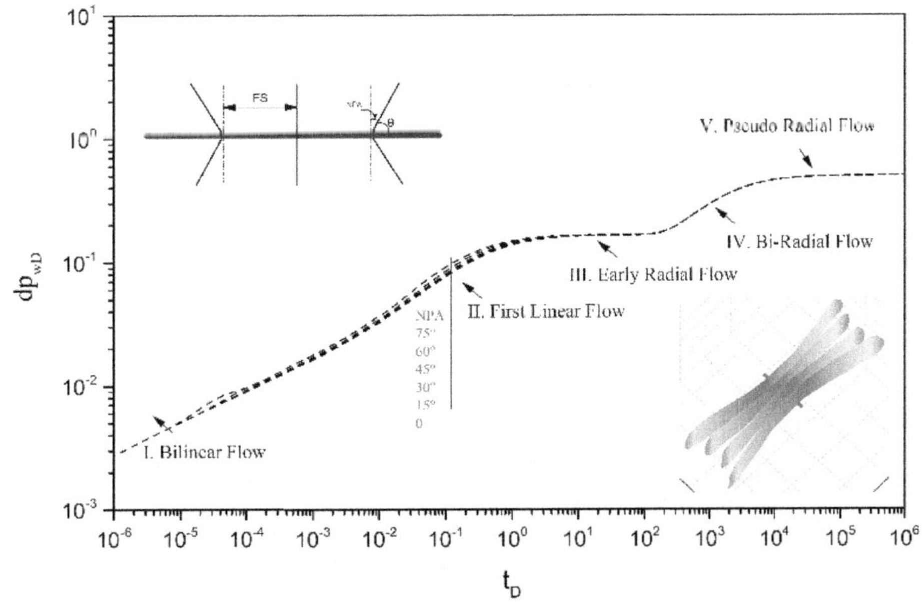

Fig. 9
The curve of pressure derivative for a horizontal well intercepted by 3 non-planar fractures ($C_{fDa} = 5$, FS $= 50$) with different NPA.

period, i.e., bilinear flow period and first linear flow period. The existence of NPA leads to the pressure derivative curve deviating from the characteristic straight line. The magnitude of curve deviation becomes more and more significant as the NPA increases. It can also be observed that as the NPA increases, the end of the bilinear flow period occurs earlier and the formation linear flow will not be identifiable for the case of fracture conductivity $C_{fDa}=5$.

4.5 Well Test Analysis of Multi-Wing Fractured Well Using Synthetic Data

The synthetic data has been generated using ECLIPSE. The simulated drawdown test of a water well for the case of four-wing orthogonal fractures is analyzed using the type curve match. The data used to simulate the test are listed in Table 1. Table 2 gives the pressure drawdown data after the well fracturing treatment.

Fig. 10 shows the matching result of the synthetic data and type curves using the fracture wing model presented in this chapter. It was found that the data was matched best when $C_{fDa}=10$. Using the obtained matching points, we can estimate the formation permeability and fracture half-length.

Matching results for new type curves: $C_{fDa}=10$ and the fracture wing number $M=4$.

$$[t_D]_{MP}=0.1,\quad [p_{wD}]_{MP}=0.37$$

$$[t]_{MP}=1.8\,\text{h},\quad [\Delta p]_{MP}=2.64\,\text{psi}$$

Sample calculations
Permeability:

From the pressure matching point relation for permeability, k, given as

$$k=141.2\frac{q\mu B}{h}\frac{[p_{wD}]_{MP}}{[\Delta p]_{MP}}$$

We can obtain the permeability value, k, for this case

$$k=141.2\frac{(100)(1)(1)}{100}\frac{0.37}{2.64}=19.79\,\text{md}$$

Table 1 Reservoir, fluid properties and production data of a well

Name of Parameters	Basic Data
Estimated net pay thickness (h)	100 ft
Average porosity (ϕ)	0.2
Volume factor (B)	1.00RB/STB
Viscosity (μ)	1 cP
Total compressibility (c_t)	$3\times10^{-6}\,\text{psi}^{-1}$
Rate (q)	100 STB/D

Table 2 Pressure drawdown test data

Time (h)	Δp (psi)	$t^*\ \Delta p'$ (psi)
hr	psi	psi
0.009486	0.4443	0.240601
0.021429	0.5435	0.156721
0.036463	0.6416	0.193901
0.05539	0.7241	0.223347
0.079219	0.814	0.258734
0.109217	0.8999	0.287213
0.146982	0.9922	0.330633
0.194525	1.0918	0.378556
0.254379	1.2012	0.442701
0.32973	1.3271	0.498432
0.424592	1.4585	0.559357
0.544016	1.6094	0.633211
0.694361	1.7725	0.707816
0.883635	1.9556	0.799198
1.121917	2.1592	0.897515
1.421896	2.3862	1.007202
1.799547	2.6392	1.126872
2.274981	2.9204	1.262924
2.873518	3.2354	1.412478
3.62703	3.585	1.570031
4.575647	3.9722	1.737925
5.769884	4.3989	1.909967
7.27334	4.8652	2.08529
9.16608	5.3726	2.258099
11.5489	5.9185	2.42109
14.54869	6.501	2.573603
18.3252	7.1172	2.711341
23.07954	7.7632	2.834089
29.06491	8.436	2.940995
36.60004	9.1313	3.033357
46.0862	9.8467	3.111622
58.02858	10.5781	3.177092
73.06314	11.3237	3.233439
91.99053	12.0811	3.280892
115.8187	12.8486	3.319891
145.8166	13.624	3.350878

Fracture half length:

Solving the time matching point relation for fracture half length, x_f, we obtain

$$x_f = 0.01624\sqrt{\frac{k}{\varphi \mu c_t}\frac{[t]_{MP}}{[t_D]_{MP}}}$$

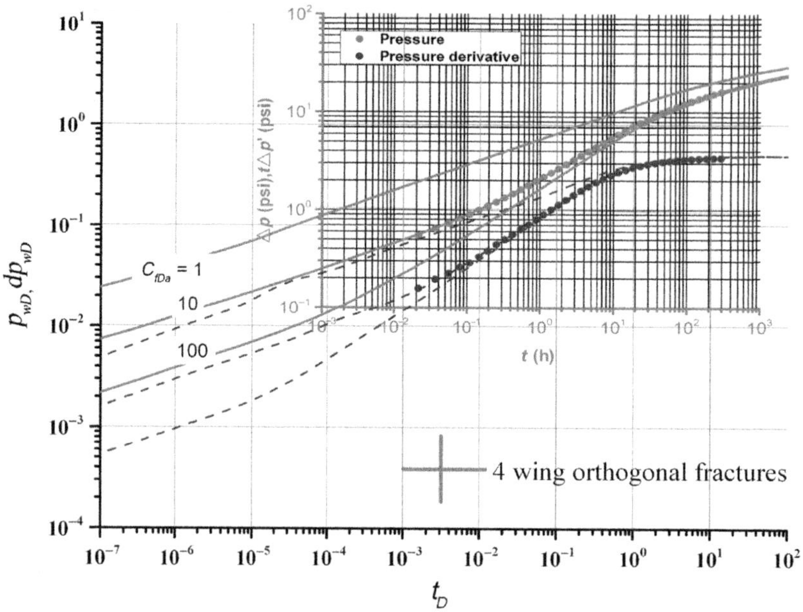

Fig. 10
Type curve analysis of 4-wing orthogonal fractures.

Table 3 Summary of calculative results

Name of Parameters	New Type Curves	Numerical Simulation Inputted
Fracture wings	4	4
Dimensionless conductivity, C_{fD}	10	10
Fracture half-length, x_f	395.7 ft	N/A
Reservoir permeability, k	19.79 md	20 md
Fracture wing value, L_{f1}	N/A	392 ft
Fracture wing value, L_{f2}	N/A	392 ft
Fracture wing value, L_{f3}	N/A	392 ft
Fracture wing value, L_{f4}	N/A	392 ft

$$x_f = 0.01624\sqrt{\frac{19.79}{(0.2)(1)\left(3\times10^{-6}\right)}\frac{1.8}{0.1}} = 395.7\text{ft}$$

Table 3 shows a comparison between the type curves and inputted parameters in the simulator, which indicates that the new type curves are in perfect agreement with the simulation data.

As can be seen from Table 3, we can obtain the equivalent length of multiple fracture wings. It is difficult to obtain the parameters of each fracture wing by type curve match.

5 Conclusion

For the low permeability reservoirs, it was found that hydraulic fractures always exhibit complex geometry patterns by direct observation on fractured cores, geological evidence, and laboratory experiments.

1. A general, rigorous, computationally stable, and accurate semi-analytical solution is presented to compute pressure responses of the fractured well in the Laplace domain.
2. Based on the fracture wing model, many complex fracture patterns can be generated by the combinations of fracture wings with different lengths and azimuths. As long as the discretized segment of each wing is fixed, we can easily calculate the coefficient matrix for fluid flow in the fracture using a tabulated sheet.
3. Three applications have been used to reveal the fluid flow in the complex fractures in low permeability reservoirs. It has been shown that the method is very flexible in the sense that it has the ability to consider the fracture conductivity, facture asymmetry, number of the wings, and fracture azimuth.

Appendix A. Dimensionless Definitions

The dimensionless reservoir and fracture pressure are given as

$$p_D = \frac{2\pi kh(p_i - p)}{\mu BQ}, \quad p_{fD} = \frac{2\pi kh\left(p_i - p_f\right)}{\mu BQ} \tag{A.1}$$

The dimensionless time is

$$t_D = \frac{k}{\varphi \mu c_t \overline{L}_f^2} t \tag{A.2}$$

In the fracture model, the dimensionless fracture wing conductivity, C_{fD}, is

$$C_{fD} = \frac{k_f w_f}{k L_f} \tag{A.3}$$

Note that conductivity is defined with the wing length of each fracture wing and may be different from each other.

The dimensionless fracture wing coordinate in the 1D polar coordinates, r_D, is:

$$r_D = r/L_f \in [0, 1] \tag{A.4}$$

and the dimensionless flow rate strength, $\tilde{q}_{fD}$, is defined as

$$\tilde{q}_{fD} = \tilde{q}_f L_f / Q \tag{A.5}$$

where $\tilde{q}_f$ is the flow rate of per unit fracture wing length. The dimensionless flow contribution of the fracture wing to wellbore rate is given as

$$q_{fwD} = q_{fw} / Q \tag{A.6}$$

Q is the sum of the flow rate of each fracture wing in the wellbore

$$Q = \sum_{i=1}^{M} q_{fwi} \tag{A.7}$$

Here, M is the total fracture wing number.

If the fracture wing has been divided into N segments, the wellbore flow rate of the fracture wing is

$$q_{fw} = \sum_{j=1}^{N} q_{fj} \tag{A.8}$$

and the dimensionless wellbore flow rate of the fracture wing is

$$q_{fwD} = \sum_{j=1}^{N} q_{fDj} \tag{A.9}$$

Combining Eq. (A.8) into Eq. (A.7) yields

$$Q = \sum_{i=1}^{M} \sum_{j=1}^{N_i} q_{fij} \tag{A.10}$$

and other dimensionless definitions in the reservoir model are

$$x_D = x/\overline{L}_f, y_D = y/\overline{L}_f, L_{fD} = L_f/\overline{L}_f, w_{fD} = w_f/\overline{L}_f \tag{A.11}$$

where L_f is the wing length of fracture wing and $\overline{L}_f$ is the reference length

$$\overline{L}_f = \frac{1}{M} \sum_{i=1}^{M} L_{fi} \tag{A.12}$$

Accordingly, the apparent fracture conductivity can also be defined

$$C_{fDa} = \frac{k_f w_f}{k \overline{L}_f} \tag{A.13}$$

Appendix B. Derivation of the Fluid Flow Equation of a Fracture Wing

As shown in Fig. 4, the fracture wing is considered as a homogeneous, finite, slab, porous medium of height, h, wing length, L_f, and width, w_f. Fluid enters the fracture at a rate $\widetilde{q}_f$ per unit of fracture length. The flow inside the fracture is assumed to be incompressible and can be described in the one-dimensional coordinate originating from the wellbore [2]

$$\frac{\partial^2 p_f}{\partial r^2}+\frac{\mu}{k}\left(\frac{\widetilde{q}_f}{w_f h}\right)=0,\quad 0\leq r\leq L_f \tag{B.1}$$

with initial condition

$$p_f(r,t=0)=p_i,\quad 0\leq r\leq L_f \tag{B.2}$$

and boundary conditions

$$\left(\frac{k_f h w_f}{\mu}\right)\cdot\left(\frac{\partial p_f}{\partial r}\right)_{r=0}=q_{fw} \tag{B.3}$$

$$\left(\frac{\partial p_f}{\partial r}\right)_{r=L_f}=0 \tag{B.4}$$

Using the dimensionless definition, we can obtain the following dimensionless equation:

Fluid flow equation

$$\frac{\partial^2 p_{fD}}{\partial r_D^2}-\frac{2\pi}{C_{fD}}\widetilde{q}_{fD}=0 \tag{B.5}$$

Initial condition

$$p_{fD}(r_D,t_D=0)=0,\quad 0\leq r_D\leq 1 \tag{B.6}$$

Boundary conditions

$$\left(\frac{\partial p_{fD}}{\partial r_D}\right)_{r_D=0}=-\frac{2\pi}{C_{fD}}q_{fwD} \tag{B.7}$$

$$\left(\frac{\partial p_{fD}}{\partial r_D}\right)_{r_D=1}=0 \tag{B.8}$$

$$p_{wD}=p_{fD}(0) \tag{B.9}$$

Integrating Eq. (B.5) from 0 to v with boundary condition (B.7) and (B.8), we can obtain the following equation

$$\frac{\partial p_{fD}(r_D)}{\partial r_D} = \frac{2\pi}{C_{fD}} \int_0^v \tilde{q}_{fD}(u)du - \frac{2\pi}{C_{fD}} q_{fwD} \tag{B.10}$$

Further integrating Eq. (B.10) from 0 to r_D with boundary condition (B.9):

$$p_{wD} - p_{fD}(r_D) = \frac{2\pi}{C_{fD}} \left(q_{fwD} r_D - \int_0^{r_D} \int_0^v \tilde{q}_{fD}(u)dudv \right) \tag{B.11}$$

Eq. (B.11) can be written in Laplace domain with initial condition (B.6)

$$\overline{p}_{wD} - \overline{p}_{fD}(r_D) = \frac{2\pi}{C_{fD}} \left(\overline{q}_{fwD} r_D - \int_0^{r_D} \int_0^v \overline{\tilde{q}}_{fD}(u)dudv \right) \tag{B.12}$$

Appendix C. Unit Conversion Factors

SI Metric Conversion Factors

bbl × 0.1589874	m^3
cP × 0.001	Pa s
ft × 0.3048	m
ft^2 × 0.0929	m^2
psi × 6.894757	kPa

Nomenclature

Dimensionless Variables: Real Domain

C_{fD} Dimensionless fracture conductivity defined in this chapter
C_{fDa} Dimensionless apparent fracture conductivity defined by Cinco-Ley
t_D Dimensionless time
p_D Dimensionless pressure
p_{uD} Dimensionless pressure of uniform-flux solution
p_{wD} Dimensionless well bottom pressure.
p_{fD} Dimensionless fracture pressure
dp_D Dimensionless pressure derivative, $p_D/d \ln t_D$
$\tilde{q}_{fD}$ Dimensionless flow rate strength in the fracture
$\tilde{q}_D$ Dimensionless flow rate strength in the reservoir
q_{fwD} Dimensionless wellbore flow rate of the fracture wing.
r_D Dimensionless coordinate in the 1D polar coordinates
Δr_D Dimensionless discretized step in the 1D polar coordinates
x_D Dimensionless coordinate in the x direction

y_D	Dimensionless coordinate in the y direction
x_{wD}	Dimensionless wellbore coordinate in the x direction
y_{wD}	Dimensionless wellbore coordinate in the y direction
L_{fD}	Dimensionless fracture wing length
α	Fracture asymmetry factor
θ	Intersection angle of fracture wing and x axis
M	Total fracture wing number
N	Discretized segments of the fracture wing
FS	Dimensionless fracture spacing of the multifractured horizontal well

Dimensionless Variables: Laplace Domain

s	Time variable in Laplace domain
$\overline{p}_D$	Dimensionless pressure p_D in Laplace domain
$\overline{p}_{wD}$	Dimensionless bottom pressure p_{wD} in Laplace domain
$\overline{p}_{fD}$	Dimensionless fracture pressure p_{fD} in Laplace domain
$\bar{\tilde{q}}_{fD}$	Dimensionless fracture-flow-rate strength $\tilde{q}_{fD}$ in Laplace domain
$\bar{\tilde{q}}_D$	Dimensionless reservoir-flow-rate strength $\tilde{q}_D$ in Laplace domain
$\overline{p}_{uD}$	Dimensionless pressure of uniform-flux solution in Laplace domain

Field Variables

c_t	Total compressibility (1/psi)
k	Formation permeability (md)
k_f	Fracture wing permeability (md)
p	Pressure (psi)
p_i	Initial formation pressure (psi)
p_f	Fracture pressure (psi)
p_w	Wellbore pressure (psi)
Q	Total rate of all wings in wellbore (STB/d)
$\tilde{q}_f$	Flow rate of per unit fracture length from formation, i.e., flow rate strength (STB/d/m)
q_{fw}	Flow rate of a fracture wing in the wellbore (STB/d)
μ	Fluid viscosity (cP)
h	Formation thickness (ft)
ϕ	Porosity
r	Radial distance at polar coordinates (ft)
r_w	Wellbore radial (ft)
x	Coordinate in the x direction (ft)
y	Coordinate in the y direction (ft)
x_f	Fracture half length defined by Cinco-Ley et al. [2]

L_f Fracture wing length (ft)
$\overline{L}_f$ Average fracture wing length (ft)
w_f Width of the fracture (ft)
t Time variable (h or d)
NPA Nonplanar angle (degrees)
B Volume factor (RB/STB)

Special Functions

$K_0(x)$ Modified Bessel function (second kind, zero order)

Special Subscripts

f Fracture property
D Dimensionless
w Wellbore property
u Uniform-flux solution
m Fracture wing
ij the jth segment of the ith wing
(ij,km) the action on the jth segment of the ith wing taken by the mth segment of the kth wing

Acknowledgment

This work was supported by the National Natural Science Foundation of China (No. 51674227), and National Key S&T Special Projects (No. 2016ZX05047-004).

References

[1] A.C. Gringarten, H.J. Ramey Jr., The use of source and Green's function in solving unsteady-flow problems in reservoirs, Soc. Pet. Eng. J. 13 (1973) 285–296.
[2] H.L. Cinco, V.F. Samaniego, N. Dominguez, Transient pressure behavior for a well with a finite-conductivity vertical fracture, Soc. Pet. Eng. J. 18 (1978) 283–264.
[3] H.L. Cinco, Transient pressure analysis for fractured wells, J. Pet. Technol. 33 (1981) 1749–1766.
[4] H.L. Cinco, H.Z. Meng, Pressure transient analysis of wells with finite conductivity vertical fractures in double porosity reservoirs, in: SPE Annual Technical Conference and Exhibition, Houston, Texas, 2–5 October, 1988.
[5] L. Larsen, T.M. Hegre, Pressure transient analysis of multi-fractured horizontal wells, in: SPE Annual Technical Conference and Exhibition, New Orleans, Louisiana, 25–28 September, 1994.
[6] D.G. Crosby, M.M. Rahman, M.K. Rahman, et al., Single and multiple transverse fracture initiation from horizontal wells, J. Pet. Sci. Eng. 35 (2002) 191–204.
[7] J. Wan, K. Aziz, Semi-analytical well model of horizontal wells with multiple hydraulic fractures, Soc. Pet. Eng. J. 7 (2002) 437–445.
[8] F. Medeiros, B. Kurtoglu, E. Ozkan, et al., Pressure-transient performances of hydraulically fractured horizontal wells in locally and globally naturally fractured formations, in: International Petroleum Technology Conference, Dubai, UAE, 4–6 December, 2007.
[9] M. Brown, E. Ozkan, R. Raghavan, et al., Practical solutions for pressure- transient responses of fractured horizontal wells in unconventional reservoirs, SPE Reser. Eval. Eng. 14 (2011) 663–676.

[10] P.B. Crawford, B.L. Landrum, Effect of unsymmetrical vertical fractures on production capacity, Soc. Pet. Eng. Trans. AIME 204 (1955) 251–254.
[11] C.O. Bennet, et al., Influence of fracture heterogeneity and wing length on the response of vertically fractured wells, Soc. Pet. Eng. J. 23 (1983) 219–230.
[12] T.N. Narasimhan, W.A. Palen, A purely numerical approach to a well intercepting a vertical fracture, in: SPE California Regional Meeting, Ventura, California, 18–20 April, 1979.
[13] F. Rodriguez, H.L. Cinco, V.F. Samaniego, Evaluation of fracture asymmetry of finite-conductivity fractured wells, SPE Form. Eval. 8 (1992) 233–239.
[14] C.E.S. Resurreicão, R. Fernando, Transient rate behavior of finite-conductivity asymmetrically fractured wells producing at constant pressure, in: SPE Annual Technical Conference and Exhibition, Dallas, Texas, 6–9 October, 1991.
[15] S. Berumen, D. Tiab, F. Rodriguez, Constant rate solutions for a fractured well with an asymmetric fracture, J. Pet. Sci. Eng. 25 (2000) 49–58.
[16] D. Tiab, Analysis of derivative data of hydraulically fractured wells by the Tiab's direct synthesis technique, J. Pet. Sci. Eng. 49 (2005) 1–21.
[17] M.K. Fisher, C.A. Wright, B.M. Davidson, et al., Integrating fracture mapping technologies to optimize stimulations in the Barnett Shale, in: SPE Annual Technical Conference and Exhibition, San Antonio, Texas, 29 September-2 October, 2002.
[18] M.K. Fisher, J.R. Heinze, C.D. Harris, et al., Optimizing horizontal completion techniques in the Barnett Shale using microseismic fracture mapping, in: SPE Annual Technical Conference and Exhibition, Houston, Texas, 26–29 September, 2004.
[19] M.J. Mayerhofer, E.P. Lolon, J.E. Youngblood, et al., Integration of microseismic fracture mapping results with numerical fracture network production modeling in the Barnett Shale, in: SPE Annual Technical Conference and Exhibition, San Antonio, Texas, USA, 24–27 September, 2006.
[20] L.N. Germanovich, L. Ring, D.K. Astakhov, et al., Hydraulic fracture with multiple segments II. Effect of interaction, Int. J. Rock Mech. Min. Sci. Geomech. Abstr. 3 (1997) 472.
[21] L.N. Germanovich, D.K. Astakhov, M.J. Mayerhofer, et al., Hydraulic fracture with multiple segments I. Observations a model formulation, Int. J. Rock Mech. Min. Sci. 34 (1997) 97.
[22] J.R. Sierra, J.D. Kaura, D. Gualtieri, et al., DTS monitoring data of hydraulic fracturing: experiences and lessons learned, in: SPE Annual Technical Conference and Exhibition, Denver, Colorado, USA, 21–24 September, 2008.
[23] M.M. Molenaar, D.J. Hill, First downhole application of distributed acoustic sensing for hydraulic-fracturing monitoring and diagnostics, SPE Drill. Completion 27 (2012) 32–38.
[24] G.S. Xu, S.W. Wong, Interaction of multiple non-planar hydraulic fractures in horizontal wells, in: International Petroleum Technology Conference, Beijing, China, 26–28 March, 2013.
[25] M. Geilikman, G.S. Xu, S.W. Wong, Interaction of multiple hydraulic fractures in horizontal wells, in: SPE Unconventional Gas Conference and Exhibition, Muscat, Oman, 28–30 January, 2013.
[26] E. Ozkan, R. Raghavan, New solutions for welltest analysis problems: Part 1-analytical considerations, SPE Form. Eval. 6 (1991) 359–368.
[27] H. Stehfest, Algorithm 368: Numerical inversion of Laplace transform, Commun. ACM 13 (1970) 47–49.

CHAPTER 10

Heat Transfer in Enhanced Geothermal Systems: Thermal-Hydro-Mechanical Coupled Modeling

Quan Gan*, Derek Elsworth†, Jianchao Cai‡

Department of Petroleum Geology & Geology, School of Geosciences, University of Aberdeen, United Kingdom, †Department of Energy and Mineral Engineering, EMS Energy Institute and G3 Center, Pennsylvania State University, University Park, PA, United States, ‡Hubei Subsurface Multi-scale Imaging Key Laboratory, Institute of Geophysics and Geomatics, China University of Geosciences, Wuhan, P.R. China

Chapter Outline

1 Introduction

Harvesting geothermal energy from deep fractured low-permeability formations has become a feasible method to ease the demand on fossil energy. Predicting mass rates and temperatures of fluid production and assessing induced seismicity are intimately connected topics that require an intimate and complete understanding of subsurface coupled thermal-hydrological-mechanical-chemical (THMC) processes [1, 2]. The associated thermal drawdown response of the rock mass results from the circulation of a heat-transfer fluid [3–5]. Thermal drawdown in the fractured porous medium may be determined by accommodating the essential components of the reservoir—heat transfer from the reservoir matrix to the fluid by

Petrophysical Characterization and Fluids Transport in Unconventional Reservoirs. https://doi.org/10.1016/B978-0-12-816698-7.00010-3

conduction and then advection across the reservoir—for which a variety of analytical approaches are available [6–14]. These approaches are based on the assumptions of locally one-dimensional heat conduction in an infinite medium, and that the fluid flow in the fractured medium instantaneously reaches local thermal equilibrium [15].

To accommodate more general flow geometries, including the inclusion of heterogeneity, a variety of numerical methods are available to represent the response. Such models also accommodate the key process of heat transfer in the subsurface in accommodating dual porosity to describe heat exchange between the porous fracture and the low porosity rock [16–19]. The principal heat transfer processes include first heat conduction between the matrix and fluid within the fractures then the advection of that heat across and then out from the reservoir. Depending on the fluid velocity in the fracture, the temperature gradient between the circulating fluid and the adjacent rock varies significantly. In general, the amount of heat energy transferred from the rock is controlled by the heat transfer area, the temperature difference between rock and fluid and the velocity of the circulating fluid [20].

In addition to exploring thermal drawdown in the reservoir, and its influence on effluent fluid temperatures, the spatial distribution of the reservoir temperature is also influenced by rates of fluid circulation. This thermal drawdown of the rock is also capable of inducing thermal stresses, which in turn are implicated in the potential for induced seismicity and increase the likelihood of late stage fault reactivation [21]. The role of thermal stresses in reservoirs has been explored with respect to the evolution of permeability [5] and of stresses [22]. Thermally driven stress changes in geothermal reservoirs are known to be potentially significant. In this work, we specifically explore rates of stress generation and their propagation through the reservoir as controlled by thermal capacitance of a dual porosity system, and ultimately on the potential to develop unstable slip. Induced thermal stresses cause the unloading of the fault by reducing maximum in situ stress, thereby reducing shear strength and therefore enabling slip reactivation as a potentially seismic event. The severity of the reactivation event appears to be directly related to the spatial gradient of rock temperature that develops in the reservoir [5]. A uniform reduction in temperature will have a muted change in thermal stresses. Hence, in this chapter, we explore the impact of fluid circulation rates on the heterogeneity of thermal drawdown that may develop within the reservoir and its potential impact on the timing and magnitude of induced seismicity.

The following study develops a dimensionless semi-analytical model that incorporates the reservoir scale, fracture spacing, and injection mass flow rate to determine thresholds for the evolution of uniform or shock-front distributions of thermal drawdown within the rock comprising the reservoir. The semi-analytical model is derived based on the balance of heat conduction within the fractured medium and the Warren-Root fracture model. Key variables are presented that may then be used to assess the propagation of stress fronts through the reservoir and from that define the likelihood, timing, and magnitude of late stage events that might occur on reactivated faults.

2 Analytical Solution

To assess and elucidate the fundamental heat transfer processes within the fractured-porous medium, a basic model is presented in Fig. 1 for the following analytical study. The two-dimensional reservoir is characterized by an orthogonal fracture network with uniform fracture spacing s (m) in both x and z directions and with a uniform fracture aperture b (m). The fractures are the sole conduits for fluid circulation within the reservoir. This parallel fracture model (PFM) has been validated to effectively characterize heat recovery from an arrangement of prismatic blocks, which are thermally isolated from the geological host medium [8]. Here we relax the restrictions imposed by prior analytical solutions by imposing the following assumptions:

1. The initial temperature of the reservoir and interstitial fluid is uniform at T_0 with cold water injected at a constant rate and constant temperature T_{inj}.
2. Heat conduction occurs primarily in z directions along the fractures with thermal conductivity K_r and with heat transfer in the vertical direction neglected. We assume that the majority of heat transfer occurs as heat conduction, perpendicular to the direction from

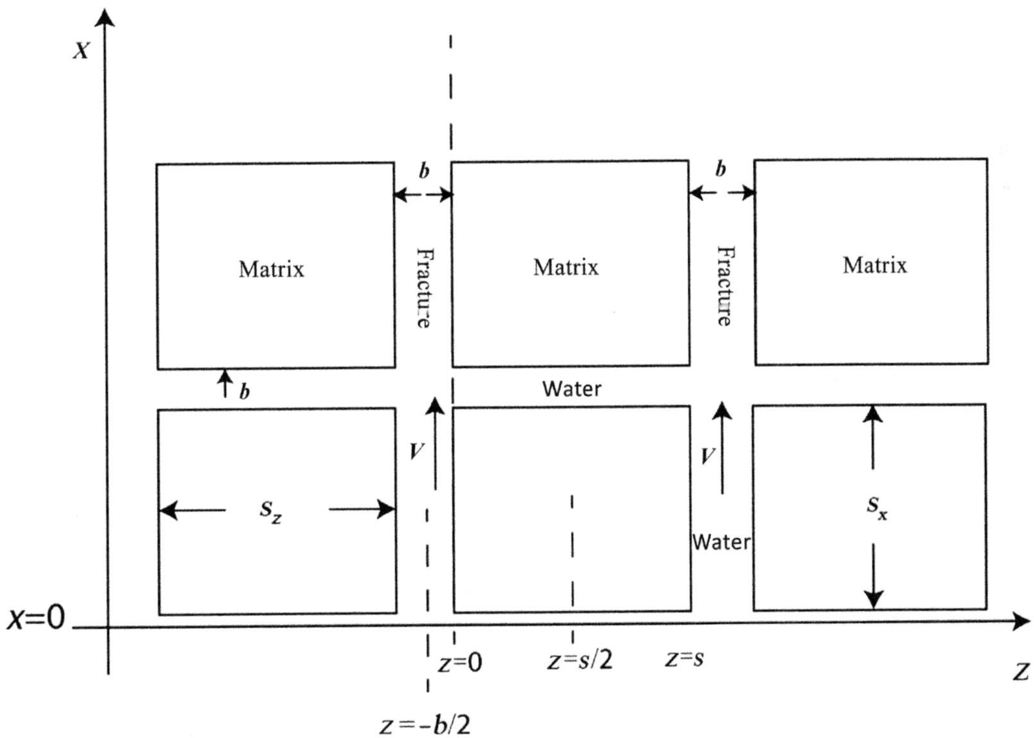

Fig.1
Schematic diagram of analytical heat conduction within fractured medium, the identical fractures with aperture b are equally separated at the spacing s.

injection well to production well. There is no heat transfer by radiation within the fractures. The diffusion of heat in the rock matrix occurs only in the direction orthogonal to the fracture plane.

3. The density and heat capacity for both the rock and fluid are constant. Also, the thermal conductivity of the rock matrix is assumed constant.

The differential equations governing heat transfer in the fracture are based on the balance of heat energy in the control volume of fractures, defined as

$$\rho_w c_w \frac{\partial T_w(x,t)}{\partial t} = -\upsilon \rho_w c_w \frac{\partial T_w(x,t)}{\partial x} + \frac{2K_r}{b} \frac{\partial T_r(x,z,t)}{\partial z}\bigg|_{z=b/2} \tag{1}$$

where υ is the fluid velocity (m/s), $T_w(x,t)$ is the temperature of water, $T_m(x,z,t)$ is the temperature of the rock matrix, b is the fracture aperture (m), c_w is the heat capacity of water (J/kg/°C), ρ_w is the density of water, and K_r is the thermal conductivity of the rock (J/s/m/°C).

The temperature of the rock matrix is governed by the one-dimensional heat conduction equation:

$$\frac{\partial^2 T_r(x,z,t)}{\partial z^2} = \frac{\rho_r c_r}{K_r} \frac{\partial T_r(x,z,t)}{\partial t} \tag{2}$$

where ρ_r is the density of the rock matrix, and c_r is the heat capacity of the rock.

The thermal drawdown response may be determined under a variety of different reservoir configurations using a unified dimensionless analytical model. We redefine the governing equations in terms of the nondimensional variables of dimensionless flow rate, Q_D, mean temperatures of the rock, T_{Dr}, and water, T_{Dw}, time, t_D,and length scales, x_D and z_D, as

$$Q_D = \frac{\rho_w c_w}{K_r} \left(\frac{q_1}{L^*}\right) s \tag{3}$$

$$q_1 = \frac{q}{\rho_w \frac{W}{s} H} = \frac{qs}{\rho_w HW} \tag{4}$$

where q is the injection rate (kg/s), q_1 is the volumetric flow rate per fracture per unit thickness (m^2/s), the terms H and W are the height and width of the reservoir respectively, and L^* is the distance between the injector and the target point. Substituting Eq. (4) into Eq. (3) defines the dimensionless flow rate as

$$Q_D = \frac{qs^2 c_w}{K_r H L^* W} \tag{5}$$

Dimensionless time t_D is defined as

$$t_D = \frac{\rho_w c_w \rho_w c_w}{K_r \ \rho_r c_r} \left(\frac{q_1}{L^*}\right)^2 t = \frac{t}{K_r \rho_r c_r} \left(\frac{qsc_w}{HWL^*}\right)^2 \tag{6}$$

The dimensionless rock temperature T_{Dr} is defined as

$$T_{Dr} = \frac{T_{inj} - T_r}{T_{inj} - T_0} \tag{7}$$

The dimensionless outlet water temperature T_{Dw} is defined as

$$T_{Dw} = \frac{T_{inj} - T_w}{T_{inj} - T_0} \tag{8}$$

The dimensionless length scales x_D and z_D are defined separately as

$$x_D = \frac{x}{L^*} \tag{9}$$

$$z_D = \frac{z}{s} \tag{10}$$

$$\eta_w = \frac{L^* K_r}{\upsilon \rho_w c_w s b} = \frac{1}{Q_D} \tag{11}$$

These defined nondimensional parameters, Q_D, t_D, T_{Dr}, T_{Dw}, length scales x_D, z_D, and η_w are used to transform and simplify the governing equations. By assuming that heat storage term in the fracture is negligible, the corresponding dimensional governing equations of Eqs. (1) and (2) could be represented in dimensionless form as

$$\frac{\partial T_{Dw}(x,t)}{\partial x_D} = 2\eta_w \frac{\partial T_{Dr}(x,z,t)}{\partial z_D}\bigg|_{z=b/2} \tag{12}$$

$$\frac{\partial^2 T_{Dr}(x,z,t)}{\partial {z_D}^2} = \eta_w^2 \frac{\partial T_{Dr}(x,z,t)}{\partial t_D} \tag{13}$$

3 Numerical Model Description

The aim of this work was to codify the evolution of rock temperature and water temperature at the outlet as a function of fracture geometry and fluid transmission characteristics and applied flow rates. The heat transfer problem was approached using semi-analytical and numerical methods and this was then applied to define the propagation of stress fronts and their impact on seismicity. Calculations were completed using a numerical simulator that couples the analysis of mass and energy transport in porous fractured media (TOUGH) with mechanical deformation (FLAC3D) [1, 16, 23]. These analyses are completed using the nondimensional parameters noted in Eq. (5) through Eq. (8).

The reservoir geometry included a two-dimensional rectangular (1500 m × 600 m × 15 m) reservoir, containing three sets of orthogonal fractures (see the mesh in Fig. 2). A strike-slip

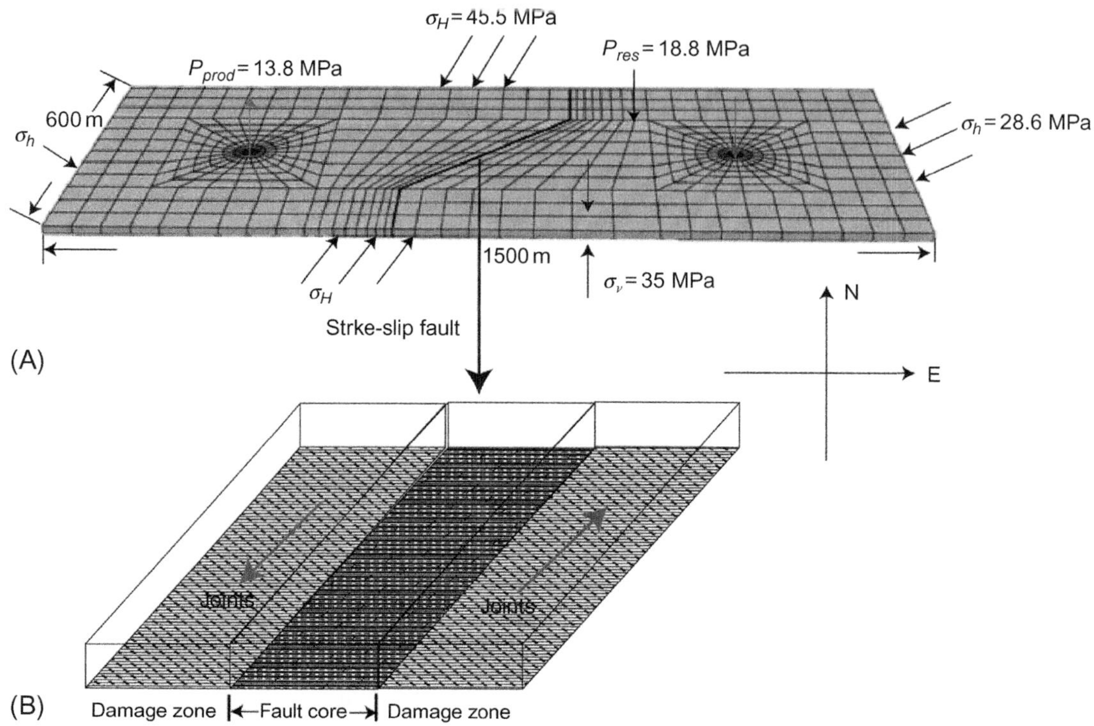

Fig. 2
(A) Model geometry and applied stress boundary condition, initial condition, and (B) strike-slip fault geometry in model.

fault was located in the center of the reservoir flanked by equidistant injection and withdrawal wells. The initial distribution of reservoir temperature was uniform with an initial rock and fluid temperature of 250°C, an initial reservoir pressure of 18.8 MPa, and with cold water injected under constant mass flow rate with a constant enthalpy of 2.0×10^5 J/kg (equivalent to 43°C), and with the production well operated under a constant pressure of 13.8 MPa. A minimum in situ stress of 28.8 MPa was imposed in the W-E direction and the maximum horizontal stress of 45 MPa was applied in the N-S direction. The model boundaries were set as no-flow boundaries with applied constant stresses.

The inserted strike-slip fault is finely discretized to represent the anticipated mechanical and transport characteristics of a fault with a low permeability core (thickness 0.8 m) flanked by higher permeability damage zones (thickness 1.2 m). The fault acts as a flow conduit along its axis but as a barrier/impediment for the propagation of the fluid front from the injection to the recovery well. A ubiquitous-joint constitutive model is applied to represent the elastic-plastic behavior of the fault. The fractures comprising the fracture network and host rock have the same initial angle of internal friction of 30 degrees (see Table 1) [24, 25]. The friction angle of the fault joint is 30 degrees. We adopted a linear strain-softening relationship that implies that the friction angle decreases with an increase in the plastic strain.

Table 1 Material properties input in the model for host rock and fault [21]

Parameters	Intact Rock	Damage Zone	Core Zone
Young's modulus (GPa)	8	1.5	1.5
Shear modulus (GPa)	5.5	1	1
Permeability (m^2)	1×10^{-16}	5×10^{-13}	1×10^{-17}
Poisson's ratio	0.2	0.2	0.2
Friction angle	30	30	30
Fracture porosity	0.8	0.8	0.8
Fracture permeability (m^2)	1×10^{-13}	5×10^{-13}	1×10^{-17}
Cohesion (MPa)	8	0	0
Tensile strength (MPa)	10	0	0

Fluid and heat transport are accommodated by representing the reservoir as an equivalent dual permeability continuum. This dual porosity model accommodates the local thermal disequilibrium in heat exchange between rock matrix and fluid in fractures. The dual permeability continuum is represented by orthogonal fracture sets spaced equally in the three principal directions and with uniform initial aperture and with a functionally impermeable matrix.

4 Thermal Drawdown Behavior

Since the thermal stresses that are induced adjacent to the fault zone may cause seismic fault reactivation, we followed the temperature change in the rock using the dimensionless parameters noted earlier. This analysis focused on predicting the evolution of temperature in the rock (as opposed to fluid temperature as in many prior models [7, 11]), and captured the intrinsic relationship between the evolution of dimensionless fault rock temperature T_{Dr} and dimensionless time t_D under various dimensionless flow rates Q_D. Moreover, complementary evaluations of water outflow temperature condition the utility of the reservoir for heat mining where magnitudes of flow rates, Q_D, are selected to minimize seismic risk.

The evolution of dimensionless fault rock temperature T_{Dr} vs. dimensionless time t_D/Q_D was determined for different dimensionless flow rates Q_D as shown in Fig. 3. This is for a fracture network permeability of $1.0 \times 10^{-13}\,m^2$ but with different fracture spacing s to allow the full parameter space of Q_D to be explored. It was apparent that the thermal drawdown response of the rock was exclusively controlled by the magnitude of Q_D. Table 2 presents the various Q_D values under different combinations of injection rate and fracture spacing. Here the main assumption for the expression of Q_D was that the direction of heat conduction in the matrix was orthogonal to the direction of the transverse fractures.

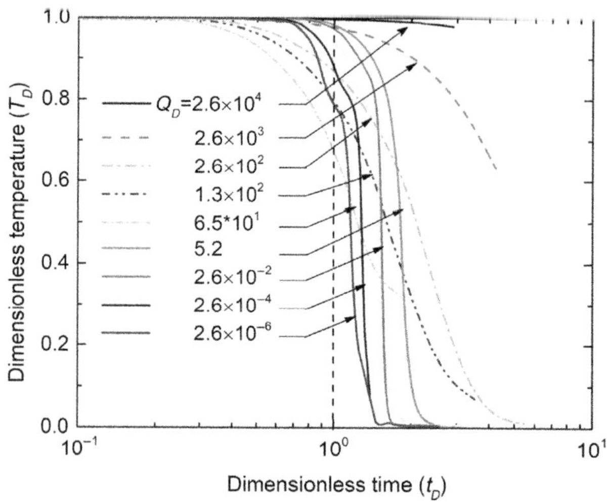

Fig. 3
Dimensionless fault temperature T_D evolution vs. dimensionless time t_D/Q_D under Q_D, values varied respectively from 2.6×10^{-6} to 2.6×10^4.

Table 2 Q_D values for the fault under different reservoir configurations by variant injection rate and fracture spacing combinations

Injection Rate (kg/s)	Fracture Spacing (m)	Q_D
100	1000	2.6×10^4
10	1000	2.6×10^3
100	100	2.6×10^2
50	100	1.3×10^2
25	100	6.5×10^1
200	10	5.2
100	1	2.6×10^{-2}
100	0.1	2.6×10^{-4}
100	0.01	2.6×10^{-6}

Fig. 3 shows that the drawdown gradients of dimensionless temperature with time became infinite as represented by a steep (vertical) drawdown response around $t_D/Q_D \sim 1$, when the Q_D value decreased below 2.6×10^{-4}. Solutions were limited to this magnitude due to advection-dominant instabilities in TOUGH (Peclet number $> 2.6 \times 10^3$) with the dashed line in Fig. 3 extrapolating this response. Furthermore, the drawdown response became asymptotic to $t_D/Q_D \sim 1$ as Q_D approached 2.6×10^{-6} defining the lower bounding condition where a plug thermal front migrates through the reservoir and consequently yields the highest thermal gradients in space. Conversely, for high Q_D, the gradient of dimensionless temperature drawdown became progressively flatter as the flow rate was increased. When Q_D reached 2.6×10^4, the curves became asymptotic to a horizontal line anchored at $T_D \sim 1$. This represents the case where water flow through the reservoir is sufficiently rapid that heat transfer from the rock to the fluid is small. This results in a uniform temperature distribution in the fluid and uniform thermal drawdown in the reservoir.

Thus, the bounding distributions of temperature in the reservoir and how they change with time are conditioned by this nondimensional flow variable, Q_D. At high flow rates ($Q_D > 2.6 \times 10^4$), there is little heat transfer from the rock to the water, and the thermal drawdown in the reservoir is uniform as shown in Fig. 5. Conversely, when the flow rate is low ($Q_D < 5.2$), the chilled front in the rock propagates through the reservoir.

An alternative way to represent the transient response of the temperature at the fault within the reservoir is by using the dimensionless timing t_D alone. This dimensionless relationship $T_D - t_D$ returns a new response of the thermal drawdown in the rock (Fig. 4). The individual drawdown curves were parallel for $Q_D < 5.2$. The factor Q_D has an impact in determining the sequence of the parallel curves with a steep decline in time. The curve with the lowest $Q_D = 2.6 \times 10^{-6}$ indicates the earliest drawdown in dimensionless time as $t_D = 10^{-5}$. For this Q_D-controlled thermal drawdown system, as the Q_D value grows one order of magnitude when $Q_D < 5.2$, the corresponding timing of drawdown is equivalently elevated by one order magnitude. When the $Q_D > 5.2$, the gradient of the rock temperature curves decreased gradually as the Q_D value increased, which represents the case of more significant heat transfer by heat advection. By means of this dimensionless type curve, the timing of thermal drawdown for reservoirs under different configurations may be rigorously explored.

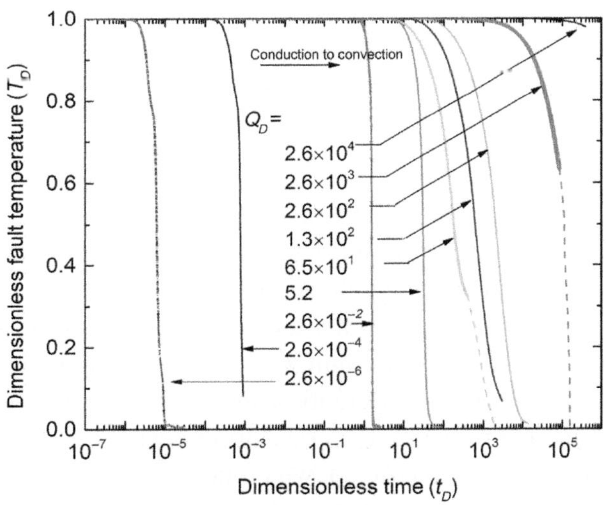

Fig. 4
Dimensionless fault temperature T_D at the fault vs. dimensionless time t_D under Q_D varied respectively from 2.6×10^{-6} to 2.6×10^4.

5 Thermal Front Propagation

The migration of the thermal front within the reservoir depicts the efficiency of heat recovery from the surrounding geologic medium and controls the form of the thermal gradient at the front. Furthermore, the propagation of the thermal front under different nondimensional flow rates Q_D alters the timing of its arrival and therefore the timing of any induced seismicity associated with that thermal stress. Importantly, two bounding behaviors are noted, based on the magnitude of the nondimensional injection rate, Q_D. When Q_D is sufficiently large, the advective heat transfer within the fast-flowing fluid is much more efficient than that due to conductive heat transfer to the fluid. In this instance, the thermal front propagates from the injection well toward the production well without the presence of a distinct uniform thermal front—indeed, the water temperature is near uniform throughout the reservoir. Conversely, for small Q_D, the short dimensionless conduction length in the blocks results in more efficient transfer by heat conduction and results in a distinct front but one that displaces at a slower rate v_T—a velocity $v_T = \frac{\rho_w c_w}{\rho_r c_r} v_w$, defined by the product of the ratio of the thermal capacities of water and rock and the fluid velocity v_w in the fracture.

Fig. 5 shows the displacement of the thermal front within the rock as nondimensional flow rates Q_D are varied. The injection well is to the left ($x = 0$ m) with the production well to the right ($x = 900$ m). The fault is intermediate (*dashed arrow*) between the injection and production wells. If the Q_D value was larger than 2.6×10^2 (Fig. 5C), the rock temperature across the entire reservoir declined uniformly. The flat and uniform drawdown curve corresponds to the larger magnitudes of Q_D (Fig. 5A). Uniform longitudinal thermal drawdown of the rock resulted in the early breakthrough of cold water (Fig. 5A). Conversely, when the Q_D values were smaller than 2.6×10^{-2}, the thermal front propagated as a thermal shock. Figure 5F with the smallest Q_D value of 2.6×10^{-6} shows the slowest rate of propagation of the thermal front, which identifies the case for the most delayed breakthrough. This represents the most desired scenario in retaining the highest outlet temperatures although flow rates may not be sufficiently high to be economically viable as a geothermal reservoir.

The thermal front propagation under various magnitudes of Q_D substantially changed the evolution of the stress state around the fault. Fig. 7 illustrates the thermal evolution of rock temperature under the two bounding magnitudes of Q_D as 2.6×10^4 and 2.6×10^{-3}. This contrasts with Fig. 6 for the evolution of water temperature distribution. For $Q_D = 2.6 \times 10^{-3}$, the rock and fluid were in thermal equilibrium with no significant temperature difference. A distinct thermal front developed to differentiate the cool and hot region. Conversely, for $Q_D = 2.6 \times 10^4$, no distinct and observable thermal front developed. Fig. 7C shows the evolution of the Coulomb stress ratio (τ/σ_{neff}), which indicated that the stress state of the fault evolved significantly differently due to the bounding styles of thermal propagation in the reservoir at $Q_D > 2.6 \times 10^4$ and $Q_D < 2.6 \times 10^{-3}$. These induced thermal stresses may trigger fault reactivation when the fluid pressures alone are insufficient.

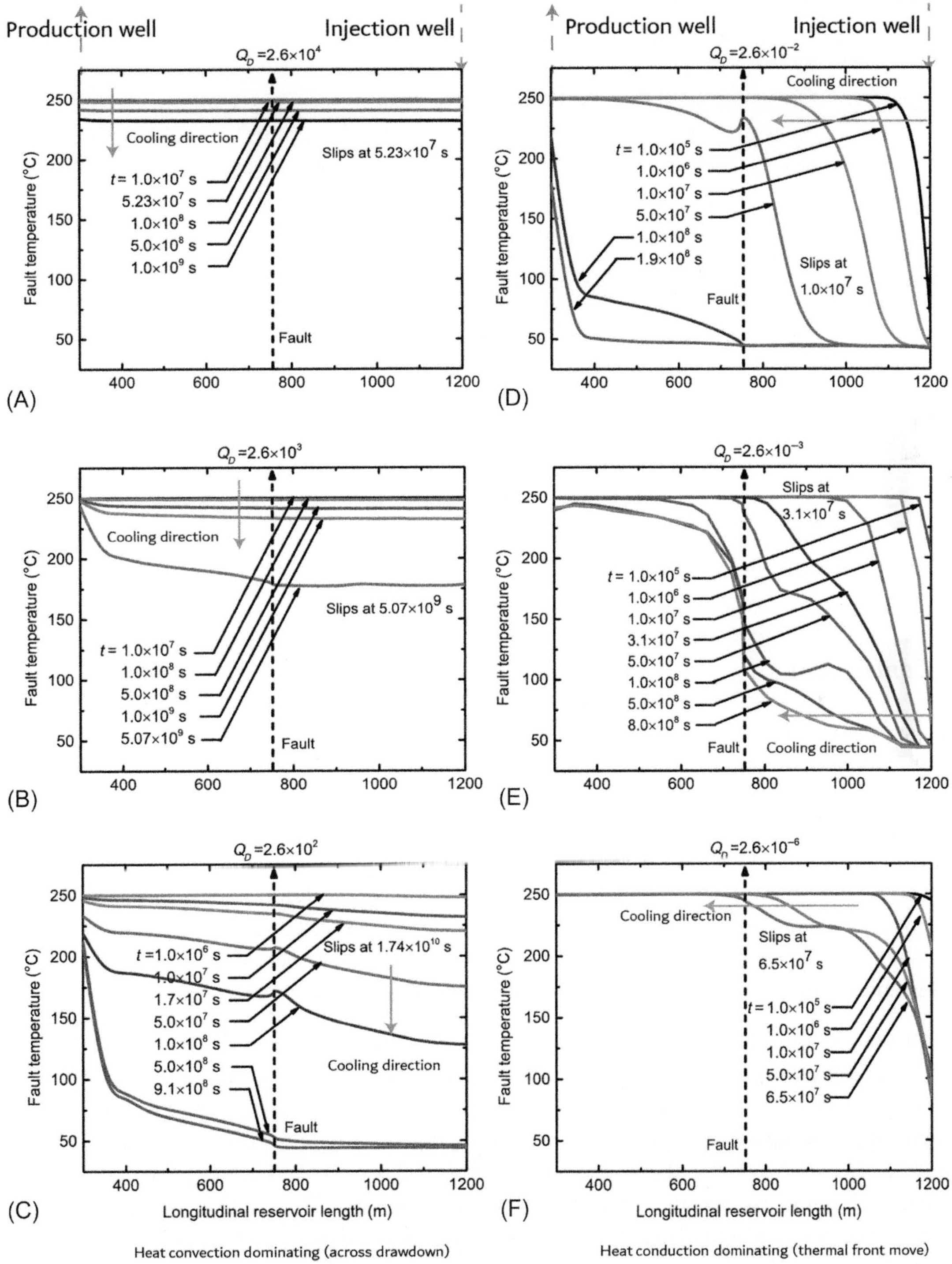

Fig. 5

Thermal front propagation from the injection well (right axis) toward the production well (left axis) under different Q_D values, the figures from (A) to (F) respectively represent the Q_D values varied from 2.6×10^{-6} to 2.6×10^4.

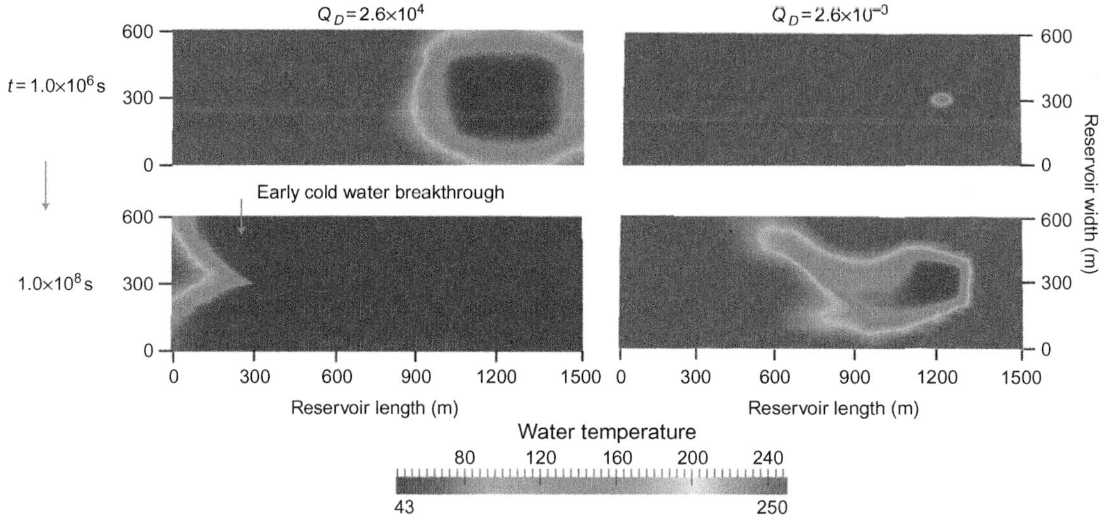

Fig. 6
Comparison of water temperature evolution in the reservoir under two bounding Q_D values, the left figures represent the condition of $Q_D = 2.6 \times 10^4$, the right figures represent the condition of $Q_D = 2.6 \times 10^{-3}$.

6 Conclusions

The foregoing defines the evolution of heat transfer in a fractured geothermal reservoir characterized by nondimensional parameters. The evolution of water temperature and mean temperature in the rock is conditioned by dimensionless parameters representing the flow rate Q_D and time t_D. The dimensionless flow rate Q_D is conditioned by the in-situ fracture spacing, prescribed mass injection rate and reservoir geometry and influences the thermal drawdown response of the reservoir. Sensitivity tests have explored the effect of fracture spacing, fault permeability, and injection temperature [19], Based on this dimensionless model, this work captures the thermal drawdown response of the host rock at different reservoir scales and for different spacing and permeabilities of fracture networks.

The primary control parameter Q_D transforms dimensionless timing and temperature data into two bounding asymptotic behaviors. These two bounding asymptotic behaviors refer to the situations of heat transfer dominated by heat conduction where the drawdown gradients of dimensionless temperatures with time become asymptotically and infinitely steep (vertical) around $t_D/Q_D \sim 1$, when the Q_D value decreased below 2.6×10^{-4}. When Q_D reached 2.6×10^4, the curves became asymptotic to a horizontal line anchored at $T_D \sim 1$. This represents the case where the velocity of fluid in the reservoir is sufficiently rapid that heat transfer from the rock to the fluid is conduction limited and small. This results in an early cold water breakthrough at the outlet and uniform thermal drawdown in the rock. The magnitude of Q_D has an impact in

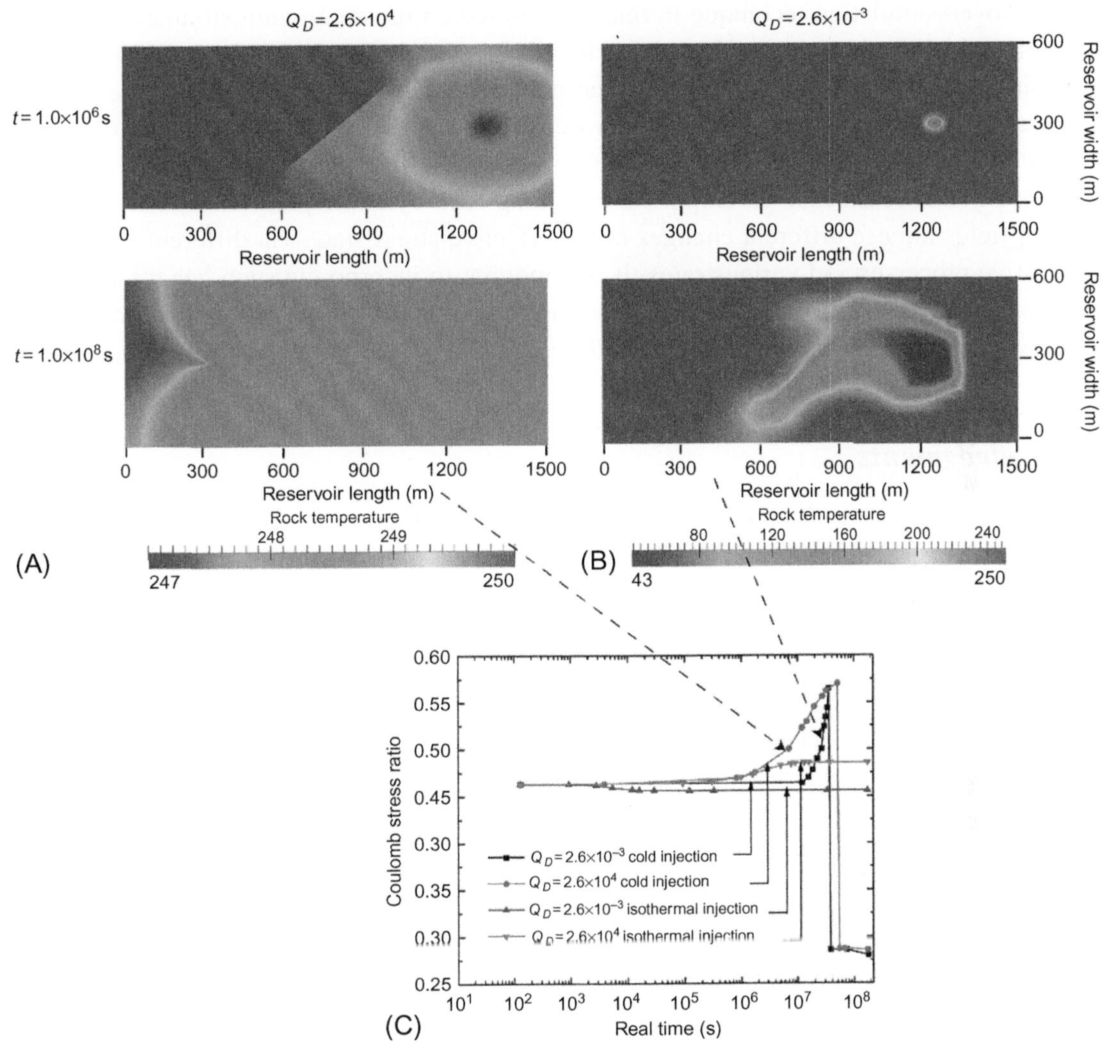

Fig. 7

Contour of rock temperature distribution at $t = 1.0 \times 10^6$ s, 1.0×10^7 s, 1.0×10^8 s respectively for $Q_D = 2.6 \times 10^4$ (A) and $Q_D = 2.6 \times 10^{-3}$ (B). (C) represents the evolution of the Coulomb stress ratio under $Q_D = 2.6 \times 10^4$ and $Q_D = 2.6 \times 10^{-3}$ with isothermal or nonisothermal injection conditions.

determining the sequence of the parallel steep curves (dimensionless t_D). When the Q_D value increased by one order of magnitude, the corresponding dimensionless timing t_D for thermal depletion of water or rock was approximately elevated by one order magnitude for $Q_D < 5.2$. The situation with a lower magnitude Q_D (<5.2 in this model configuration) yielded a uniform propagation of the thermal front, while the case with larger magnitude Q_D ($>2.6 \times 10^2$) produced a more uniform thermal distribution across the whole reservoir without a distinct thermal front.

The employed simulation technique in this work relied on the continuum simulator Tough-FLAC3D, which enables representation of the mechanical response of the fractured porous subsurface due to cold fluid injection and thermal recovery. This model satisfies the requirement of representing the coupled elastic-plastic behavior of fractured rocks. The strategy of THM coupling allows this model to be used in the field of geothermal, shale gas, and conventional fossil fuel development, with long term dynamic prediction. Different physical fields impose different changes in the effective stress state, via different constitutive processes and various rates. It is important to incorporate such behaviors. In this work, it can be identified that the late thermal drawdown induced cooling on the host rocks, which in turn changed the timing and magnitude of thermal induced earthquakes.

Acknowledgments

This work is the result of partial support from the US Department of Energy under project DOE-DE-EE0002761. The source of support is gratefully acknowledged.

References

[1] J. Taron, D. Elsworth, Coupled mechanical and chemical processes in engineered geothermal reservoirs with dynamic permeability, Int. J. Rock Mech. Min. Sci. 47 (2010) 1339–1348.
[2] Q. Gan, D. Elsworth, A continuum model for coupled stress and fluid flow in discrete fracture networks, Geomech. Geophys. Geo-Energy Geo-Resour. 2 (2016) 43–61.
[3] G. Bodvarsson, On the temperature of water flowing through fractures, J. Geophys. Res. 74 (1969) 1987–1992.
[4] G.S. Bödvarsson, C.F. Tsang, Injection and thermal breakthrough in fractured geothermal reservoirs, J. Geophys. Res. Solid Earth 87 (1982) 1031–1048.
[5] Q. Gan, D. Elsworth, Production optimization in fractured geothermal reservoirs by coupled discrete fracture network modeling, Geothermics 62 (2016) 131–142.
[6] D. Elsworth, Thermal recovery from a multiple stimulated HDR reservoir, Geothermics 18 (1989) 761–774.
[7] D. Elsworth, A comparative evaluation of the parallel flow and spherical reservoir models of HDR geothermal systems, J. Volcanol. Geotherm. Res. 44 (1990) 283–293.
[8] D. Elsworth, Theory of thermal recovery from a spherically stimulated hot dry rock reservoir, J. Geophys. Res. Solid Earth 94 (1989) 1927–1934.
[9] S. Ganguly, M.M. Kumar, Analytical solutions for transient temperature distribution in a geothermal reservoir due to cold water injection, Hydrogeol. J. 22 (2014) 351–369.
[10] A.C. Gringarten, P.A. Witherspoon, Extraction of heat from multiple-fractured dry hot rock, Geothermics 2 (1973) 119–122.
[11] A.C. Gringarten, P.A. Witherspoon, Y. Ohnishi, Theory of heat extraction from fractured hot dry rock, J. Geophys. Res. 80 (1975) 1120–1124.
[12] K. Pruess, Y.S. Wu, A new semi-analytical method for numerical simulation of fluid and heat flow in fractured reservoirs, SPE Adv. Technol. Ser. 1 (1993) 63–72.
[13] K. Pruess, Heat transfer in fractured geothermal reservoirs with boiling, Water Resour. Res. 19 (1983) 201–208.
[14] A. Ghassemi, S. Tarasovs, H.-D. Cheng, An integral equation solution for three-dimensional heat extraction from planar fracture in hot dry rock, Int. J. Numer. Anal. Methods Geomech. 27 (2003) 989–1004.

[15] A.R. Shaik, S.S. Rahman, N.H. Tran, et al., Numerical simulation of fluid-rock coupling heat transfer in naturally fractured geothermal system, Appl. Therm. Eng. 31 (2011) 1600–1606.
[16] T. Xu, E. Sonnenthal, N. Spycher, et al., TOUGHREACT—a simulation program for non-isothermal multiphase reactive geochemical transport in variably saturated geologic media: applications to geothermal injectivity and CO_2 geological sequestration, Comput. Geosci. 32 (2006) 145–165.
[17] T. Xu, J.A. Apps, K. Pruess, Reactive geochemical transport simulation to study mineral trapping for CO_2 disposal in deep arenaceous formations, J. Geophys. Res. Solid Earth 108 (2003) 2071.
[18] T. Xu, E. Sonnenthal, N. Spycher, et al., TOUGHREACT User's Guide: A Simulation Program for Non-isothermal Multiphase Reactive Geochemical Transport in Variably Saturated Geologic Media, Lawrence Berkeley National Laboratory Report LBNL-55460, 2004.
[19] D. Elsworth, J. Xiang, A reduced degree of freedom model for permeability enhancement in blocky rock, Geothermics 18 (1989) 691–709.
[20] J.P. Holman, Heat Transfer, ninth ed., McGraw-Hill, New York, 2002.
[21] Q. Gan, D. Elsworth, Analysis of fluid injection-induced fault reactivation and seismic slip in geothermal reservoirs, J. Geophys. Res. Solid Earth 119 (2014) 3340–3353.
[22] P. Segall, S.D. Fitzgerald, A note on induced stress changes in hydrocarbon and geothermal reservoirs, Tectonophysics 289 (1998) 117–128.
[23] J. Taron, D. Elsworth, Thermal–hydrologic–mechanical–chemical processes in the evolution of engineered geothermal reservoirs, Int. J. Rock Mech. Min. Sci. 46 (2009) 855–864.
[24] J. Byerlee, Friction of rocks, Pure Appl. Geophys. 116 (1978) 615–626.
[25] M.A. Biot, General theory of three-dimensional consolidation, J. Appl. Phys. 12 (1941) 155–164.

CHAPTER 11

Pore-Scale Modeling and Simulation in Shale Gas Formations

Yang Ning*, **Bikai Jin**†, **Yu Liang**‡, **Guan Qin***

*Department of Petroleum Engineering, Cullen College of Engineering, The University of Houston, Houston, TX, United States, †Harold Vance Department of Petroleum Engineering, College of Engineering, Texas A&M University, College Station, TX, United States, ‡Hildebrand Department of Petroleum and Geosystems Engineering, Cockrell School of Engineering, The University of Texas at Austin, Austin, TX, United States

Chapter Outline

1 Overview of the Transport and Storage Mechanisms of Shale Gas in Organic Nanopores

Being both source rock and reservoir rock, shale gas reservoirs differ from most of the conventional reservoirs in the aspect of geological features, and contain multiscale pore systems in organic matter and nonorganic matter [1]. A significant amount of shale gas is in the adsorbed state stored in organic matter that contains numerous nanoscale pores [2, 3].

Petrophysical Characterization and Fluids Transport in Unconventional Reservoirs. https://doi.org/10.1016/B978-0-12-816698-7.00011-5

The nanopore systems of gas shales are an important measure and control of natural gas storage capacity and transport to fracture networks. The coexistence of free-state gas and adsorbed-state gas results in great challenges in understanding the storage and transport mechanisms at the pore-scale. Thus, pore-scale modeling is essential for the measurement of the petrophysical properties of gas shales, the prediction of long-term production, and the estimation of shale gas reserves. Fig. 1 (data obtained from [4]) shows the pore size distribution (PSD) of Lower Silurian Longmaxi shale samples based on low-pressure adsorption experiments and it indicates that the majority of pores have a pore diameter smaller than 10nm [4]. PSD may vary largely for different shale formations, but the general consensus is that a significant proportion of total pores are nanometer-scaled pores [5].

In nanopores, natural gas transport is characterized by a dimensionless number, Knudsen number (K_n). The Knudsen number is defined as the ratio of the molecular mean free path (λ) to a representative physical length of interest (L) that is usually the representative diameter of nanopores. Based on the Knudsen number, gas flows are categorized by four flow regimes, which are continuum flow regime ($K_n < 0.001$), slip flow regime ($0.001 < K_n < 0.1$), transitional flow regime ($0.1 < K_n < 10$), and free molecular flow regime ($K_n > 10$). Due to the extreme confinement effect, natural gas transport in nanopores falls into the slip or transitional flow regime depending on the size of nanopores, reservoir pressure, stress, temperature, etc. [6]. In either flow regime, the continuum assumption of gas flow is no longer valid and macroscopic equations based on the continuum assumption may fail to describe the shale gas transport. Consequently, Darcy's equation may not be adequate for describing natural gas flow in shale formations.

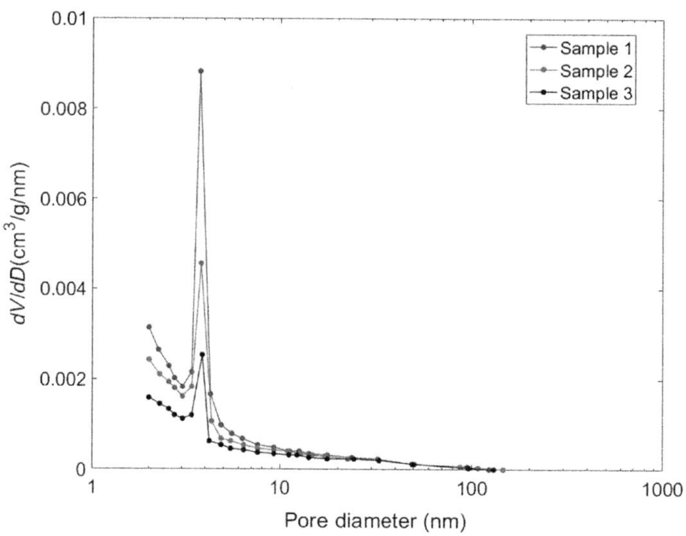

Fig. 1
Pore size distribution of Lower Silurian Longmaxi shale samples.

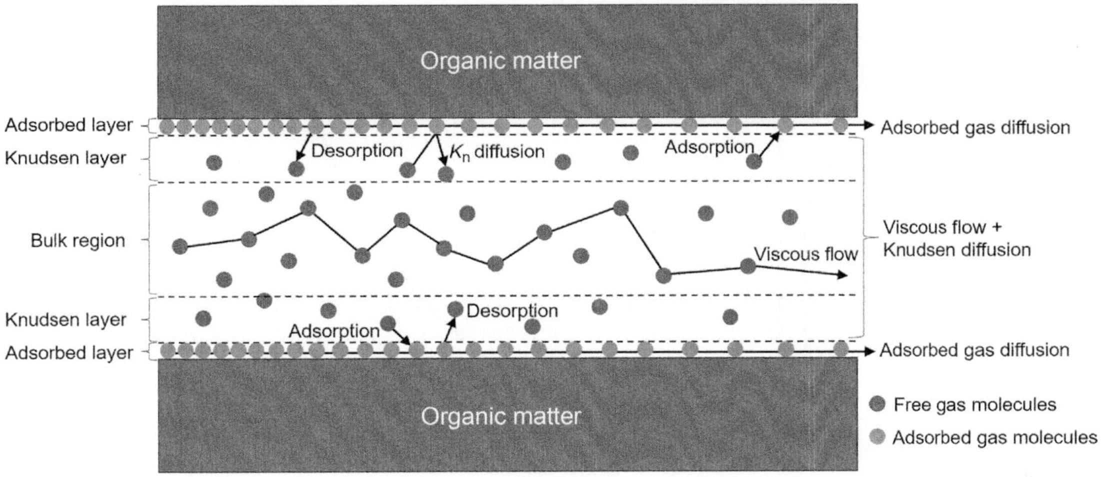

Fig. 2
Schematic diagram of shale gas transport and adsorption in an organic nanopore.

In Fig. 2, we illustrate the transport and adsorption/desorption processes of gas molecules in an organic nanopore. The single-component gas flow in an organic nanopore consists of multiple transport mechanisms: viscous flow, Knudsen diffusion, and surface diffusion, etc. [7–9]. Viscous flow is also known as continuous flow when the distance between gas molecules is small and the collisions between gas molecules dominate. Knudsen diffusion occurs when the characteristic length is comparable to the mean free path of gas molecules and the collision between gas molecules and solid surface dominate. Surface diffusion is a complex phenomenon compared to viscous flow and Knudsen diffusion. Surface diffusion is a migration process caused by the concentration gradient of the adsorbed gas molecules on the solid surface and more detailed theories on this topic can be found in [10]. Surface diffusion is of great importance because it may contribute much of the mass flux of natural gas, especially in organic nanopores. Another complexity is the unclear fluid phase behaviors in shale nanopores. Depending on the type of pore material, the interactions between gas molecules and pore surface significantly complicate the phase behavior in nanopores.

For correct characterization of various petrophysical properties, one feasible way is to apply an appropriate pore-scale modeling approach, which can help to gain insights to the complex system consisting of multiple mechanisms in shale nanopores. In this section, two well-known mechanisms in shale nanopores, confinement and adsorption, are discussed.

1.1 Confinement Effect

Gas molecules in an extremely confined space encounter solid surfaces frequently and the mean free path of gas molecules becomes a similar order of magnitude with the characteristic length. The Navier-Stokes equations may not be adequate to describe the flow behavior in shale

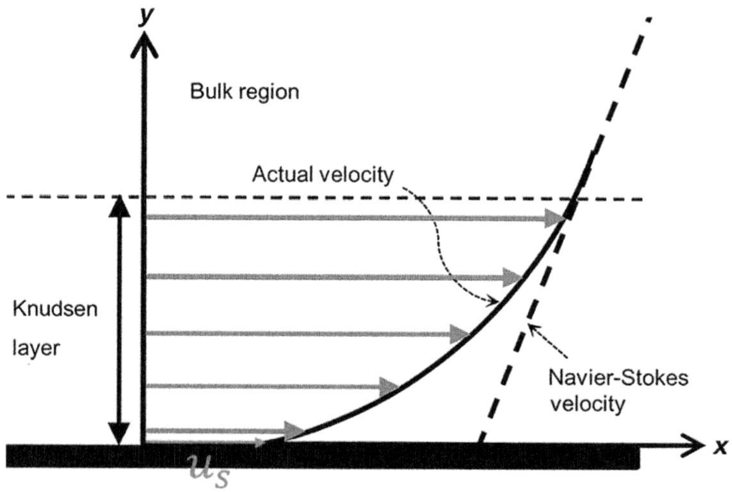

Fig. 3
Illustration of the Knudsen layer.

nanopores due to the failure of capturing the Knudsen layer. The Knudsen layer, shown in Fig. 3, is a kinetic layer that is always formed when gas flows over a solid surface. In an unconfined space, the Knudsen layer is usually negligible because the thickness of a Knudsen layer is much smaller than the characteristic length. However, in nanoscaled shale pores, it has the same order of magnitude as the mean free path of gas molecules and may take up a large portion of the total pore space. Within the Knudsen layer, since the collisions of gas molecules are not sufficient to reach the quasithermodynamic-equilibrium condition, the standard Navier-Stokes equations fail to describe gas slippage. In Fig. 3, it can be seen that the actual slip velocity u_s is smaller than the slip velocity obtained from the Navier-Stokes equations with a first-order slip boundary condition. In fact, there are many approaches that can successfully capture the flow behavior within the Knudsen layer, but we focus on the effective mean free path approach with higher-order slip boundary conditions in this section because they are common.

1.1.1 Effective mean free path/effective viscosity

Gas flow behavior in shale nanopores is significantly different from the gas flow in the bulk region due to the confinement effect. The mean free path of gas molecules becomes smaller in nanopores because gas molecules collide with solid surfaces more frequently. Therefore, an effective mean free path is introduced to include the confinement effect, which can be written as a function of the Knudsen number:

$$\lambda_e = \lambda_\infty \Psi\left(K_n, \vec{x}\right) \tag{1}$$

where λ_∞ is the mean free path in an unconfined system defined as $\lambda_\infty = \frac{k_B T}{\sqrt{2}\pi p d^2}$, in which k_B is the Boltzmann constant, T is temperature, p is pressure, and d is the diameter of a gas molecule. For simplicity, let us take a gas system between two parallel plates as an example,

Ψ can be derived through the probability distribution function of the free path of the gas molecules [11, 12] as

$$\Psi(z) = \frac{1}{2}\left[\emptyset\left(\frac{z}{\lambda_\infty}\right) + \emptyset\left(\frac{H-z}{\lambda_\infty}\right)\right] \tag{2}$$

where the parallel plates are located at $z=0$ and $z=H$. The function $\emptyset$ is defined as $\emptyset(y)=1+(y-1)e^{-y}-y^2E_i(y)$, where $E_i(y)=\int_1^\infty t^{-1}e^{-yt}dt$. Fig. 4 shows the nondimensionalized effective mean free path (λ_e/λ_∞) against the system location for Knudsen numbers between 0.01 and 10 that cover from the free molecular regime to the slip flow regime. One should note that $\lambda_e \approx \lambda_\infty$ for the continuum regime ($K_n < 0.001$). For $K_n \leq 0.1$, λ_e is close to λ_∞ in the central region, but it becomes smaller near the walls. λ_e becomes even smaller as the Knudsen number increases, indicating that the confinement effect becomes more pronounced throughout the entire channel. Note that the exact expression of Ψ is rather complicated because it requires a numerical integration and depends on the location, and it could become impossible for complex pore geometries. To address this complexity, Guo et al. [13] approximated Eq. (2) with

$$\Psi(K_n) = \frac{2}{\pi}\arctan\left(\sqrt{2}K_n^{-\frac{3}{4}}\right) \tag{3}$$

In Eq. (3), Ψ becomes a function that solely depends on K_n and it can be easily applied to practical problems. Eq. (3) has been shown to fit Eq. (2) quite well over a wide range of K_n [13].

According to the kinetic theory of gases, the gas viscosity is linearly dependent on the mean free path of gas molecules. Thus, we can also write the effective viscosity of gases in a confined system as

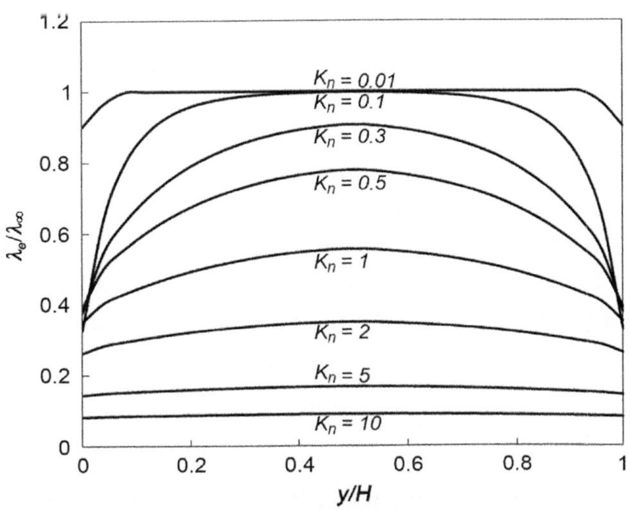

Fig. 4
Effective mean free path at different Knudsen numbers.

$$\mu_e = \mu_\infty \Psi(K_n) \tag{4}$$

where μ_∞ is the bulk viscosity in an unconfined system, and $\Psi(K_n)$ has the same form with Eq. (3).

1.1.2 Slip boundary condition

As illustrated in Fig. 3, the actual velocity within the Knudsen layer is smaller than the macroscopic slip velocity extrapolated linearly from the Navier-Stokes solutions. In addition, Karniadakis et al. [14] concluded that the first-order slip boundary condition should only be used for gas flows with $K_n \leq 0.1$. As K_n increases, the higher-order slip boundary condition must be included in the Navier-Stokes equations in order to capture the correct slip velocity. The second-order slip boundary condition is written as

$$u_s = A_1 \lambda_e \frac{\partial u}{\partial y}\bigg|_{\text{wall}} - A_2 \lambda_e^2 \frac{\partial^2 u}{\partial y^2}\bigg|_{\text{wall}} \tag{5}$$

where A_1 and A_2 are the first-order and second-order slip coefficients, respectively. The values of the slip coefficients by different researchers can be found in [14]. For example, Loyalka et al. [15, 16] derived $A_1 = \frac{2-\sigma}{\sigma}(1-0.1817\sigma)$ by modifying the Maxwell's argument in the kinetic theory, where σ is the tangential momentum accommodation coefficient that describes the tangential momentum exchange of gas molecules with a solid surface.

1.2 Adsorption and Surface Diffusion

Methane adsorption in shale formations is one of the most significant properties that impact the predictions of gas-in-place and production. The mechanism of methane adsorption is controlled by mineralogical compositions and the microstructure of shale. Among various compositions, kerogen pores and slit-like spaces in clay minerals are more favorable to adsorb methane molecules. The methane adsorption characteristics in shale have been studied extensively by experiments and molecular simulations [1, 17, 18]. Fig. 5 gives an example of the excess methane adsorption isotherms of Lower Silurian Longmaxi shale samples at 60°C. Adsorption models have been developed based on different adsorption theories to fit the adsorptions isotherms [19], which include monolayer adsorption models (e.g., Langmuir adsorption), micropore filling models, (Dubinin-Radushkevich model and Dubinin-Astakhov model), and so on. The most commonly used adsorption model is the Langmuir model expressed as

$$n_{ab} = \frac{n_L p}{p_L + p} \tag{6}$$

where n_{ab} is the absolute adsorption capacity, n_L is the maximum adsorption capacity on monolayers, p is the equilibrium pressure, and p_L is the Langmuir pressure that is the pressure at half of the maximum capacity adsorbed by methane. The access (observed) adsorption capacity

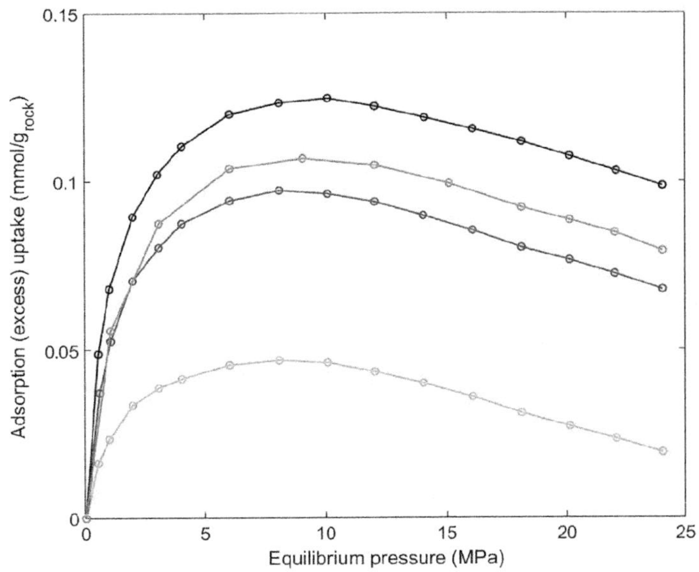

Fig. 5

The excess adsorption isotherms of Lower Silurian Longmaxi shale samples at 60°C. *Modified from S. Zhou, H. Xue, Y. Ning, et al., Experimental study of supercritical methane adsorption in Longmaxi shale: insights into the density of adsorbed methane, Fuel 211 (2018) 140–148.*

is $n_{ex} = n_{ab}\left(1 - \frac{\rho_g}{\rho_a}\right)$, where ρ_g and ρ_a are the bulk gas density and the adsorbed gas density, respectively.

Surface diffusion describes a dynamic migration process of gas molecules in the adsorbed state and can be viewed as a diffusion process driven by the concentration gradient of adsorbed molecules. While it has been claimed that a significant proportion of total mass flux is contributed by surface diffusion, it is still one of the least understood phenomena in the shale gas development. Surface diffusion in shale nanopores is a very complex, physical phenomenon mainly controlled by interactions between methane molecules and the surface adsorbents and by the surface geometric complexity. Medveď and Černý reviewed various theories of surface diffusion in porous media in [10]. The commonly used ones in shale gas formations are the hopping model and the Fickian-type model. In the theory of the hopping model, gas molecules are randomly hopping between adsorption sites on the surface. When an adsorbed gas molecule acquires the minimum activation energy, it hops to a neighboring, available adsorption site. The Fickian-type model assumes that the fluxes of free-state and adsorbed-state gas molecules contribute to the total flux independently, i.e., $J_{\text{total}} = J_{\text{bulk}} + J_{\text{surface}}$. Through experimental measurements, J_{surface} can be determined by J_{total} (measured) $- J_{\text{bulk}}$ (theoretical or measured). Interested readers may refer to the review works [10, 20–22].

In the following sections, we focus on the discussion of pore-scale modeling approaches at two scales, i.e., microscopic and mesoscopic, and simulation results in characterizing petrophysical and fluid properties in shale formations.

2 Molecular Simulation (Microscopic Approach)

As reported in [4], the PSD in the shale system is usually in the range from a few to several hundred nanometers. In a nanometer-scaled pore, molecular interactions need to be investigated to explain the fluid behavior and transport. Also, since nanopores are dominant in shale rock, the fluid-pore contact surface is relatively higher than that in the bulk condition and fluid thermodynamic properties are thus different. To build a deep understanding of the shale nanopore system, extensive research has been presented in recent years by conducting molecular simulations [23, 24]. Molecular simulation can help us understand the microstructure of a system and determine the fluid system properties, providing reliable results and a detailed description of the system behavior, which conventional methods cannot achieve. In this section, some basic knowledge about molecular simulation is presented. Two major molecular simulation techniques are discussed based on some typical simulation results in shale formations.

2.1 Basics

2.1.1 Molecular interactions

Based on the assumption of classical mechanics, molecular simulation only considers intermolecular interactions (nonbonded term) and intramolecular interactions (bonded terms) in the potential energy computation, and the quantum effect is usually ignored [25].

$$U\left(\boldsymbol{r}^N\right) = U_{nonbonded}\left(\boldsymbol{r}^N\right) + U_{bonded}\left(\boldsymbol{r}^N\right). \tag{7}$$

The calculation of nonbonded potential energy $U_{nonbonded}(\boldsymbol{r}^N)$ is similar to the typical many-body problem, which is prohibitively expensive in terms of computational costs, even if only the explicit three-body term is included in the simulation. Thus, the simple pairwise model is commonly used since it is sufficient to represent the energy potential in most of the studies [26]. The Lennard-Jones (LJ) is the most widely applied continuous potential form:

$$U_{nonbonded}\left(\boldsymbol{r}^N\right) = 4\varepsilon_{ij}\left[\left(\frac{\sigma_{ij}}{r_{ij}}\right)^{12} - \left(\frac{\sigma_{ij}}{r_{ij}}\right)^{6}\right] \tag{8}$$

where ε_{ij} is the potential well depth between particle i and j, and σ_{ij} is the specific distance at $U=0$. Since the absolute value decreases quite fast (Fig. 6A) and the calculation of more interactions is needed at a larger distance, the LJ potential is usually truncated at some cutoff distance and one tail correction is included in the computation to compensate for the energy

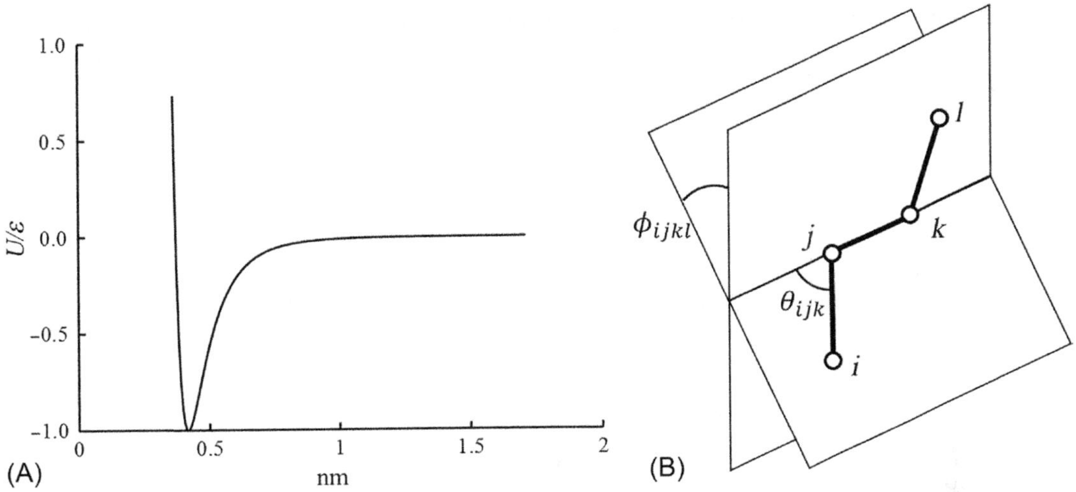

Fig. 6

(A) Schematic diagram of the Lennard-Jones potential and (B) Long chain molecular model.

beyond the cutoff distance [25]. In addition, the Coulomb potential is generally added in the nonbonded term if electrostatic charges are considered:

$$U_{coulomb}\left(r_{ij}\right)=\frac{q_1q_2}{4\pi\varepsilon_0 r_{ij}} \tag{9}$$

where q_1, q_2 are charges and ε_0 is the vacuum permittivity.

For the intramolecular interactions, the general form of $U_{bonded}(\boldsymbol{r}^N)$ consists of three terms:

$$U_{bonded}\left(\boldsymbol{r}^N\right)=\frac{1}{2}\sum_{bond}K_b\left(r_{ij}-r_0\right)^2+\frac{1}{2}\sum_{bending}K_\theta\left(\theta_{ijk}-\theta_0\right)^2+\frac{1}{2}\sum_{torsion}\sum_{n}K_\phi^n\left(1+\cos\left(n\phi_{ijkl}-\phi_0\right)\right) \tag{10}$$

As illustrated in Fig. 6B, the "bond" term refers to the vibration potential between adjacent atoms separated in the length of the equilibrium bond, and it is usually presented in the harmonic form. The "bending" potential denotes the potential energy between two bonds, which involves three consecutive atoms and it is also formulated in the quadratic form with an equilibrium angle. The "torsion" energy is defined based on the dihedral angle, which is formed of three successive bonds. Based on the definitions, the bending and torsion potentials have the periodic effect. When the molecular structure is complex (e.g., consider the chirality or maintain the planarity), another "improper" torsion potential should be included in the simulation.

Based on different force field settings, the above potentials can be specified with various parameters. Most of the studies in the oil and gas industry are conducted using long chain molecules. There are two major types of force field used in this research: explicit-atom (or all-atom) and united-atom force fields. Every atom is explicitly expressed in the explicit-atom force field but not in the united-atom force field, in which the nonpolar hydrogen atoms are grouped with adjacent atoms to form the pseudoatoms, such as the TraPPE force field [27]. Please refer to [28–31] for a comparison of these two approaches through applications.

2.1.2 Periodic boundary condition

Molecular simulation aims to build a bridge between the macroscopic system and the microscopic cell. Thus, we can accurately estimate system properties based on the simulation in a unit cell. Most of the studies were conducted in a system with molecules in the range of dozens to thousands, which is still an extremely small number compared with the content of a thermodynamic system. The unit cell size is usually kept in the magnitude of nanometers to reduce the computation cost. To represent a large system, it is necessary to apply the periodic boundary condition in the unit cell [32, 33]. In general, this condition is conducted by surrounding the cell with duplicates of itself, as shown in Fig. 7. Based on this setting, the cubic box is the common choice for the cell. When an atom leaves the original box, another image will enter this box based on the minimal image convention [32, 33]. Attention must be paid to the tradeoff between the force field cutoff range and box size.

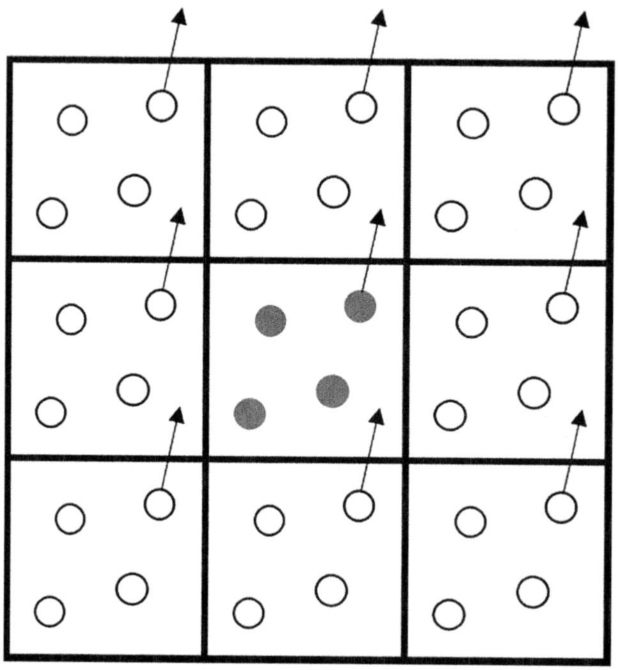

Fig. 7
Schematic diagram of periodic boundary condition. As one particle leaves the simulation box, another image particle will move in as the replacement.

2.1.3 Ensemble

An ensemble is a set of system states that have the same thermodynamic parameters at the macroscopic scale. For cases based on different ensembles, the simulation target is also different. Several ensembles are described below:

- The canonical ensemble, also called NVT: the system has fixed molecular number (N), volume (V), and temperature (T).
- Grand canonical ensemble, also called μVT: the system has fixed chemical potential (μ), volume (V), and temperature (T).
- Microcanonical ensemble, also called NVE: the system has fixed molecular number (N), volume (V), and energy (E).
- The isobaric-isothermal ensemble, also called NPT: the system has fixed molecular number (N), pressure (P), and temperature (T).

2.2 Monte Carlo Molecular Simulation

The Monte Carlo (MC) molecular simulation is the application of the Monte Carlo method in the molecular simulation. Based on classical statistical mechanics, the state of a system can be represented as a partition function according to the Boltzmann probabilities. For example, the partition function for a system with a volume V and a particle number N at the temperature T can be expressed as

$$Q = c \int d\boldsymbol{p}^N d\boldsymbol{r}^N \exp\left[-H\left(\boldsymbol{r}^N, \boldsymbol{p}^N\right)/k_B T\right] \tag{11}$$

where k_B is the Boltzmann's constant, H is the system Hamiltonian, and $\boldsymbol{p}^N$ and $\boldsymbol{r}^N$ are momenta and coordinates of the N particles, respectively. Usually, an average of the observation A can be calculated based on the partition function:

$$\langle A \rangle = \frac{\int d\boldsymbol{p}^N d\boldsymbol{r}^N A\left(\boldsymbol{r}^N, \boldsymbol{p}^N\right) \exp\left[-H\left(\boldsymbol{r}^N, \boldsymbol{p}^N\right)/k_B T\right]}{\int d\boldsymbol{p}^N d\boldsymbol{r}^N \exp\left[-H\left(\boldsymbol{r}^N, \boldsymbol{p}^N\right)/k_B T\right]} \tag{12}$$

It is a daunting challenge to calculate the partition function directly since analytical solutions are only available in some specific cases. To improve the computation efficiency, Metropolis et al. [34] introduced a modified Monte Carlo scheme to sample the configuration space. In this method, only the relative but not the absolute probability of a system state in the configuration space is required in the computation of average observations. This Metropolis method is one kind of Markov chain process, in which the new state depends only on the current state but not the previous ones. The general Metropolis Monte Carlo scheme is designed as follows:

1. Select a system state o randomly with energy U_o.
2. Generate a new state n with energy U_n through a trial.
3. Accept this trial move with the probability $\mathrm{acc}(o \rightarrow n) = \min(1, \exp[-\beta(U_n - U_o)])$.
4. Repeat the above processes to reach the system equilibrium.

Different methods are developed based on different ensembles, such as Gibbs ensemble Monte Carlo (GEMC) [35, 36], Grand canonical Monte Carlo (GCMC) [37, 38], and gauge-cell Monte Carlo [39–42]. Note that we only focus on the Monte Carlo methods to study the static properties of a system. The dynamic Monte Carlo method [43], which calculates the time evolution related system properties, is not covered here.

The prediction of phase equilibrium is widely conducted by Monte Carlo simulations in bulk and confined systems [23]. Kamath and Potoff [44] determined the pressure-composition diagrams of binary mixtures based on the GCMC method and one ternary phase diagram by using the GEMC method. The calculated phase diagrams in bulk have an excellent agreement with the experiment data (Fig. 8A). For confined fluids, the phase diagrams can be simulated by gauge-cell and GCMC methods. Jiang et al. [45] produced the temperature-density diagrams of n-alkanes in a carbon nanotube (Fig. 8B), which has similar results to that presented by Dr. Jin's group [46] and Dr. Nasrabadi's group [41], that the critical temperature and pressure are decreased while the critical density is increased (Fig. 8C). Also, the phase equilibrium diagrams are investigated by considering the effects of PSD [47] and the boundary material model [48].

The Monte Carlo simulation has the advantage of describing the system behaviors, such as adsorption and density distribution. Severson and Snurr [49] calculated the single component adsorption of alkanes, such as pentane in Fig. 9A), in carbon slit pores based on GCMC method.

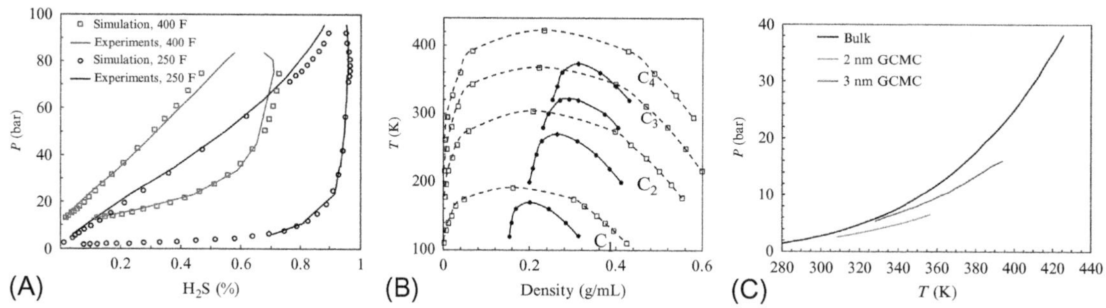

Fig. 8

(A) Pressure-composition diagrams of H_2S + heptane. (B) Temperature-density diagrams for pure n-alkanes. (C) Pressure-temperature diagram of nC4. *Part (A) reproduced from G. Kamath, J.J. Potoff, Monte Carlo predictions for the phase behavior of H_2S + n-alkane, H_2S + CO_2, CO_2 + CH_4 and H_2S + CO_2 + CH_4 mixtures, Fluid Phase Equilib. 246 (2006) 71–78; part (B) reproduced from J. Jiang, S. I. Sandler, B. Smit, Capillary phase transitions of* n*-alkanes in a carbon nanotube, Nano Lett. 4 (2004) 241–244; part (C) reproduced from Z. Jin, A. Firoozabadi, Phase behavior and flow in shale nanopores from molecular simulations, Fluid Phase Equilib. 430 (2016) 156–168.*

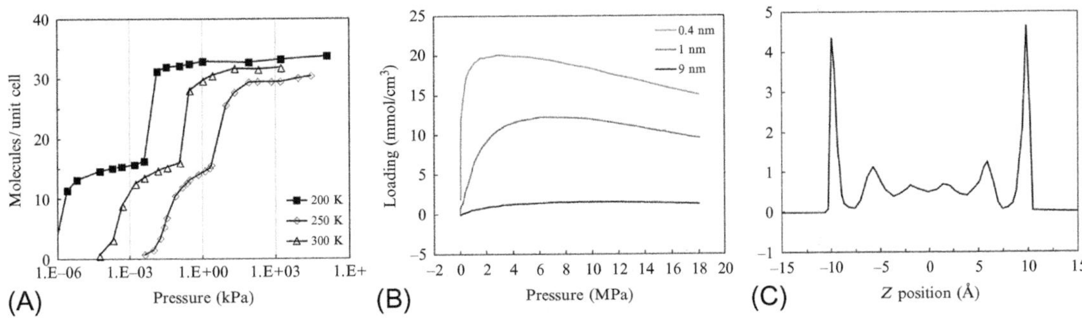

Fig. 9

(A) Pentane isotherms in a 19.2 Å slit. (B) Excess adsorption isotherms in a silt at 298 K. (C) Entane density profile at 300 K and 10,000 kPa in a 27.1 Å slit. *Parts (A & C) modified from B.L. Severson, R.Q. Snurr, Monte Carlo simulation of* n-*alkane adsorption isotherms in carbon slit pores, J. Chem. Phys. 126 (2007) 134708; part (B) reproduced from K. Mosher, J. He, Y. Liu, et al., Molecular simulation of methane adsorption in micro- and mesoporous carbons with applications to coal and gas shale systems, Int. J. Coal Geol. 109 (2013) 36–44*

Mosher et al. [50] showed the excess methane adsorption isotherm as a function of pore size, in which the general trend varies with pore sizes (Fig. 9B). Both studies reported the density profile along the pore diameter with peaks near the pore boundary surface, which agrees with other researches (Fig. 9C) [47, 51]. For more application results, please refer to the review works, e.g., [52, 53].

2.3 Molecular Dynamics Simulation

Molecular dynamics (MD) is a simulation methodology in which the movement of system particles is calculated in a certain period of time and the system evolution is investigated. It can be used in the studies of equilibrium and dynamic properties of a system. Usually, Newton's equations of motion are used to capture the trajectories of particles in the system and the potential energy is computed based on the force fields. For systems under the ergodic hypothesis, the averages of a MD simulation can represent the thermodynamic properties of a system at the macroscopic scale. The simulation procedure is usually constructed as follows:

1. Initialize the system under the criterion of zero total momentum.
2. Compute the forces for each particle.
3. Integrate Newton's equation of motion.
4. Repeat steps 2 and 3 for a desired length of time.

The Verlet algorithm is commonly used in the integration of Newton's equation of motion. It derives the velocity from the particle trajectory instead of computing the new position from the velocity.

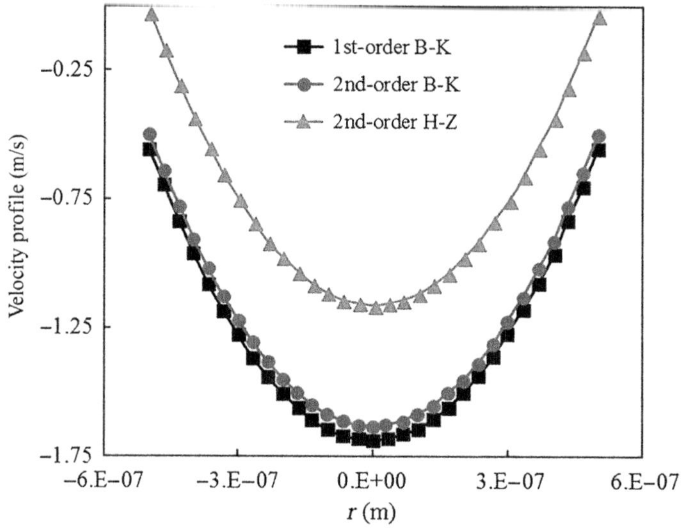

Fig. 10

Velocity profiles of first- and second-order Beskok-Karniadakis slip models and Zhang et al. model at $K_n = 0.1$. *Reproduced from C. Niu, Y. Hao, D. Li, et al., Second-order gas-permeability correlation of shale during slip flow, SPE J. 19 (2014) 786–792.*

MD simulation is a process similar to real experiments. Thus, it is an ideal tool to study the fluid transport at the nanometer-scale. Niu et al. [54] used MD simulation to show the gas-accumulation effect along the methane flow through nanoscale kaolinite channels and proposed a second-order permeability correlation shown in Fig. 10. Also, the slip velocity of methane flow in nanopores is calculated by considering the difference between kerogen and quartz surfaces by Okamoto et al. [55]. Wang et al. [56, 57] reported their molecular studies of flow transport through inorganic and organic nanopores (Fig. 11). It was observed that there was a layering structure of confined octane and the bulk-liquid-like properties at the center of slits. Sun et al. [58] performed the MD simulation to study the oil aggregation and displacement inside a cylindrical silica nanopore and investigated the distribution of heavy oil components inside the silica nanopores.

As designed, the MD method is suitable to capture the dynamic process of system evolution. Zhong et al. studied the adsorption behavior of oil components on the water-wet silica surface and proposed a two-step adsorption process [59]. Bui et al. [60] used the MD method to simulate the two-phase flow in kerogen pores, in which the nonionic surfactant and the terpene solvent were carried by water in the form of a microemulsion. Fig. 12 shows the density distribution profiles of the oil and water phases with dissolved components (surfactant and solvent) for a steady-state flow on a capillary wall (located on the left).

The MD method is not limited to the study of fluid properties, but also the shale system modeling. Hu et al. [61] presented a practical model with a proper description of the kerogen

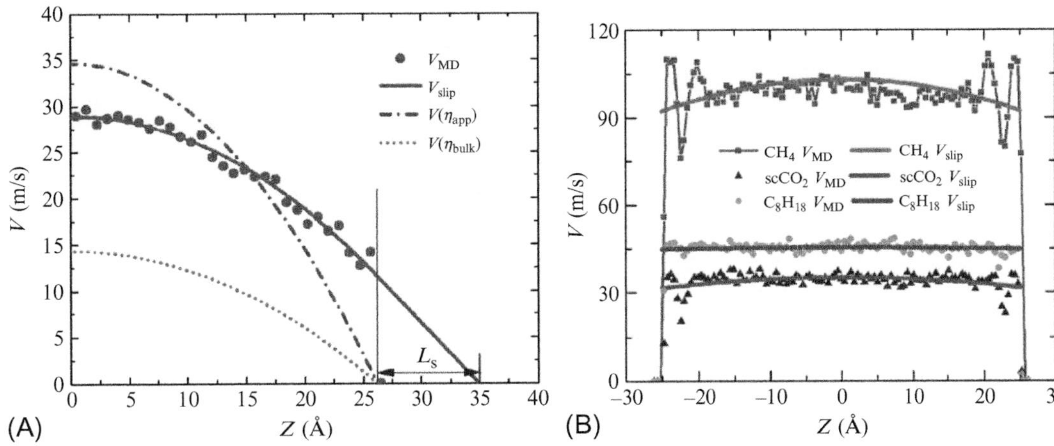

Fig. 11

(A) Velocity profiles of *n*-octane in a 5.24 nm quartz nanopore. (B) Comparison of velocity profiles for CH_4, $scCO_2$, and *n*-C_8H_{18} transport in graphene slits. *Part (A) reused with permission from S. Wang, F. Javadpour, Q. Feng, Molecular dynamics simulations of oil transport through inorganic nanopores in shale, Fuel 171 (2016) 74–86; Part (B) reused with permission from S. Wang, F. Javadpour, Q. Feng, Fast mass transport of oil and supercritical carbon dioxide through organic nanopores in shale, Fuel 181 (2016) 741–758.*

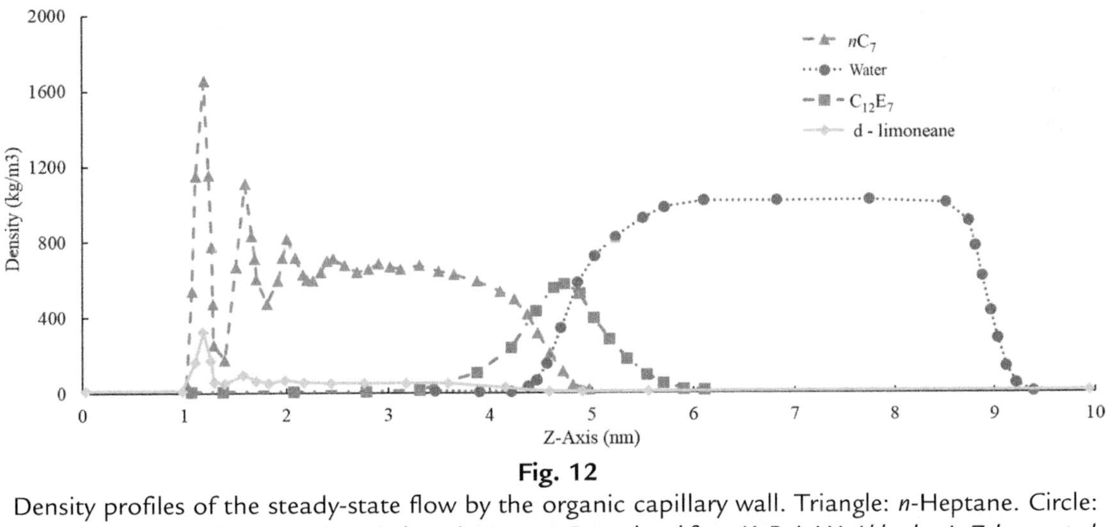

Fig. 12

Density profiles of the steady-state flow by the organic capillary wall. Triangle: *n*-Heptane. Circle: Water. Square: Surfactant. Diamond: D-Limonene. *Reproduced from K. Bui, I.Y. Akkutlu, A. Zelenev, et al., Insights into mobilization of shale oil by use of microemulsion, SPE J. 21 (2016) 613–620*

pore system considering the wettability, surface roughness, tortuous path, and material disorder. Collell et al. [62] constructed a molecular model to represent the organic content in type II shales by mimicking the bulk organic matter in typical reservoir conditions. Fig. 13A shows an amorphous 3D structure with eight type II kerogen molecules with the formula $C_{175}H_{102}O_9N_4S_2$, representing one kind of overmatured organic matter, and Fig. 13B

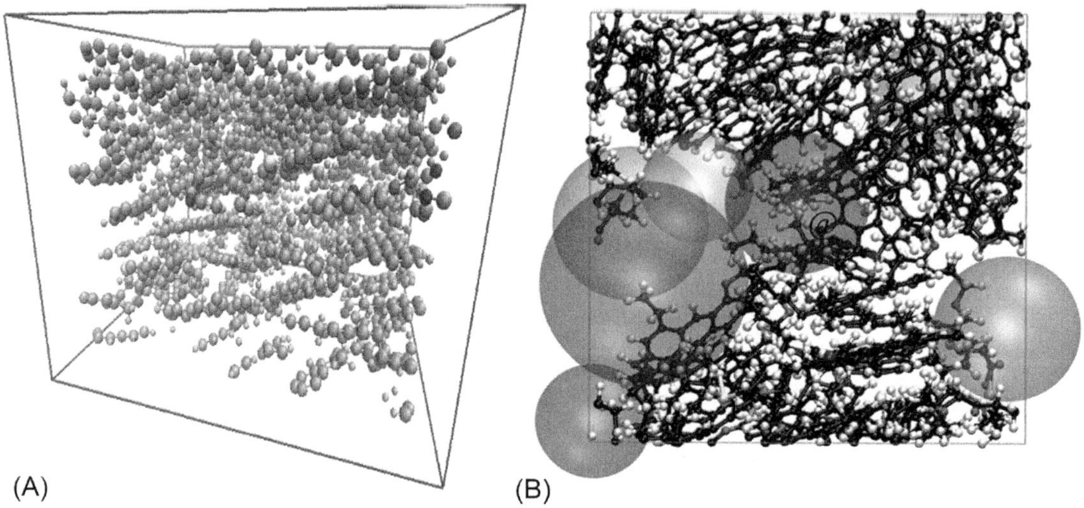

Fig. 13

(A) An illustration of a type II kerogen structure.(B) 3D kerogen structure with pores materialized by spheres. *Part (B) reused with permission from J. Collell, G. Galliero, P. Ungerer, et al., Transport of multicomponent hydrocarbon mixtures in shale organic matter by molecular simulations, J. Phys. Chem. C 119 (2015) 22587–22595.*

shows a 3D kerogen structure with pores materialized by spheres. This model is also used in the studies of methane release, methane transport [18, 63, 64], and phase equilibrium [48]. By considering the amorphous microporous structure of the kerogen nodules, Collell et al. [63] conducted the MD simulations to study the permeation ("apparent permeability") of hydrocarbon fluids through the kerogen molecular model. As shown in Fig. 14, the permeation mechanisms are purely diffusive in this simulation, and the results have good agreement with the Onsager's autocorrelation coefficients.

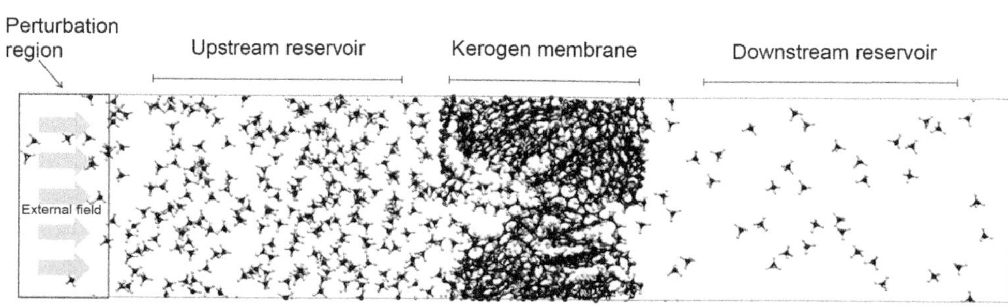

Fig. 14

Transport of multicomponent hydrocarbon mixtures in shale organic matter by molecular simulations [63]. *Reused with permission from J. Collell, G. Galliero, P. Ungerer, et al., Transport of multicomponent hydrocarbon mixtures in shale organic matter by molecular simulations, J. Phys. Chem. C 119 (2015) 22587–22595.*

Since the MD method is very flexible and useful for a wide range of research, e.g., diffusion, adsorption, transport, and system modeling, we have only presented a couple of examples. In addition, several molecular simulation methods are usually conducted together to provide more accurate results. Interested readers may refer to the literature. Molecular simulation is powerful and reliable to characterize petrophysical and fluid properties of shale at the nanometer-scale, but two limitations hinder this approach from wider application in the oil and gas industry. First, it requires massive computational resources to recover a representative property in shale formations. Second, molecular simulation is limited by the time- and length-scales. The typical MD simulation time, even on a powerful modern computer, is within 1–100ns, and the size of a computational cell is in the order of tens of nanometers. Knowing the limitations of molecular simulations, we discuss one widely used pore-scale modeling approach at the mesoscopic scale in the next section, i.e., the lattice Boltzmann method (LBM). The LBM has become a routine tool to characterize pore-scale petrophysical properties for many types of reservoir rocks.

3 Lattice Boltzmann Method (Mesoscopic Approach)

3.1 Lattice Boltzmann Models for Gaseous Flows at the Micro-/Nanoscale

Generally, it is well accepted that the Boltzmann equation is adequate to model gas flows of all flow regimes, ranging from the continuum flow regime ($K_n < 0.001$) to the free-molecular flow regime ($K_n > 10$). Among the numerical approaches based on the Boltzmann equation, the LBM is a widely used one for modeling pore-scale fluid flows and it is a simple, discretized form of the continuous Boltzmann equation. The standard LBM is a second-order approximation of the continuous Boltzmann equation and it recovers the Navier-Stokes equation through Chapman-Enskog multiscale expansion analysis. The standard LBM has shown great success in characterizing permeability, relative permeability, and capillary pressure for many conventional formations. However, the rarefaction effect becomes important for gaseous flows in nanometer-scaled shale pores as the Knudsen number is finite, and the continuum assumption may break down. The standard LBM that recovers the Navier-Stokes equations fails to work for gaseous flows at the micro/nanoscale. In this section, we will focus on the standard LBM, some necessary modifications for micro-/nanoscale gaseous flows, and applications of LBM in shale formations.

3.1.1 LBM with BGK approximation

The lattice Boltzmann equation consists of two fundamental steps: collision and streaming. The standard LBM with the Bhatnagar–Gross–Krook (BGK) collision model can be expressed as

$$f_\alpha(\mathbf{x}+\mathbf{e}_\alpha\delta_t, t+\delta_t) - f_\alpha(\mathbf{x}, t) = -\frac{1}{\tau}\left[f_\alpha(\mathbf{x}, t) - f_\alpha^{eq}(\mathbf{x}, t)\right] + \delta_t F_\alpha \tag{13}$$

where f_α is the density distribution function associated with gas molecules moving at the discrete particle velocity $\mathbf{e}_\alpha$. Fig. 15 illustrates typical lattice structures in 2D and 3D. $\mathbf{e}_\alpha$ is given as

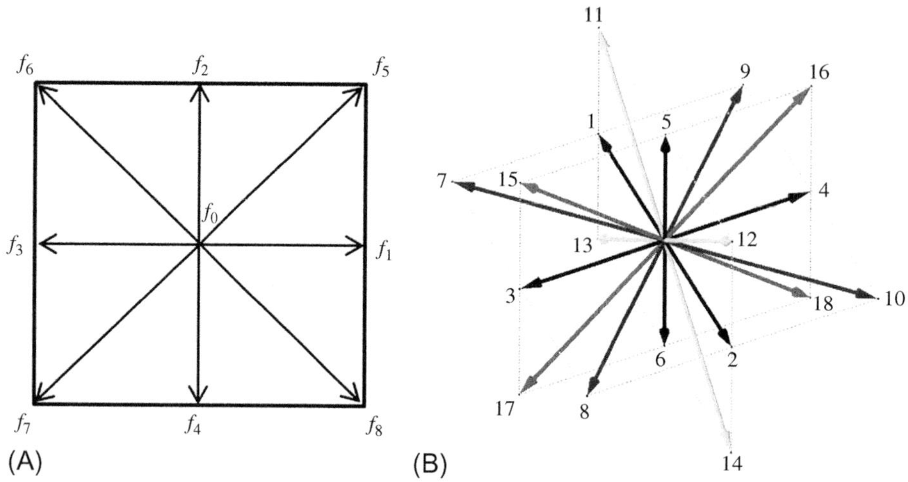

Fig. 15
Typical lattice structures (A) 2D lattice structure D2Q9 and (B) 3D lattice structure D3Q19.

$$\vec{e}_\alpha = \begin{cases} (0,0) & \alpha=0 \\ (1,0);(0,1);(-1,0);(0,-1) & \alpha=1;2;3;4 \\ (1,1);(-1,1);(-1,-1);(1,-1) & \alpha=5;6;7;8 \end{cases} \tag{14}$$

for 2D, and

$$\vec{e}_\alpha = \begin{cases} (0,0) & \alpha=0 \\ (\pm 1,0,0);(0,\pm 1,0);(0,0,\pm 1) & \alpha=1,2,3,4,5,6 \\ (\pm 1,\pm 1,0);(\pm 1,0,\pm 1);(0,\pm 1,\pm 1) & \alpha=7,8,9,10,11,12,13,14,15,16,17,18 \end{cases} \tag{15}$$

for 3D.

The left-hand side of Eq. (13) is the streaming term that describes the propagations of the fluid particle distribution functions on the lattices. The right-hand side contains the collision and source terms. The collision term describes the evolution of the distribution function to its equilibrium state and the source term $\delta_t \vec{F}_\alpha$ is a discrete form of forcing induced by body or external forces. f_α^{eq} is a discrete form of the equilibrium (Maxwell-Boltzmann) distribution function that can be written as

$$f_\alpha^{eq} = \omega_\alpha \rho \left[1 + \frac{\mathbf{e}_\alpha \cdot \mathbf{u}}{c_s^2} + \frac{(\mathbf{e}_\alpha \cdot \mathbf{u})^2}{2c_s^4} - \frac{\mathbf{u} \cdot \mathbf{u}}{2c_s^2} \right] \tag{16}$$

We only cover the 2D case in this chapter and readers are referred to literature for the 3D case. In Eq. (16), w_α is the weighting factor with $w_0 = \frac{4}{9}$, $w_i = \frac{1}{9}(i=1,2,3,4)$, and $w_i = \frac{1}{36}(i=5,6,7,8)$; ρ is the fluid density; $\mathbf{u}$ is the fluid velocity; and $c_s = c/\sqrt{3}$ is the speed of sound. The discrete forcing term F_α is

$$F_\alpha = \left(1 - \frac{1}{2\tau}\right)\omega_\alpha \left[\frac{\mathbf{e}_\alpha - \mathbf{u}}{c_s^2} + \frac{\mathbf{e}_\alpha \cdot \mathbf{u}}{c_s^4}\mathbf{e}_\alpha\right] \cdot \mathbf{F} \tag{17}$$

where $\mathbf{F}$ is the continuous force term. τ is the relaxation time that describes the rate of a fluid system driving to its equilibrium state and it is related to the fluid viscosity through

$$\nu = c_s^2\left(\tau - \frac{1}{2}\right)\delta_t \tag{18}$$

The macroscopic quantities, the fluid density, velocity, and the pressure can be obtained as

$$\rho = \sum_\alpha f_\alpha,\ \vec{u} = \sum\nolimits_\alpha \vec{e}_\alpha f_\alpha + \frac{\delta_t}{2}\vec{F},\ \text{and} p = c_s^2 \rho \tag{19}$$

Through the Chapman-Enskog expansion, the Navier-Stokes equations can be derived from the LBM:

$$\frac{\partial \rho}{\partial t} + \nabla \cdot (\rho \mathbf{u}) = 0$$

$$\frac{\partial(\rho \mathbf{u})}{\partial t} + \nabla \cdot (\rho \mathbf{u}\mathbf{u}) = -\nabla p + \nabla \cdot \left[\mu\left(\nabla \mathbf{u} + \nabla \mathbf{u}^T\right)\right] \tag{20}$$

We have only introduced the LBM with the BGK model for single phase flows, and other Lattice Boltzmann models, such as the multi-relaxation-time model, the two-relaxation-time model, the entropy model, the cascaded model, and multi-component multiphase models, can be found in [65–68].

3.1.2 Effective relaxation time

To model gas flows in nanoscale shale pores, two possible approaches are available: the higher-order LBM and the effective relaxation time method. The higher-order LBM utilize more discrete particle velocities to approximate the continuous Boltzmann equations with a higher-order accuracy. However, the higher-order models increase the computational cost dramatically and the knowledge is still limited on what the order of the discrete velocity set should be for a specific microflow. The other approach is to make use of the effective relaxation time, which has shown reliable results and practicability [69–72]. The effective relaxation time is defined as

$$\tau_e = \tau \Psi(Kn) \tag{21}$$

where $\Psi(K_n)$ has the same form with Eq. (3). Fig. 16 shows the effective relaxation time as a function of the Knudsen number. Later, Li et al. [73] adopted the idea of a Bosanquet-type effective viscosity and derived the effective relaxation time as

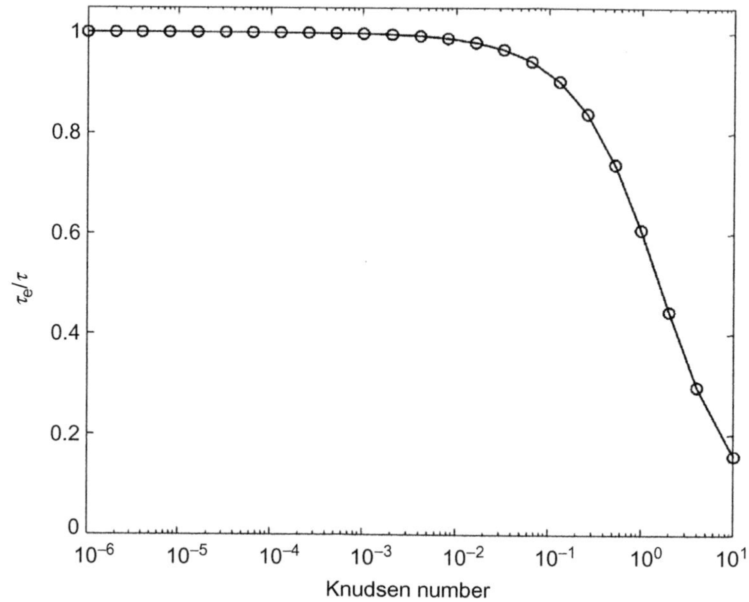

Fig. 16
Nondimensionalized effective relaxation time vs. Knudsen number.

$$\tau_e = \frac{1}{2} + \sqrt{\frac{6}{\pi}}\left(\frac{NK_n}{1 + aK_n}\right) \tag{22}$$

where N is the lattice number along with the direction of characteristic length, and a is the rarefaction coefficient. Sun and Chan [74] suggested $a = 2$ based on the results of the direct simulation Monte Carlo (DSMC) and the linearized Boltzmann equation, while a varies in other researchers' works [10, 70].

3.1.3 Boundary conditions in LBM

Fig. 17A illustrates three common boundary conditions in LBM, i.e., bounce-back (BB), specular reflection (SR), and Maxwell diffuse reflection (MR). BB is the most widely used boundary condition, ensuring the nonslip velocity for a continuum flow regime, namely $K_n < 0.001$. However, the BB scheme is not suitable for gas flow in nanoscale pores where

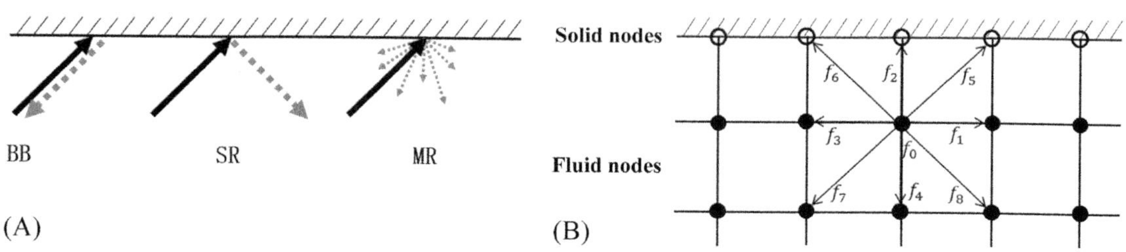

Fig. 17
(A) Conceptual illustration of BB, SR, and MR boundaries. (B) Schematic diagram of LBM fluid-solid nodes.

the slip velocity within the Knudsen layer plays an important role. To model the slip flow boundary condition, SR, MR, and the combination of them can be used for gas flows. SR is usually used for slip flows on a perfectly smooth surface, and the MR boundary assumes that the particle velocities leaving the surface follow the Maxwell distribution. In this section, taking a 2D boundary domain as an example shown in Fig. 17B, we list five types of boundary schemes in LBM as below.

- BB boundary scheme:

$$f_4 = f_2; f_7 = f_5; f_8 = f_6 \tag{23}$$

- SR boundary scheme:

$$f_4 = f_2; f_7 = f_6; f_8 = f_5 \tag{24}$$

- Maxwell diffuse reflection (MR) boundary scheme:

$$f_\alpha(\mathbf{x}, t) = K f_\alpha^{eq}(\rho_w, u_w), \ (\mathbf{e}_\alpha - \mathbf{u}_w)\cdot\mathbf{n} > 0 \tag{25}$$

where $K = \dfrac{\sum_{\boldsymbol{\varepsilon}'_\alpha\cdot\mathbf{n}<0} |\boldsymbol{\varepsilon}'_\alpha \cdot \mathbf{n}| f'_\alpha}{\sum_{\boldsymbol{\varepsilon}'_\alpha\cdot\mathbf{n}<0} |\boldsymbol{\varepsilon}'_\alpha \cdot \mathbf{n}| f_\alpha^{eq}(\rho_w, u_w)}$, $\boldsymbol{\varepsilon}'_\alpha = (\mathbf{e}_\alpha - \mathbf{u}_w)$, in which the subscripts w and α denote the wall and particle velocity directions, respectively, and f_α' is the particle distribution function after streaming.

- Combined BB with SR (CBBSR) boundary scheme:

$$f_\alpha = \gamma f_\alpha^{BB} + (1-\gamma) f_\alpha^{SR}, \tag{26}$$

- Combined specular with Maxwell diffusive reflection (CSMR) boundary scheme:

$$f_\alpha = (1-\beta) f_\alpha^{SR} + \beta f_\alpha^{MR}. \tag{27}$$

where γ and β are the combination coefficients.

Fig. 18 presents two examples of the slip velocity of gas flow in shale nanopores with the CBBSR scheme. In Fig. 18A, we can see that the slip velocity is significantly affected by the combination coefficient, so a sophisticated γ value is required. Fig. 18B shows the LBM slip velocity is in good agreement with the MD results for different K_n with appropriate γ values (see [75] for more details). Zheng et al. [76] derived the combination coefficients for the CBBSR and CSMR schemes, and showed that the two schemes are mathematically equivalent for prescribing the slip velocity on the surface. The combination coefficient γ in the CBBSR scheme and β in the CSMR scheme and the comparison results between the two schemes can be found in [76].

3.1.4 Adsorbed layer

To capture the adsorbed layer and the gas nonideality by LBM, Fathi and Akkutlu [77] first applied the pseudopotential approach and introduced the adsorptive force into LBM. The adsorptive force has the following form:

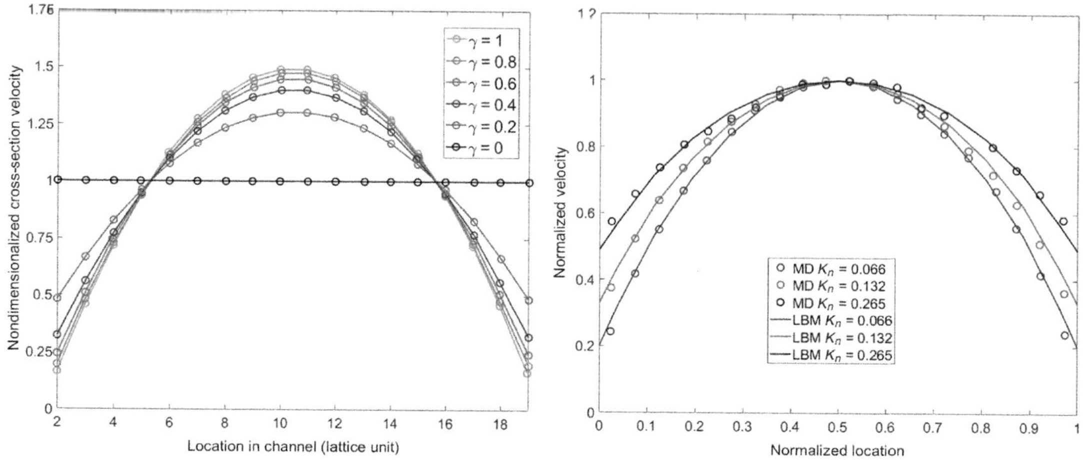

Fig. 18

(A) Cross-sectional velocity profile in a channel ($K_n = 3.2 \times 10^{-2}$) for different slip coefficients γ. (B) Comparison results between LBM and MD for different K_n. *Part (A) reproduced from X. Zhang, L. Xiao, X. Shan, et al., Lattice Boltzmann simulation of shale gas transport in organic nano-pores, Sci. Rep. 4 (2015) 4843; part (B) reproduced from Y. Ning, Y. Jiang, H. Liu, et al., Numerical modeling of slippage and adsorption effects on gas transport in shale formations using the lattice Boltzmann method, J. Nat. Gas Sci. Eng. 26 (2015) 345–355.*

$$\vec{F}_{\text{ads}}(\mathbf{x}, t) = G_{\text{ads}}\Phi(\mathbf{x}, t)\sum_{\alpha} w_\alpha s(\mathbf{x} + \mathbf{e}_\alpha\delta_t, t)\mathbf{e}_\alpha \tag{28}$$

where G_{ads} is the adsorption strength, $\Phi(\mathbf{x}, t)$ is the interaction potential function, and s is a switch function (1 for solid-nearest nodes and 0 for elsewhere). Ning et al. [75] extended this approach to a multi-relaxation-time model and proposed a form for the adsorption strength $G_{\text{ads}} = \rho_{\max}[1 - \exp(-\rho/\rho_{\max})]$, and Yu et al. [72] obtained $\rho_{\max}$ by using the MD simulations. Zhao et al. [71] determined the adsorption parameter by matching the LBM results with the GCMC simulation results on an identical microflow problem. Fig. 19A shows the density profile with and without considering the adsorption, showing that this approach can successfully capture the methane adsorption on solid surfaces.

Ren et al. [78] derived the adsorbed-gas velocity based on the Langmuir isotherm model and applied the CBBSR boundary condition to the LBM for nonideal gases. The adsorbed-gas layer is treated as a moving wall for the CBBSR boundary condition and it has velocity:

$$u_{\text{surface}} = -D_{\text{surface}}\frac{\rho_S M}{\rho_{\text{ads}} V_{\text{std}}}\frac{q_L p_L}{(p_L + p)^2}\nabla p \tag{29}$$

where D_{surface} is the surface diffusivity, ρ_S is the density of kerogen, ρ_{ads} is the density of adsorbed gas, V_{std} is the gas volume per mole at the standard temperature and pressure, q_L and p_L are the Langmuir volume and pressure, respectively. Note that these parameters are required from experiments. We can thus calculate the free-gas transport using the LBM with the

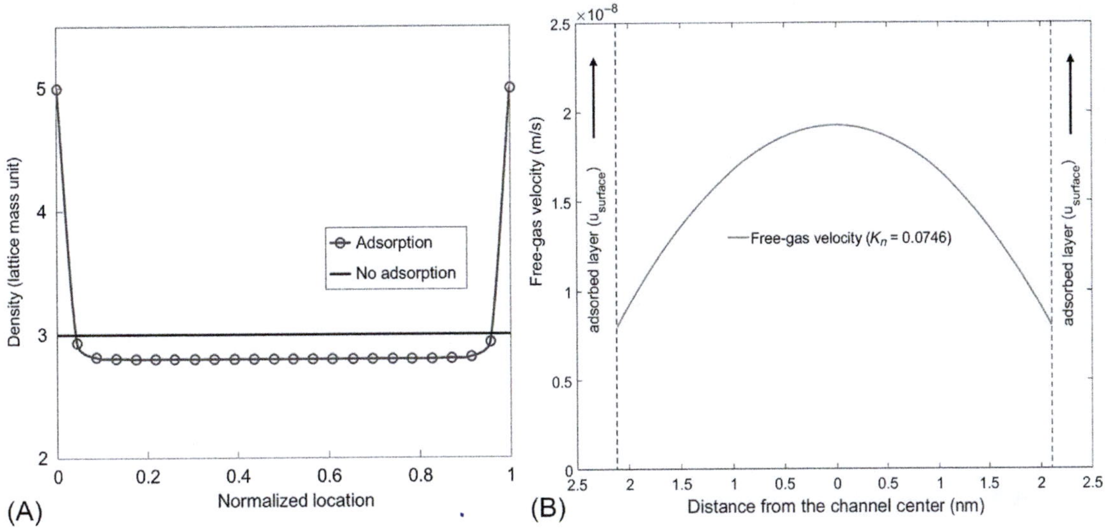

Fig. 19

(A) Density profiles considering with and without adsorption. (B) Free-gas velocity profile and adsorbed gas transport. *Part (B) modified from J. Ren, P. Guo, Z. Guo, et al., A lattice Boltzmann model for simulating gas flow in kerogen pores, Transp. Porous Media 106 (2015) 285–301.*

adsorbed layer as a moving wall with the velocity of u_{surface}. The total mass flux can be achieved by summing up the free-gas and the adsorbed-gas mass fluxes shown in Fig. 19B. Later, Cudjoe and Barati [79] applied a similar approach to evaluate CO_2 sequestration in organic-rich shales using lattice Boltzmann simulation. Up to date research on the topic of the adsorption and surface diffusion by LBM is still lacking but has demonstrated some potential applications.

3.2 Pore-Scale LBM Characterization of Petrophysical Properties of Shale Rock

As more advanced imaging and computing resources are available, LBM has become a popular and reliable tool to perform direct simulations in reconstructed digital pore structures using highly efficient parallelization procedures [80, 81]. Furthermore, LBM's mesoscopic nature provides a viable way to include appropriate microscopic dynamics in the existing models. Kelly et al. [82] presented a workflow on digital rock physics to characterize single phase permeability from FIB-SEM images of shale. The workflow consists of two major components, image analysis and flow simulation by LBM (Fig. 20). Chen et al. [83] applied nanometer-scale SEM in visualizing the microscopic pores within shale kerogen and estimated permeability and tortuosity using LBM simulations. Chen et al. [81] presented a multi-GPU implementation of 3D LBM on hybrid CPU-GPUs cluster with a 1000+ times speedup and estimated the intrinsic permeability of shale kerogen based on a FIB-SEM digital rock (500^3 voxels) shown in Fig. 21. Chen et al. [84] proposed a dusty gas model based

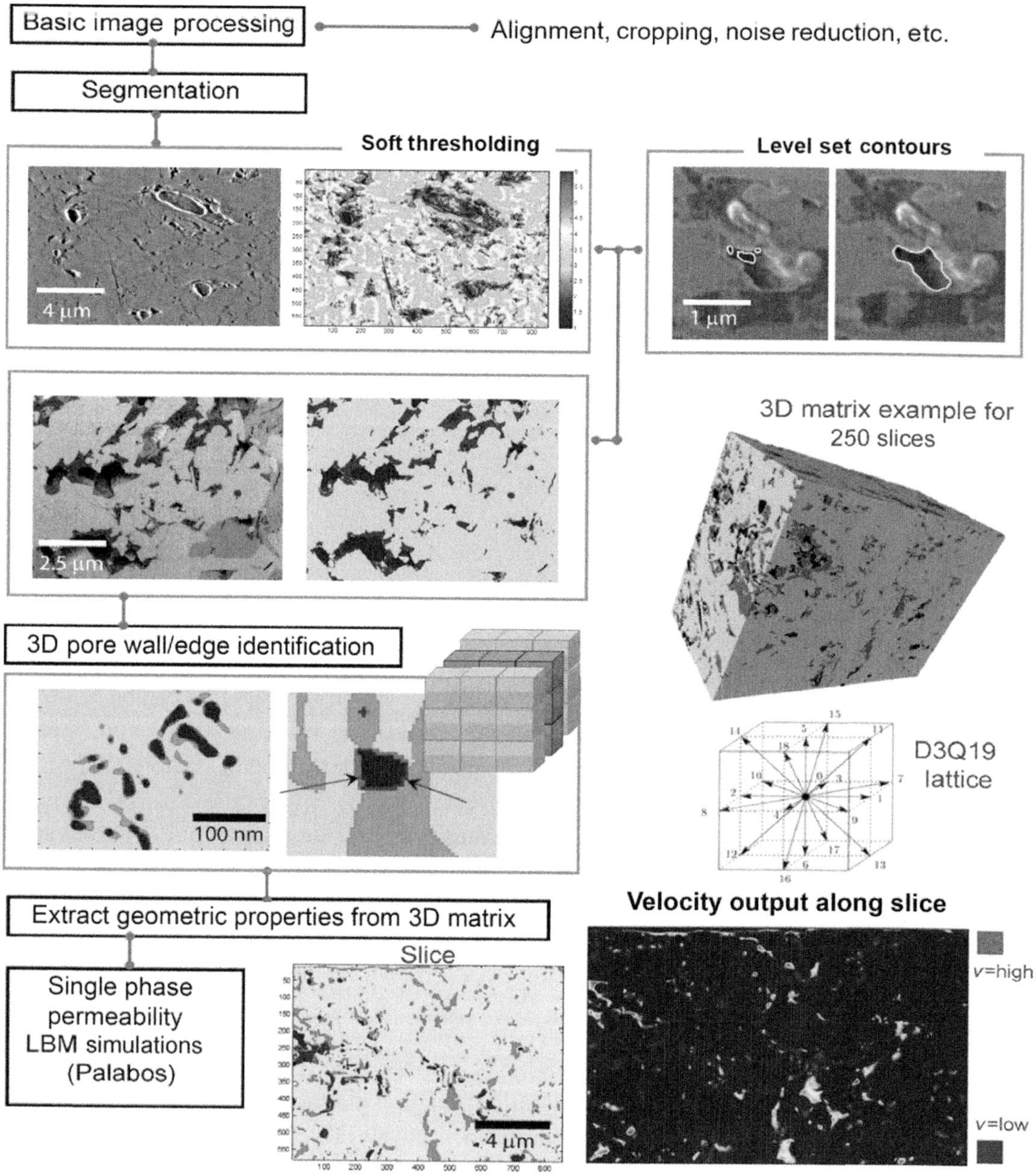

Fig. 20

Workflow to characterize petrophysical properties using LBM on reconstructed FIB-SEM shale digital rocks [84]. *Reused with permission from S. Kelly, H. El-Sobky, C. Torres-Verdín, et al., Assessing the utility of FIB-SEM images for shale digital rock physics, Adv. Water Resour. 95 (2016) 302–316.*

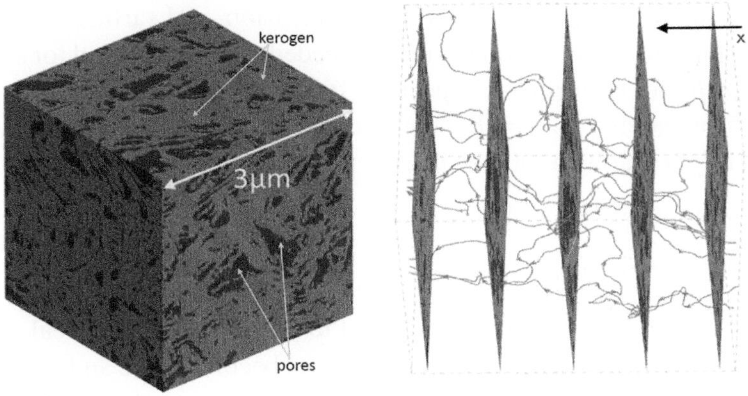

Domain size	500 × 500 × 500
Image resolution (length of a lattice)	6 nm
Temperature	300 K
Viscosity	1.252×10^{-5} pa · s
Inlet pressure	6069310.982pa
Outlet pressure	606844.868pa
Porosity	0.31
Tortuosity	1.8
Permeability	8.905 μD

Fig. 21

(A) FIB-SEM digital rock of kerogen. (B) Streamlines based on LBM flow simulation. (C) Simulation conditions and results. *Modified from T. Chen, Y. Ning, A. Amritkar, et al., Multi-GPU solution to the lattice Boltzmann method: an application in multiscale digital rock simulation for shale formation, Concurr. Comput. Pract. Exp. (2018) e4530.*

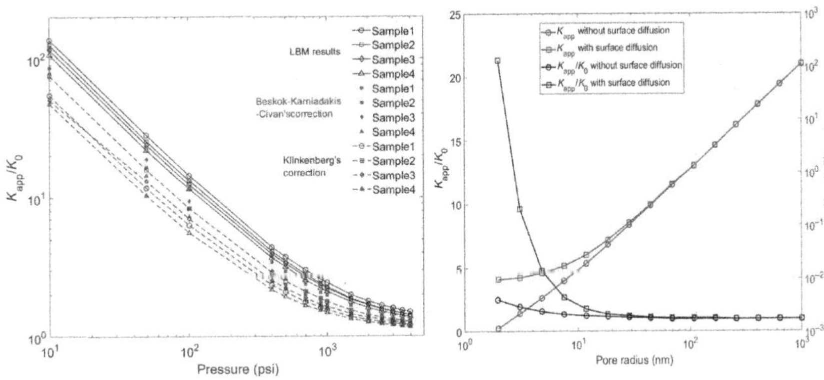

Fig. 22

(A) The correction factor in shale apparent permeability function ($f_c = K_{app}/K_0$) [84]. (B) Apparent permeability estimation considering surface diffusion. *Part (B) reproduced from J. Wang, L. Chen, Q. Kang, et al., Apparent permeability prediction of organic shale with generalized lattice Boltzmann model considering surface diffusion effect, Fuel 181 (2016) 478–490.*

LBM and calculated the apparent permeability of shale shown in Fig. 22A [84]. The intrinsic permeability and the effective Knudsen diffusivity using LBM are estimated. Surface diffusivity is not considered in their work. Wang et al. [85] adopted the dusty gas model and generalized Maxwell-Stefan model to calculate the apparent permeability accounting for multiple transport mechanisms including viscous flow, Knudsen diffusion, and surface diffusion. They found that the surface diffusion in organic matter needed to be considered for an accurate determination of the apparent permeability, especially for very small pores at a low pressure. As seen in Fig. 22B, the apparent permeability considering the surface diffusion can be an order of magnitude higher than that without considering the surface diffusion.

4 Summary

Understanding the transport and storage mechanisms is essential to estimate petrophysical properties in shale formation. Pore-scale modeling and simulation has become more and more important for the estimation of petrophysical properties, the prediction of long-term hydrocarbon production, and reserve estimation. Due to the confined space and adsorption effects, transport, storage, and phase behavior of natural gas in nanoscale shale pores, these pores behave significantly different to that in conventional rock pores. In this chapter, we first presented an overview of the gas transport and adsorption mechanisms in an organic nanopore. The confinement effect and surface diffusion have been briefly discussed. Second, we discussed molecular simulation in terms of its basic theory and modeling. Two approaches in molecular simulations have been presented, i.e., Monte Carlo simulation and Molecular Dynamics simulation. The applications of molecular simulation on characterizing transport and thermodynamic properties have been discussed. Molecular simulation is very flexible and useful for research fields, e.g., diffusion, adsorption, transport, and system modeling. Third, we discussed a mesoscopic modeling approach, i.e., the lattice Boltzmann method (LBM). The LBM has become a routine tool to characterize pore-scale petrophysical properties for many types of reservoirs rocks. However, the standard LBMs are not adequate in describing the transport behavior of natural gas in shale formations, so we focused on the effective relaxation time approach and several feasible boundary conditions. Lastly, a few examples of the estimation of intrinsic and apparent permeability of shale were presented.

Acknowledgments

This research is supported in part by the US National Science Foundation through NSF Grant DMS-1209124, Society of Petroleum Engineers-Gulf Coast Section through its distinguish professorship endowment, Institute of Petroleum Exploration & Development – Langfang Branch, PetroChina Company Limited, CNPC Chuanqing Drilling Engineering Company Limited, and Sinopec Tech Houston LLC – Research Centre.

References

[1] C.R. Clarkson, N. Solano, R.M. Bustin, et al., Pore structure characterization of north American shale gas reservoirs using USANS/SANS, gas adsorption, and mercury intrusion, Fuel 103 (2013) 606–616.

[2] S.M. Kang, E. Fathi, R.J. Ambrose, et al., Carbon dioxide storage capacity of organic-rich shales, SPE J. 16 (2011) 842–855.

[3] R. Heller, M. Zoback, Adsorption of methane and carbon dioxide on gas shale and pure mineral samples, J. Unconv. Oil Gas Resour. 8 (2014) 14–24.

[4] S. Zhou, Y. Ning, H. Wang, et al., Investigation of methane adsorption mechanism on Longmaxi shale by combining the micropore filling and monolayer coverage theories, Adv. Geo-Energy Res. 2 (2018) 269–281.

[5] G.R. Chalmers, R.M. Bustin, I.M. Power, Characterization of gas shale pore systems by porosimetry, pycnometry, surface area, and field emission scanning electron microscopy/transmission electron microscopy image analyses: examples from the Barnett, Woodford, Haynesville, Marcellus, and Doig units, AAPG Bull. 96 (2012) 1099–1119.

[6] J. Li, T. Yu, X. Liang, et al., Insights on the gas permeability change in porous shale, Adv. Geo-Energy Res. 1 (2017) 69–73.

[7] B. Jia, J. Tsau, R. Barati, A workflow to estimate shale gas permeability variations during the production process, Fuel 220 (2018) 879–889.

[8] B. Jia, J. Tsau, R. Barati, Role of molecular diffusion in heterogeneous, naturally fractured shale reservoirs during CO_2 huff-n-puff, J. Pet. Sci. Eng. 164 (2018) 31–42.

[9] Y. Ning, S. He, G. Qin, An integrated workflow to predict macro-scale transport properties in gas shales by coupling molecular dynamics simulation with lattice Boltzmann method, Microsc. Microanal. 23 (Suppl. 1) (2017) 2126–2127.

[10] I. Medveď, R. Černý, Surface diffusion in porous media: a critical review, Microporous Mesoporous Mater. 142 (2011) 405–422.

[11] D.W. Stops, The mean free path of gas molecules in the transition regime, J. Phys. D. Appl. Phys. 3 (1970) 685–696.

[12] Z.L. Guo, B.C. Shi, C.G. Zheng, An extended Navier-Stokes formulation for gas flows in the Knudsen layer near a wall, EPL 80 (2007) 24001.

[13] Z. Guo, T.S. Zhao, Y. Shi, Physical symmetry, spatial accuracy, and relaxation time of the lattice Boltzmann equation for microgas flows, J. Appl. Phys. 99 (2006) 074903.

[14] G. Karniadakis, A. Beskok, N. Aluru, Microflows and Manoflows: Fundamentals and Simulation, Springer Science & Business Media, 2006.

[15] S.K. Loyalka, Velocity slip coefficient and the diffusion slip velocity for a multicomponent gas mixture, Phys. Fluids 14 (1971) 2599–2604.

[16] S.K. Loyalka, N. Petrellis, T.S. Storvick, Some numerical results for the BGK model: thermal creep and viscous slip problems with arbitrary accommodation at the surface, Phys. Fluids 18 (1975) 1094–1099.

[17] S. Zhou, H. Xue, Y. Ning, et al., Experimental study of supercritical methane adsorption in Longmaxi shale: insights into the density of adsorbed methane, Fuel 211 (2018) 140–148.

[18] S. He, Y. Ning, T. Chen, et al., Estimation of transport diffusivity of natural gas in organic matter using molecular dynamics simulation, SPE-180198, in: Proceedings of SPE Low Perm Symposium, Denver, Colorado, USA, 5–6 May, 2016.

[19] B. Jia, J. Tsau, R. Barati, Different flow behaviors of low-pressure and high-pressure carbon dioxide in shales, SPE J. (2018) SPE-191121-PA.

[20] D.D. Do, H.D. Do, Surface diffusion of hydrocarbons in activated carbon: comparison between constant molar flow, differential permeation and differential adsorption bed methods, Adsorption 7 (2001) 189–209.

[21] K. Wu, X. Li, C. Wang, et al., Model for surface diffusion of adsorbed gas in nanopores of shale gas reservoirs, Ind. Eng. Chem. Res. 54 (2015) 3225–3236.

[22] Y. Ning, S. He, H. Liu, et al., Permeability prediction considering surface diffusion for gas shales by lattice Boltzmann simulations on multi-scale reconstructed digital rocks, IPTC-18845, in: Proceedings of International Petroleum Technology Conference, Bangkok, Thailand, 14–16 November, 2016.

[23] P. Ungerer, C. Nieto-Draghi, V. Lachet, et al., Molecular simulation applied to fluid properties in the oil and gas industry, Mol. Simul. 33 (2007) 287–304.

[24] P. Ungerer, B. Tavitian, A. Boutin, Applications of Molecular Simulation in the Oil and Gas Industry: Monte Carlo Methods, Editions Technip, 2005.

[25] M.P. Allen, D.J. Tildesley, Computer Simulation of Liquids, Oxford University Press, 2017.

[26] G.C. Maitland, Intermolecular Forces: Their Origin and Determination, Clarendon Press, 1981.
[27] M.G. Martin, J.I. Siepmann, Transferable potentials for phase equilibria. 1. United-atom description of *n*-alkanes, J. Phys. Chem. B 102 (1998) 2569–2577.
[28] K. Bolton, W.L. Hase, W.F. Schneider, et al., Comparison of explicit and united atom models for alkane chains physisorbed on α-Al_2O_3 (0001), J. Phys. Chem. B 103 (1999) 3885–3895.
[29] C. Chen, P. Depa, V.G. Sakai, et al., A comparison of united atom, explicit atom, and coarse-grained simulation models for poly (ethylene oxide), J. Chem. Phys. 124 (2006) 234901.
[30] D.Y. Yoon, G.D. Smith, T. Matsuda, A comparison of a united atom and an explicit atom model in simulations of polymethylene, J. Chem. Phys. 98 (1993) 10037–10043.
[31] J.P. Ewen, C. Gattinoni, F.M. Thakkar, et al., A comparison of classical force-fields for molecular dynamics simulations of lubricants, Materials (Basel) 9 (2016) 1–17.
[32] D.N. Theodorou, U.W. Suter, Geometrical considerations in model systems with periodic boundaries, J. Chem. Phys. 82 (1985) 955–966.
[33] S. Pfalzner, P. Gibbon, Many-Body Tree Methods in Physics, Cambridge University Press, 2005.
[34] N. Metropolis, A.W. Rosenbluth, M.N. Rosenbluth, et al., Equation of state calculations by fast computing machines, J. Chem. Phys. 21 (1953) 1087–1092.
[35] A.Z. Panagiotopoulos, Direct determination of phase coexistence properties of fluids by Monte Carlo simulation in a new ensemble, Mol. Phys. 61 (1987) 813–826.
[36] A.Z. Panagiotopoulos, N. Quirke, M. Stapleton, et al., Phase equilibria by simulation in the Gibbs ensemble: alternative derivation, generalization and application to mixture and membrane equilibria, Mol. Phys. 63 (1988) 527–545.
[37] L.A. Rowley, D. Nicholson, N.G. Parsonage, Monte Carlo grand canonical ensemble calculation in a gas-liquid transition region for 12-6 Argon, J. Comput. Phys. 17 (1975) 401–414.
[38] D. Frenkel, B. Smit, Monte Carlo Simulations in Various Ensembles, in: Understanding Molecular Simulation, second ed., Academic Press, San Diego, CA, 2002, , pp. 111–137.
[39] A. Neimark, A. Vishnyakov, Gauge cell method for simulation studies of phase transitions in confined systems, Phys. Rev. E 62 (2000) 4611.
[40] A. Vishnyakov, A.V. Neimark, Studies of liquid-vapor equilibria, criticality, and spinodal transitions in nanopores by the gauge cell Monte Carlo simulation method, J. Phys. Chem. B 105 (2001) 7009–7020.
[41] B. Jin, H. Nasrabadi, Phase behavior of multi-component hydrocarbon systems in nano-pores using gauge-GCMC molecular simulation, Fluid Phase Equilib. 425 (2016) 324–334.
[42] M.T. Miyahara, H. Tanaka, Determination of phase equilibria in confined systems by open pore cell Monte Carlo method, J. Chem. Phys. 138 (2013) 084709.
[43] K.A. Fichthorn, W.H. Weinberg, Theoretical foundations of dynamical Monte Carlo simulations, J. Chem. Phys. 95 (1991) 1090–1096.
[44] G. Kamath, J.J. Potoff, Monte Carlo predictions for the phase behavior of H_2S+*n*-alkane, H_2S+CO_2, CO_2+CH_4 and $H_2S+CO_2+CH_4$ mixtures, Fluid Phase Equilib. 246 (2006) 71–78.
[45] J. Jiang, S.I. Sandler, B. Smit, Capillary phase transitions of *n*-alkanes in a carbon nanotube, Nano Lett. 4 (2004) 241–244.
[46] Z. Jin, A. Firoozabadi, Phase behavior and flow in shale nanopores from molecular simulations, Fluid Phase Equilib. 430 (2016) 156–168.
[47] B. Jin, R. Bi, H. Nasrabadi, Molecular simulation of the pore size distribution effect on phase behavior of methane confined in nanopores, Fluid Phase Equilib. 452 (2017) 94–102.
[48] B. Jin, H. Nasrabadi, Phase behavior in shale organic/inorganic nanopores from molecular simulation, SPE-187307-PA, SPE Reserv. Eval. Eng. (2018).
[49] B.L. Severson, R.Q. Snurr, Monte Carlo simulation of *n*-alkane adsorption isotherms in carbon slit pores, J. Chem. Phys. 126 (2007) 134708.
[50] K. Mosher, J. He, Y. Liu, et al., Molecular simulation of methane adsorption in micro- and mesoporous carbons with applications to coal and gas shale systems, Int. J. Coal Geol. 109 (2013) 36–44.
[51] S.K. Singh, A. Sinha, G. Deo, et al., Vapor – liquid phase coexistence, critical properties, and surface tension of confined alkanes, J. Phys. Chem. C 113 (2009) 7170–7180.

[52] L.D. Gelb, K.E. Gubbins, R. Radhakrishnan, et al., Phase separation in confined systems, Rep. Prog. Phys. 62 (1999) 1573–1659.
[53] E. Barsotti, S.P. Tan, S. Saraji, et al., A review on capillary condensation in nanoporous media: implications for hydrocarbon recovery from tight reservoirs, Fuel 184 (2016) 344–361.
[54] C. Niu, Y. Hao, D. Li, et al., Second-order gas-permeability correlation of shale during slip flow, SPE J. 19 (2014) 786–792.
[55] N. Okamoto, K. Kobayashi, Y. Liang, et al., Slip velocity of methane flow in nanopores with kerogen and quartz surfaces, SPE J. 23 (2018) 102–116.
[56] S. Wang, F. Javadpour, Q. Feng, Molecular dynamics simulations of oil transport through inorganic nanopores in shale, Fuel 171 (2016) 74–86.
[57] S. Wang, F. Javadpour, Q. Feng, Fast mass transport of oil and supercritical carbon dioxide through organic nanopores in shale, Fuel 181 (2016) 741–758.
[58] J. Sun, H. Zhang, M. Hu, et al., Molecular dynamics study on oil migration inside silica nanopore, Chem. Phys. Lett. 678 (2017) 186–191.
[59] J. Zhong, P. Wang, Y. Zhang, et al., Adsorption mechanism of oil components on water-wet mineral surface: a molecular dynamics simulation study, Energy 59 (2013) 295–300.
[60] K. Bui, I.Y. Akkutlu, A. Zelenev, et al., Insights into mobilization of shale oil by use of microemulsion, SPE J. 21 (2016) 613–620.
[61] Y. Hu, D. Devegowda, A. Striolo, et al., Microscopic dynamics of water and hydrocarbon in shale-kerogen pores of potentially mixed wettability, SPE J. 20 (2015) 112–124.
[62] J. Collell, P. Ungerer, G. Galliero, et al., Molecular simulation of bulk organic matter in type II shales in the middle of the oil formation window, Energy Fuel 28 (2014) 7457–7466.
[63] J. Collell, G. Galliero, P. Ungerer, et al., Transport of multicomponent hydrocarbon mixtures in shale organic matter by molecular simulations, J. Phys. Chem. C 119 (2015) 22587–22595.
[64] S. He, Y. Ning, T. Chen, et al., Transport properties of natural gas in shale organic and inorganic nanopores using non-equilibrium molecular dynamics simulation, IPTC-18875, in: Proceedings of International Petroleum Technology Conference, Bangkok, Thailand, 14–16 November, 2016.
[65] X. Shan, H. Chen, Lattice Boltzmann model for simulating flows with multi phases and components, Phys. Rev. E 47 (1993) 1815–1819.
[66] S.S. Chikatamarla, I.V. Karlin, Entropy and Galilean invariance of lattice Boltzmann theories, Phys. Rev. Lett. 97 (2006) 190601.
[67] I. Ginzburg, F. Verhaeghe, D. dHumieres, Two-relaxation-time lattice Boltzmann scheme: about parametrization, velocity, pressure and mixed boundary conditions, Commun. Comput. Phys. 3 (2008) 427–478.
[68] Y. Ning, K.N. Premnath, D.V. Patil, Numerical study of the properties of the central moment lattice Boltzmann method, Int. J. Numer. Methods Fluids 82 (2016) 59–90.
[69] Z. Guo, C. Zheng, B. Shi, Lattice Boltzmann equation with multiple effective relaxation times for gaseous microscale flow, Phys. Rev. E 77 (2008) 1–12.
[70] Y. Ning, S. He, T. Chen, et al., Simulation of shale gas transport in 3D complex nanoscale-pore using the lattice Boltzmann method, SPE-176997, in: SPE Asia Pacific Unconventional Resources Conference and Exhibition, Brisbane, Australia, 9–11 November, 2015.
[71] J. Zhao, J. Yao, L. Zhang, et al., Pore-scale simulation of shale gas production considering the adsorption effect, Int. J. Heat Mass Transf. 103 (2016) 1098–1107.
[72] H. Yu, J. Chen, Y.B. Zhu, et al., Multiscale transport mechanism of shale gas in micro/nano-pores, Int. J. Heat Mass Transf. 111 (2017) 1172–1180.
[73] Q. Li, Y.L. He, G.H. Tang, et al., Lattice Boltzmann modeling of microchannel flows in the transition flow regime, Microfluid. Nanofluid. 10 (2011) 607–618.
[74] S. Yuhong, W.K. Chan, Analytical modeling of rarefied Poiseuille flow in microchannels, J. Vac. Sci. Technol. A 22 (2014) 383–394.
[75] Y. Ning, Y. Jiang, H. Liu, et al., Numerical modeling of slippage and adsorption effects on gas transport in shale formations using the lattice Boltzmann method, J. Nat. Gas Sci. Eng. 26 (2015) 345–355.

[76] L. Zheng, Z. Guo, B. Shi, Microscale boundary conditions of the lattice Boltzmann equation method for simulating microtube flows, Phys. Rev. E 86 (2012) 016712.
[77] E. Fathi, I.Y. Akkutlu, Lattice Boltzmann method for simulation of shale gas transport in kerogen, SPE J. 18 (2011) 27–37.
[78] J. Ren, P. Guo, Z. Guo, et al., A lattice Boltzmann model for simulating gas flow in kerogen pores, Transp. Porous Media 106 (2015) 285–301.
[79] S. Cudjoe, R. Barati, Lattice Boltzmann simulation of CO_2 transport in kerogen nanopores—an evaluation of CO_2 sequestration in organic-rich shales, J. Earth Sci. 28 (2017) 926–932.
[80] P.R. Rinaldi, E.A. Dari, M.J. Vénere, et al., A lattice-Boltzmann solver for 3D fluid simulation on GPU, Simul. Model. Pract. Theory 25 (2012) 163–171.
[81] T. Chen, Y. Ning, A. Amritkar, et al., Multi-GPU solution to the lattice Boltzmann method: an application in multiscale digital rock simulation for shale formation, Concurr. Comput. Pract. Exp. (2018) e4530.
[82] S. Kelly, H. El-Sobky, C. Torres-Verdín, et al., Assessing the utility of FIB-SEM images for shale digital rock physics, Adv. Water Resour. 95 (2016) 302–316.
[83] C. Chen, D. Hu, D. Westacott, et al., Nanometer-scale characterization of microscopic pores in shale kerogen by image analysis and pore-scale modeling, Geochem. Geophys. Geosyst. 14 (2013) 4066–4075.
[84] L. Chen, L. Zhang, Q. Kang, et al., Nanoscale simulation of shale transport properties using the lattice Boltzmann method: Permeability and diffusivity, Sci. Rep. 5 (2015) 8089.
[85] J. Wang, L. Chen, Q. Kang, et al., Apparent permeability prediction of organic shale with generalized lattice Boltzmann model considering surface diffusion effect, Fuel 181 (2016) 478–490.

Further Reading

[86] V.K. Michalis, A.N. Kalarakis, E.D. Skouras, et al., Rarefaction effects on gas viscosity in the Knudsen transition regime, Microfluid. Nanofluid. 9 (2010) 847–853.
[87] X. Zhang, L. Xiao, X. Shan, et al., Lattice Boltzmann simulation of shale gas transport in organic nano-pores, Sci. Rep. 4 (2015) 4843.

CHAPTER 12

High-Pressure Methane Adsorption in Shale

Shangwen Zhou*, **Hongyan Wang***, **Jianchao Cai**†

PetroChina Research Institute of Petroleum Exploration & Development, Beijing, P.R. China, †Hubei Subsurface Multi-scale Imaging Key Laboratory, Institute of Geophysics and Geomatics, China University of Geosciences, Wuhan, P.R. China

Chapter Outline

1 Introduction

Shale gas differs from conventional natural gas stored in sandstone and carbonate formations because shale formations are often both the generating and storing place of natural gas [1–3]. In addition, within the shale formation, shale gas exists in three different phases: (i) free gas in pores; (ii) adsorbed gas on the pore surface; and (iii) dissolved gas in liquid hydrocarbon and brine [1]. Shale gas-in-place (GIP) is estimated by summing these three components [4]. The adsorbed gas accounts for 20%–85% of the total gas amount in five major shale gas reservoirs in the United States [1]. Thus, the accurate determination of the amount of adsorbed gas, the main component of which is methane, significantly influences the final calculation of the GIP quantity and estimated ultimate recovery of the shale gas producing well [4].

Methane adsorption in shale has been extensively studied by volumetric and gravimetric isothermal adsorption experiments [5–8]. However, the maximum pressure in most of these experiments was lower than 15 MPa, which is far below the actual formation pressure [9–14]. The in situ pressure of the Longmaxi formation in the southern Sichuan Basin is approximately

Petrophysical Characterization and Fluids Transport in Unconventional Reservoirs. https://doi.org/10.1016/B978-0-12-816698-7.00012-7

26 MPa at a depth of 2000 m, with a pressure gradient of 13 MPa/km. In this regard, experiments with a high pressure range (>25 MPa) are essential for accurate estimation of the actual adsorption capacity in the shale formation [15–17].

Many adsorption models have been proposed to characterize the methane adsorption process under the formation temperature and pressure, including Langmuir [16], Langmuir-Freundlich (L-F) [13], Ono-Kondo [12], simplified local-density (SLD) [18], Dubinin-Radushkevich (D-R), and Dubinin-Astakhov (D-A) models [19]. Researchers have attempted to reveal methane adsorption mechanisms in shale using these models [17, 20]. However, these models were originally established based on a variety of assumptions. For example, the Langmuir model assumes that gas molecules are monolayer-adsorbed in adsorbents, and the surface of the solid adsorbent is homogeneous with a constant adsorption heat, which is over-simplified to describe complicated situations in shale gas reservoirs [16]. Although the Langmuir model fits some experimental isotherms quite well [9, 17], it cannot be concluded that the methane adsorption mechanism is monolayer adsorption. Therefore, its mechanism cannot be determined by those models as all the equations show a similar fitting effect [19]. In other words, it is inaccurate to explain the adsorption mechanism using a model only based on the quality of curve fitting. A suitable model should have its internal hypothesis consistent with the real characteristics of shale. Therefore, an adsorption model considering the complexity and heterogeneity of shale pore structures, which most of adsorption models have ignored, is critically needed.

In conclusion, high-pressure methane adsorption experiments, an appropriate adsorption model and its adsorption mechanism are the key issues, which will be discussed in this chapter.

2 High-Pressure Methane Adsorption Experiments in Shale

The high-pressure adsorption isotherm experiments were conducted using a Rubotherm gravimetric adsorption instrument. Its core component is the magnetic suspension balance (MSB) with high precision of 10 μg [21]. The maximum test pressure and temperature were 35 MPa and 150°C, respectively. The long time fluctuation range of temperature can be controlled within 0.2°C. The basic principle of this instrument is shown in Fig. 1.

The blank test (without shale samples) was first conducted in order to obtain the mass and volume of the adsorption cell. The whole system was pumped down to vacuum conditions, and then the measurement was conducted by dosing pure nitrogen into the adsorption cell up to 5 MPa. The apparent weight of the adsorption cell was recorded from MSB, which is the interaction between the weight of the adsorption cell and the buoyancy induced by the dosing helium. After the shale sample was put in the adsorption cell, the nonadsorbed pure helium was dosed into the system after a vacuum was applied. The apparent weight of the adsorption cell with shale was recorded from MSB. The mass of the shale sample was obtained using

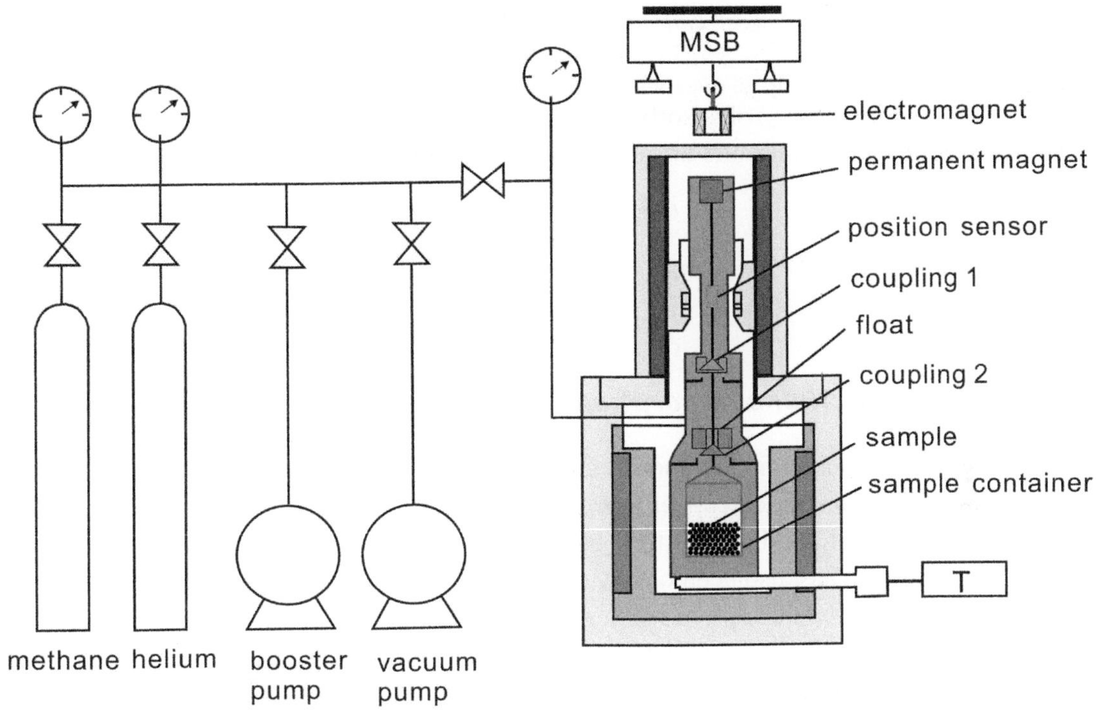

Fig. 1
The principle diagram of the high-pressure gravimetric adsorption instrument.

the above two steps. Then the adsorption cell was dosed with methane after being vacuumed. When the experimental temperature was constant (60°C), the adsorbed phase of methane was formed on the surface of nanopores in shale. The mass balance can be written as

$$m_b = m_{sc} + m_s + m_{abs} - (V_{sc} + V_s + V_a) \times \rho_g \tag{1}$$

where m_b is the MSB reading, m_{sc} is the mass of the sample container, m_s is the mass of the sample, m_{abs} is the mass of adsorbed methane, V_{sc} is the volume of the sample container, V_s is the sample volume, V_a is the volume of adsorbed phase, and ρ_g is the density of free methane at a given temperature and pressure.

From Eq. (1), the adsorption capacity can be expressed as

$$m_{abs} = m_b - m_{sc} - m_s + (V_{sc} + V_s + V_a) \times \rho_g \tag{2}$$

where all variables except for V_a on the right side of Eq. (2) can be measured by experiment. Thus, the actual adsorption capacity cannot be determined readily from the experiment. If a new parameter (m_{ex}) is defined to represent the excess adsorption, then

$$m_{ex} = m_{abs} - V_a \times \rho_g = V_a \times (\rho_a - \rho_g) \tag{3}$$

Substituting Eq. (3) into Eq. (2)

$$m_{ex} - \Delta m - m_{sc} - m_s + (V_{sc} + V_s) \times \rho_g \tag{4}$$

Obviously, the excess adsorption m_{ex} can be measured by adsorption experiment. If either the density (ρ_a) or the volume (V_a) of adsorbed methane is known [16, 21], the observed adsorption can be transformed to the absolute adsorption by the following:

$$m_{abs} = m_{ex} + V_a \times \rho_g \tag{5}$$

$$m_{abs} = m_{ex} / (1 - \rho_g/\rho_a) \tag{6}$$

3 Supercritical Methane Adsorption Characteristics

The measured excess adsorption isotherms are presented in Fig. 2. The experimental results show that the excess adsorption capacity (n_{ex}) increased to a maximum value at the approximate pressure of 10 MPa, and then began to decline with an increased pressure. This abnormal phenomenon has also been observed in some previous studies [15–17, 22], but most researchers stated that the adsorption isotherm of methane in shale monotonically increased with pressure and reached a constant value at a high pressure (i.e., type I isotherm) [9–11]. The main reason for these two different types of adsorption isotherms is whether or not the volume of the adsorbed phase is assumed to be negligible. This assumption is acceptable as long as the density of the free gas is much lower than the density of the adsorbed phase at low pressures. However, when the pressure becomes high, and the density of free-phase methane approaches the density of adsorbed methane, the volume of the adsorbed phase can no longer

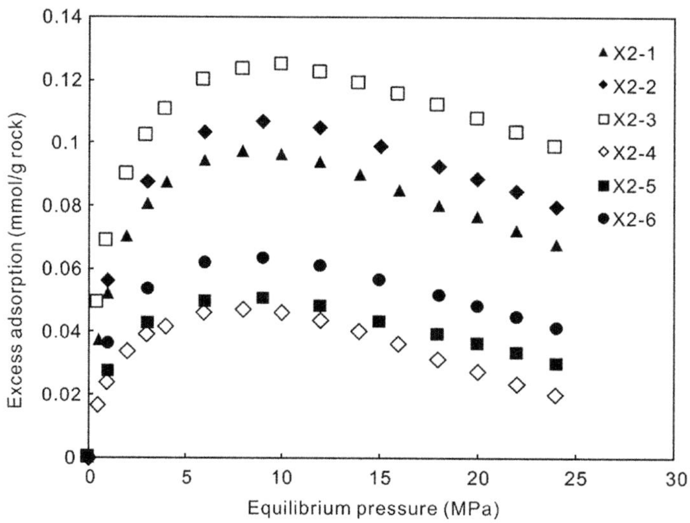

Fig. 2
The excess adsorption isotherms of the shale samples at 60°C.

be neglected. As shown in Eq. (3), the excess adsorption will approach zero when the density of free-phase methane approaches the density of adsorbed methane.

Total organic carbon (TOC) content is one of the most important parameters to evaluate a shale gas reservoir as it strongly correlates with the nanopore development and the gas adsorption capacity. $n_{ex\text{-}max}$ is positively correlated with the TOC content. These observations are consistent with previous studies [9, 11] and further illustrate that TOC is also a controlling parameter on methane adsorption as organic matter is the main contributor to the specific surface area and the micropore volume for shale samples. Adsorption in micropores is much more favorable than on relatively flat surfaces of meso-macropores due to more intense interactions of the molecule with surrounding pore walls (Fig. 3).

4 Excess Adsorption Models

A suitable model should have its internal hypothesis consistent with the real characteristics of shale. Therefore, an adsorption model considering the complexity and heterogeneity of shale pore structures, which most of adsorption models have ignored, is critically needed. In micropores, gas molecules are adsorbed in the form of pore filling [23], which is caused by the overlapping of the adsorption potential from both sides of the pore wall surfaces [24]. On the other hand, in mesopores, gas molecules are adsorbed as a single layer on the pore wall surfaces [25]. The adsorption mechanisms and capacities depend on the characteristics of pore wall surfaces in both micropores and mesopores. Therefore, pore filling in micropores and

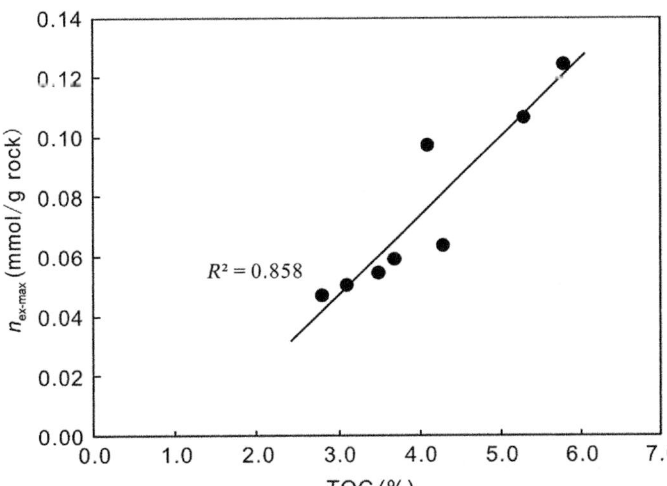

Fig. 3
The correlation between the maximum excess adsorption capacity ($n_{ex\text{-}max}$) and TOC of shale samples. R^2 represents the correlation coefficient of linear fitting.

monolayer adsorption in mesopores might occur simultaneously during the process of methane adsorption.

Based on the micropore filling theory and Polanyi adsorption potential theory, the D-R and D-A equations were developed to model Type I adsorption isotherms [26, 27]. The two equations were considered to provide an appropriate description of the adsorption phenomena occurring in the adsorbent micropores. The D-A adsorption model can be expressed by Eq. (1) and can be further simplified to the D-R equation when $k=2$ [28].

$$n_{ab} = n_0 \exp\left[-D\left[\ln\left(\frac{p_0}{p}\right)\right]^k\right] \tag{7}$$

where n_{ab} is the absolute adsorption capacity, and n_0 is the maximum absolute adsorption capacity of micropore filling. D is a parameter related to the pore structure through $D = (RT/\beta E)^k$, where E is the characteristic energy and β is the coefficient of adsorbate affinity. In Eq. (7), p_0 is the saturation vapor pressure of the adsorbate at temperature T, p is the equilibrium pressure, and k is the structural heterogeneity parameter. Notice that the parameter p_0 in the D-R and D-A equations cannot be physically defined for a gas above its critical temperature T_c [29]. The supercritical D-A absolute adsorption model can be expressed as [19]

$$n_{ab} = n_0 \exp\left[-D \ln\left(\frac{\rho_a}{\rho_g}\right)^k\right] \tag{8}$$

Consequently, the D-A-based excess adsorption model can be rewritten as

$$n_{ex} = n_0 \exp\left[-D \ln\left(\frac{\rho_a}{\rho_g}\right)^k\right]\left(1-\frac{\rho_g}{\rho_a}\right) \tag{9}$$

The monolayer coverage theory was proposed by Langmuir [30], assuming only a single layer of molecules covered solid surfaces during the adsorption process. Moreover, this theory is also based on the assumptions of a homogeneous pore surface and no interaction between neighboring molecules. The Langmuir model is the most widely used model in coal and shale because of its simplicity, effectiveness, and the reasonable explanation of its parameters [30]. The Langmuir model is expressed as

$$n_{ab} = \frac{n_L p}{p + p_L} \tag{10}$$

where n_L is the maximum absolute adsorption capacity of the monolayer adsorption, p is the equilibrium pressure, and p_L is the Langmuir pressure defined as the pressure at which the amount of adsorbed methane molecules equals half of the maximum adsorption capacity.

Considering the heterogeneity of adsorption sites in adsorbents, Sips [31] established the L-F adsorption model extended from the Langmuir equation. The L-F equation is expressed as

$$n_{ab} = \frac{n_L (bp)^m}{1 + (bp)^m} \tag{11}$$

At lower concentrations, it can be simplified to the Freundlich adsorption model. When $m = 1$, the L-F adsorption model is simplified to a Langmuir adsorption model. Similarly, the L-F-based excess adsorption model is

$$n_{ex} = \frac{n_L (bp)^m}{1 + (bp)^m} \left(1 - \frac{\rho_g}{\rho_a}\right) \tag{12}$$

Based on the above analysis, a novel adsorption model combining micropore filling and monolayer coverage theory is established. This model is derived from the integration of the D-A and L-F adsorption models and is written as

$$n_{ex} = \left\{ n_0 \exp\left[-D\left[\ln\left(\frac{\rho_a}{\rho_g}\right)\right]^k\right] + \frac{n_L (bP)^m}{1 + (bP)^m} \right\} \left(1 - \frac{\rho_g}{\rho_a}\right) \tag{13}$$

The unknown parameters (n_0, n_L, D, b, k, m, and ρ_a) can be obtained from experimental adsorption data via the least-squares fitting analysis by Eq. (13), and the fitting error can be evaluated by the RSS (residual sum of squares). To minimize the RSS error, the seven independent fitting parameters must obey the following limits: $0 < n_0 < 0.2$ mmol/g, $0 < D < 1$, $k > 1$, $0 < n_L < 0.2$ mmol/g, $b > 0$, $m > 0$. Once these unknown parameters are determined, the adsorption capacities can then be obtained for micropores and mesopores, respectively.

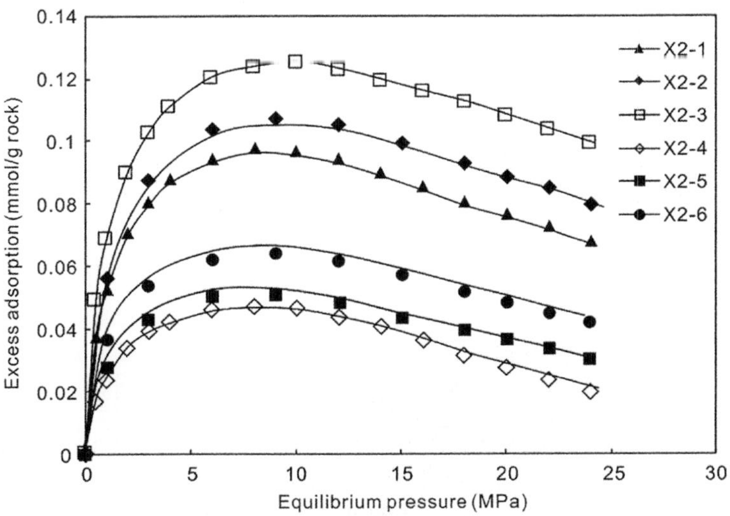

Fig. 4
The fitting results of the excess adsorption isotherms using the proposed D-A–L-F adsorption model.

Fig. 4 shows the fitting results using the new D-A–L-F model, and perfect fittings were observed. The value of n_0 ranged from 0.054 to 0.128 mmol/g, which is much larger than the n_L value that ranged from 0.008 to 0.043 mmol/g. This indicates that the maximum methane adsorption capacity in micropores is larger than that in mesopores. The most important criterion to choose a suitable model by is that it must correctly describe the internal mechanism of supercritical methane adsorption in shale. As mentioned above, the D-A model was established based on the micropore filling theory, describing the adsorption process in micropores (<2 nm); whereas, the L-F model was based on the monolayer coverage theory, describing the adsorption process in mesopores (2–50 nm). However, pores in shale have a broad size distribution, ranging from nanometers to micrometers. Therefore, in order to accurately simulate the adsorption isotherms for the broad variety of pore sizes, neither the D-A model nor the L-F model works independently according to their assumptions. The proposed D-A–L-F adsorption model combines these two theories, and it covers the complete adsorption process in micropores and mesopores. Therefore, the new model is more suitable for shale.

5 High-Pressure Methane Adsorption Mechanism

The volume of the adsorbed phase in shale can be calculated after the density of adsorbed phase and the absolute adsorption amount are determined (i.e., Eq. 3). The differences between the volume of micropores (V_{micro}), adsorbed methane (V_a), and meso-macropores (V_{BJH}) are plotted in Fig. 5. V_a was greater than the V_{micro} and smaller than the V_{BJH} for all samples (i.e., $V_{micro} < V_a < V_{BJH}$). The methane adsorption behavior in shale arises primarily from the interaction between methane molecules and pores walls, and the energy sites of micropores are more than that of meso-macropores [23], so the methane molecules are preferentially

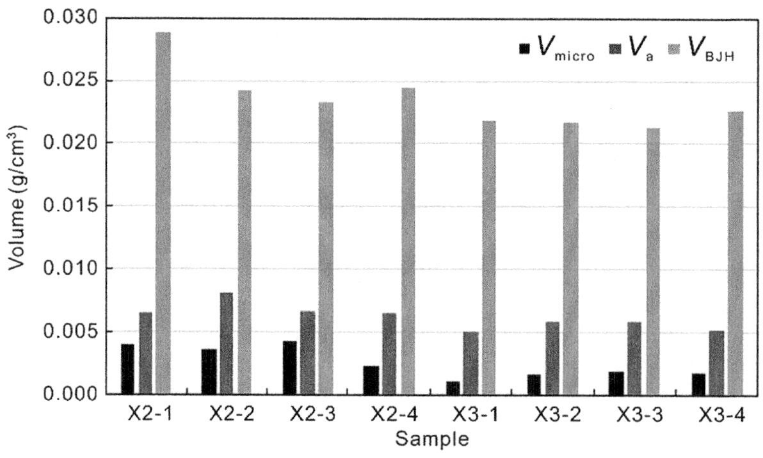

Fig. 5
The comparisons of the volume of micropores (V_{micro}), adsorbed methane (V_a), and meso-macropores (V_{BJH}) for different samples.

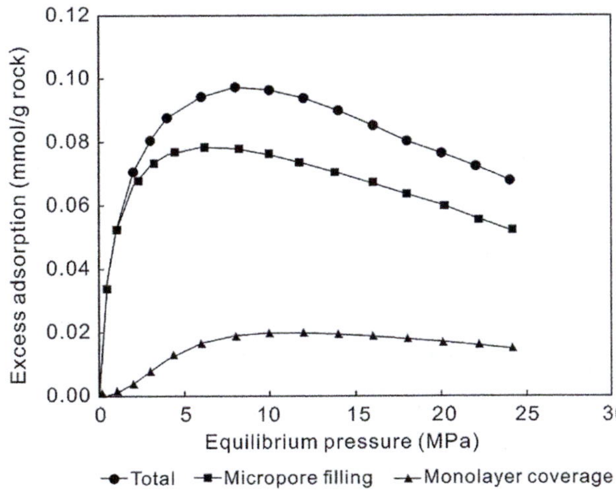

Fig. 6
The calculated absolute adsorption isotherms of methane adsorbed in micropores and mesopores through micropore filling and monolayer coverage mechanism, respectively.

adsorbed in micropores and thereafter adsorbed in meso-macropores. Therefore, the adsorbed methane is not only stored in the micropores (<2 nm) but also stored in the meso-macropores (2–200 nm) as $V_{micro} < V_a < V_{BJH}$.

The total adsorption capacity is the sum of adsorptions in micropores and mesopores that are described by the first and second terms in the proposed model (Eq. 13). Fitting the total adsorption isotherms using this model, the individual adsorption isotherms for micropores and mesopores can be obtained. Fig. 6 shows the methane adsorption isotherms in micropores and mesopores. Similar to the total adsorption capacity, the adsorption capacity in micropores also increases first and then decreases with pressure. Although both the surface area and pore volume of mesopores are larger than those of micropores, the adsorbed capacity in micropores is much larger than that in mesopores. This indicates that the methane molecules are more likely to fill micropores than adsorb onto relatively flat surfaces in mesopores because the attractive interactions between methane molecules and micropore walls are much stronger.

The proportions of methane adsorbed in micropores and mesopores can also be calculated. Fig. 7 shows that the methane adsorption in micropores accounts for 77%–97% of the total adsorption capacity, with an average of 86%, which is far greater than that in the mesopores. The primary force related to gas adsorption includes the interactions between gas molecules and pore surfaces, the interactions between pore surfaces in extremely confined pores, and the interactions between gas molecules themselves either in a free gas state or an adsorbed state [23]. If the size of a micropore is comparative with an adsorbed gas molecule, the forces from

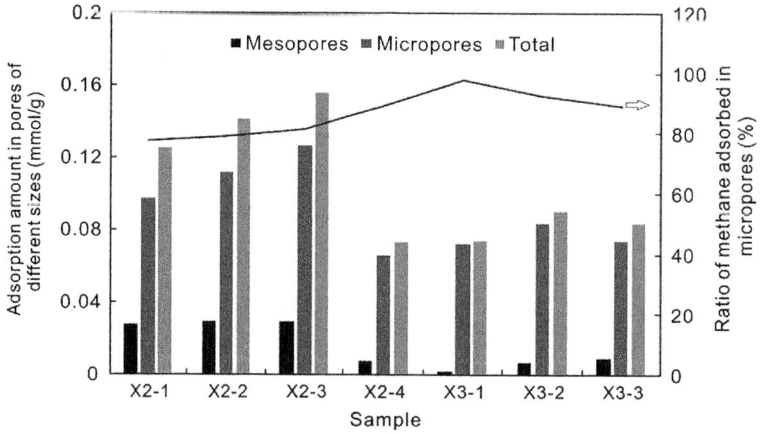

Fig. 7
Comparison between methane adsorption capacity in micropores and mesopores.

the opposing walls will be more significant as well as forces from the neighboring molecules adsorbed to the pore walls. Therefore, the affinity of micropore surfaces is much stronger than that of mesopores surfaces, leading to the result that methane molecules are primarily adsorbed in micropores.

In this study, the different adsorption mechanisms of methane for different-sized nanopores have been considered, and a new model is presented to characterize the whole adsorption process. Based on the above discussions, the methane adsorption mechanism in shale is as follows: the majority of methane molecules fill the micropores, and the remainder are monolayer-adsorbed in mesopores.

6 Conclusions

In this chapter, high-pressure methane adsorption isotherms were measured in shale samples collected from the Longmaxi formation in the northeast Chongqing area near Sichuan Basin in China. Supercritical methane adsorption characteristics and their controlling factors were analyzed. A novel D-A–L-F model was established that can fit the high-pressure isotherms quite well by combining the micropore filling and monolayer coverage theories. Based on this new model, the adsorption isotherms and capacity in micropores and mesopores can be calculated separately. Considering different adsorption mechanisms for different-sized pores, the methane adsorption mechanism in shale is summarized. This chapter helps us to have a deeper understanding of the adsorption characteristics and mechanism of shale gas under formation condition.

Acknowledgments

This work was supported by the National Science and Technology Major Project (2017ZX05035002-002) and the National Natural Science Foundation of China (41722403; 41572116).

References

[1] R.J. Ambrose, R.C. Hartman, M. Diaz-Campos, et al., New pore-scale considerations for shale gas in place calculations, SPE-131772, in: SPE Unconventional Gas Conference, Pittsburgh, Pennsylvania, USA, 23–25 February, 2010.

[2] C. Zou, D. Dong, Y. Wang, et al., Shale gas in China: characteristics, challenges and prospects (II), Pet. Explor. Dev. 42 (2015) 753–767.

[3] S. Zhou, G. Yan, H. Xue, et al., 2D and 3D nanopore characterization of gas shale in Longmaxi formation based on FIB-SEM, Mar. Pet. Geol. 73 (2016) 174–180.

[4] R.J. Ambrose, R.C. Hartman, M. Diaz-Campos, et al., Shale gas-in-place calculations part I: New pore-scale considerations, SPE J. 17 (2012) 219–229.

[5] P. Weniger, W. Kalkreuth, A. Busch, et al., High-pressure methane and carbon dioxide sorption on coal and shale samples from the Paraná Basin, Brazil, Int. J. Coal Geol. 84 (2010) 190–205.

[6] P. Chareonsuppanimit, S.A. Mohammad Jr., R.L. Robinson, et al., High-pressure adsorption of gases on shales: measurements and modelling, Int. J. Coal Geol. 95 (2012) 34–46.

[7] M. Gasparik, P. Bertier, Y. Gensterblum, et al., Geological controls on the methane storage capacity in organic-rich shales, Int. J. Coal Geol. 123 (2014) 34–51.

[8] M. Gasparik, Y. Gensterblum, A. Ghanizadeh, et al., High-pressure/high-temperature methane-sorption measurements on carbonaceous shales by the manometric method: Experimental and data evaluation considerations for improved accuracy, SPE J. 20 (2015) 790–809.

[9] T. Zhang, G.S. Ellis, S.C. Ruppel, et al., Effect of organic-matter type and thermal maturity on methane adsorption in shale-gas systems, Org. Geochem. 47 (2012) 120–131.

[10] T.F. Rexer, M.J. Benham, A.C. Aplin, et al., Methane adsorption on shale under simulated geological temperature and pressure conditions, Energy Fuel 27 (2013) 3099–3109.

[11] W. Ji, Y. Song, Z. Jiang, et al., Estimation of marine shale methane adsorption capacity based on experimental investigations of lower Silurian Longmaxi formation in the upper Yangtze platform, South China, Mar. Pet. Geol. 68 ((2015) 94–106.

[12] H. Bi, Z. Jiang, J. Li, et al., The Ono-Kondo model and an experimental study on supercritical adsorption of shale gas: a case study on Longmaxi shale in southeastern Chongqing, China, J. Nat. Gas Sci. Eng. 35 (2016) 114–121.

[13] Y. Wang, Y. Zhu, S. Liu, et al., Methane adsorption measurements and modeling for organic-rich marine shale samples, Fuel 172 (2016) 301–309.

[14] A. Li, W. Ding, X. Zhou, et al., Investigation of the methane adsorption characteristics of marine shale: A case study of lower Cambrian Qiongzhusi shale in eastern Yunnan Province, South China, Energy Fuel 31 (2017) 2625–2635.

[15] M. Gasparik, A. Ghanizadeh, P. Bertier, et al., High-pressure methane sorption isotherms of black shales from the Netherlands, Energy Fuel 26 (2012) 4995–5004.

[16] H. Tian, T. Li, T. Zhang, et al., Characterization of methane adsorption on overmature lower Silurian–upper Ordovician shales in Sichuan Basin, Southwest China: Experimental results and geological implications, Int. J. Coal Geol. 156 (2016) 36–49.

[17] P. Chareonsuppanimit, S.A. Mohammad Jr., R.L. Robinson, et al., High-pressure adsorption of gases on shales: Measurements and modeling, Int. J. Coal Geol. 95 (2012) 34–46.

[18] R. Sakurovs, S. Day, S. Weir, et al., Application of a modified Dubinin-Radushkevich equation to adsorption of gases by coals under supercritical conditions, Energy Fuel 21 (2007) 992–997.

[19] L. Zhou, S. Bai, W. Su, et al., Comparative study of the excess versus absolute adsorption of CO_2 on superactivated carbon for the near-critical region, Langmuir 19 (2003) 2683–2690.

[20] F. Dreisbach, H.W. Lösch, P. Harting, Highest pressure adsorption equilibria data: Measurement with magnetic suspension balance and analysis with a new adsorbent/adsorbate-volume, Adsorption 8 (2002) 95–109.
[21] T.F. Rexer, M.J. Benham, A.C. Aplin, et al., Methane adsorption on shale under simulated geological temperature and pressure conditions, Energy Fuel 27 (2013) 3099–3109.
[22] M.M. Dubinin, Adsorption in micropores, J. Colloid Interface Sci. 23 (1967) 487–499.
[23] K. Mosher, J. He, Y. Liu, et al., Molecular simulation of methane adsorption in micro- and mesoporous carbons with applications to coal and gas shale systems, Int. J. Coal Geol. 109 (2013) 36–44.
[24] D.D. Do, H.D. Do, Adsorption of supercritical fluids in non-porous and porous carbons: Analysis of adsorbed phase volume and density, Carbon 41 (2003) 1777–1791.
[25] K.Y. Foo, B.H. Hameed, Insights into the modeling of adsorption isotherm systems, Chem. Eng. J. 56 (2010) 2–10.
[26] R. Pini, S. Ottiger, L. Burlini, et al., Sorption of carbon dioxide, methane and nitrogen in dry coals at high pressure and moderate temperature, Int. J. Greenhouse Gas Control 4 (2010) 90–101.
[27] M.M. Dubinin, V.A. Astakhov, Development of the concepts of volume filling of micropores in the adsorption of gases and vapors by micro-porous adsorbents, Bull. Acad. Sci. USSR 20 (1971) 3–7.
[28] C.R. Clarkson, R.M. Bustin, J.H. Levy, Application of the monolayer/multilayer and adsorption potential theories to coal methane adsorption isotherms at elevated temperature and pressure, Carbon 35 (1997) 1689–1705.
[29] I. Langmuir, The adsorption of gases on plane surfaces of glass, mica and platinum, J. Am. Chem. Soc. 40 (1918) 1361–1403.
[30] X. Tang, N. Ripepi, K. Luxbacher, et al., Adsorption models for methane in shales: review, comparison, and application, Energy Fuel 31 (2017) 10787–10801.
[31] R. Sips, On the structure of a catalyst surface, J. Chem. Phys. 16 (1948) 490–495.

CHAPTER 13

Coal Permeability Modeling Considering Nonconstant Vertical Stress Condition

Yuling Tan[*,†,‡], Zhejun Pan[‡], Guanglei Cui[§]

[*]*State Key Laboratory of Geomechanics and Geotechnical Engineering, Institute of Rock and Soil Mechanics, Chinese Academy of Sciences, Wuhan, P.R. China,* [†]*University of Chinese Academy of Sciences, Beijing, P.R. China,* [‡]*CSIRO Energy Business Unit, Clayton South, VIC, Australia,* [§]*Key Laboratory of Ministry of Education on Safe Mining of Deep Metal Mines, Northeastern University, Shenyang, P.R. China*

Chapter Outline

1 *Introduction*

Coal permeability is considered one of the most important reservoir properties for gas production [1]. Compared to the permeability of conventional gas reservoirs, coal permeability has unique characteristics as it is more sensitive to stress change [2] and it is affected by the coal matrix shrinking/swelling with gas desorption/adsorption [3–5]. Permeability change can be very large during coalbed methane (CBM) production and there is evidence from the San Juan Basin in the USA that it can change up to 100 times [6]. Therefore, understanding the permeability change behavior is critical for understanding the CBM production behavior from coal reservoirs.

Analytical models that can describe the permeability increase during reservoir depletion are accessible, easy to use, and practical [6]. Over the past 30 years, there have been a great number

Petrophysical Characterization and Fluids Transport in Unconventional Reservoirs. https://doi.org/10.1016/B978-0-12-816698-7.00013-9

of coal permeability models proposed that consider both the stress sensitivity and sorption-induced coal strain change. Two types of models are popular options: the strain-based models such as the Palmer and Mansoori (P&M) model [7] and stress-based models such as the Shi and Durucan (S&D) model [1]. They are widely used in CBM reservoir simulators, including commercial CBM simulation packages. There are typically two common assumptions used in these models: constant vertical stress and the uniaxial strain [6]. Because of their simplicity and also their ability to describe the permeability change behavior in the field, the P&M and S&D models are widely used. Other permeability models and their assumptions have been reviewed by Palmer [6], who reviewed the P&M model, S&D model, ARI model [8], Cui and Bustin (C&B) model [9], and Gu and Chalaturnyk (G&C) model [10–12]; and by Pan and Connell [13], who reviewed all analytical permeability models up to 2012.

However, the two assumptions mentioned above (the constant vertical stress and uniaxial strain conditions) need to be examined. Coupled geomechanical and reservoir simulations can provide information on how vertical stress and horizontal strains change during gas production or injection in the coal reservoir. Connell [14] and Connell and Detournay [15] investigated these two assumptions by coupling the coal seam gas reservoir simulator, SIMED II, with the geomechanical simulator, FLAC^{3D}, to study the permeability change. Their simulations, which considered a range of property values relevant to the San Juan Basin, showed that the assumption of uniaxial strain introduced negligible discrepancy, while the assumption of constant vertical stress led to significant differences between the S&D model and that calculated from the SIMED-FLAC^{3D} simulation, especially early in the production cycle [14]. It was found that the pressure and sorption strain changes induced by production led to an arch of stress in the vicinity of the production well, caused the differences between the coupled simulation and S&D model, and produced significant differences in the calculated flow rate, particularly at an early time during production [14].

While coupled geomechanical and reservoir simulations are often difficult to perform, analytical permeability models are still preferred. Nevertheless, no effort has been paid to revisit the constant vertical stress in the analytical permeability models, yet this is important in calculating the permeability behavior. In this chapter, we examine the constant vertical stress assumption and then develop an analytical model that considers nonconstant vertical stress on the coal reservoir during depletion. The impact of reservoir properties on the permeability change is studied and compared with the S&D model. The impact on gas production is also studied via a series of reservoir simulation cases using SIMED II.

2 Model Development

Considering one layer of coal seam and one layer of overlaying roof rock, typically mudstone or shale, Fig. 1 shows a conceptual model of the stress and strain change during depletion. The roof rock has different mechanical properties to the coal layer. During gas production,

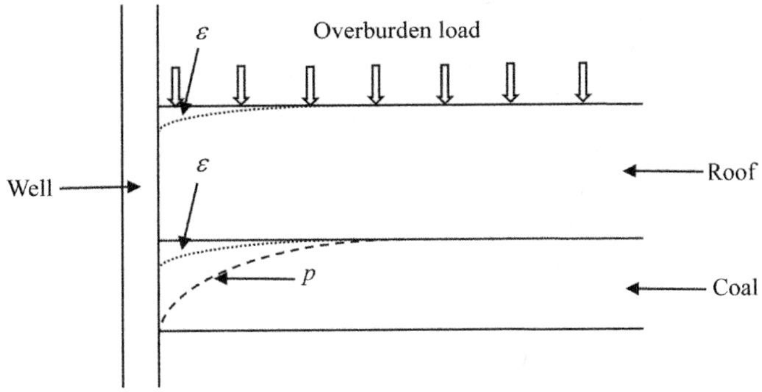

Fig. 1
Strain-stress conceptual plot considering roof impact on coal.

pressure in the coal reservoir drops, leading to the effective stress change in the coal reservoir. Moreover, gas desorption will also cause the coal matrix to shrink. The strain change in the coal reservoir will lead to the uneven distribution of strain change in the roof rock with more strain change close to the wellbore. This strain change of the roof rock will cause the stress change in the roof rock and offset part of the overburden load. Therefore, the vertical stress applied on the coal reservoir does not equal the overburden load and changes with respect to the gas pressure in the coal reservoir.

Using the stress and strain relationship considering the adsorption strain in the S&D model [1]:

$$\sigma_x = (\lambda + 2G)\varepsilon_x + \lambda\varepsilon_y + \lambda\varepsilon_z + \left(\lambda + \frac{2}{3}G\right)\varepsilon_s \tag{1}$$

$$\sigma_y = \lambda\varepsilon_x + (\lambda + 2G)\varepsilon_y + \lambda\varepsilon_z + \left(\lambda + \frac{2}{3}G\right)\varepsilon_s \tag{2}$$

$$\sigma_z = \lambda\varepsilon_x + \lambda\varepsilon_y + (\lambda + 2G)\varepsilon_z + \left(\lambda + \frac{2}{3}G\right)\varepsilon_s \tag{3}$$

where G and λ are the Lame constants of the porous rock, σ is stress, ε is strain, subscripts x, y, and z refer to the three primary directions, and ε_s is the adsorption-induced swelling strain. Note that the stress and strain are incremental changes from their original state. While still consider uniaxial strain as this assumption is a good approximation [14]:

$$\varepsilon_x = \varepsilon_y = 0 \tag{4}$$

Thus Eqs. (1)–(3) can be simplified to Eqs. (5)–(7), respectively:

$$\sigma_x = \lambda\varepsilon_z + \left(\lambda + \frac{2}{3}G\right)\varepsilon_s \tag{5}$$

$$\sigma_y = \lambda\varepsilon_z + \left(\lambda + \frac{2}{3}G\right)\varepsilon_s \tag{6}$$

$$\sigma_z = (\lambda + 2G)\varepsilon_z + \left(\lambda + \frac{2}{3}G\right)\varepsilon_s \tag{7}$$

As the deformation of the coal will also cause the roof rock to deform and considering that the roof and coal are still in contact as the deformation is small, the change of vertical stress applied on the coal can be described by

$$\Delta\sigma_z = \Delta p_{load} - \frac{h_{coal}}{h_{roof}}\Delta\varepsilon_z E_{roof} - \Delta p_{free} \tag{8}$$

where p_{load} is the overburden load, h_{coal} is the thickness of the coal layer, h_{roof} is the thickness of the roof rock, $\Delta\varepsilon_z$ is the vertical strain in the coal reservoir, E_{roof} is the Young's modulus of the roof rock, and p_{free} is the pore pressure in the coal reservoir. Using the Langmuir like equation to describe the volumetric swelling strain:

$$\varepsilon_s = \frac{\varepsilon_V p_{ads}}{P_\varepsilon + p_{ads}} \tag{9}$$

where ε_V is the maximum volumetric swelling strain, P_ε is the characteristic swelling pressure in the Langmuir like swelling strain equation, and p_{ads} is the adsorption pressure in the coal matrix.

Combining Eqs. (7)–(9), it yields

$$\Delta\sigma_z = \Delta p_{load} - \frac{h_{coal}}{h_{roof}}\Delta\varepsilon_z E_{roof} - \Delta p_{free} = \frac{E_{coal}(1-\nu_{coal})}{(1+\nu_{coal})(1-2\nu_{coal})}\Delta\varepsilon_z + \frac{E_{coal}}{3(1-2\nu_{coal})}\Delta\frac{\varepsilon_V p_{ads}}{P_\varepsilon + p_{ads}} \tag{10}$$

Solve $\Delta\varepsilon_z$ from the Eq. (10) to yield

$$\Delta\varepsilon_z = \frac{\dfrac{-E_{coal}\Delta\dfrac{\varepsilon_V p_{ads}}{P_\varepsilon + p_{ads}}}{3(1-2\nu_{coal})} - \Delta p_{free}}{\dfrac{h_{coal}}{h_{roof}}E_{roof} + \dfrac{E_{coal}(1-\nu_{coal})}{(1+\nu_{coal})(1-2\nu_{coal})}} \tag{11}$$

where ν_{coal} is coal's Poisson's ratio.

Thus the horizontal stress change can be obtained by applying Eq. (11) into Eqs. (5) and (6):

$$\Delta\sigma_x = \Delta\sigma_y = \frac{E_{coal}\nu_{coal}}{(1+\nu_{coal})(1-2\nu_{coal})}\frac{\dfrac{-E_{coal}\Delta\dfrac{\varepsilon_V p_{ads}}{P_\varepsilon + p_{ads}}}{3(1-2\nu_{coal})} - \Delta p_{free}}{\dfrac{h_{coal}}{h_{roof}}E_{roof} + \dfrac{E_{coal}(1-\nu_{coal})}{(1+\nu_{coal})(1-2\nu_{coal})}} + \frac{E_{coal}}{3(1-2\nu_{coal})}\Delta\frac{\varepsilon_V p_{ads}}{P_\varepsilon + p_{ads}} \tag{12}$$

Rearrange Eq. (12) to yield

$$\Delta\sigma_x = \Delta\sigma_y = \frac{\nu_{coal}\left(-\Delta p_{free}\right)}{\frac{h_{coal}E_{roof}}{h_{roof}E_{coal}}(1+\nu_{coal})(1-2\nu_{coal})+(1-\nu_{coal})} + \left(\frac{(1+\nu_{coal})\frac{h_{coal}E_{roof}}{h_{roof}E_{coal}}+1}{(1+\nu_{coal})(1-2\nu_{coal})\frac{h_{coal}E_{roof}}{h_{roof}E_{coal}}+(1-\nu_{coal})}\right)\frac{E_{coal}}{3}\Delta\frac{\varepsilon_V p_{ads}}{P_\varepsilon + p_{ads}} \tag{13}$$

The permeability change can then be described by [1]:

$$k = k_0 e^{-3c_f\Delta\sigma_x} \tag{14}$$

where k is the permeability, k_0 is the original permeability, and c_f is the cleat/fracture compressibility.

3 Model Validation

To validate the model, the data from Mavor and Vaughn [16] were used. These data were also used in the model validation of the P&M model and S&D model. The data were from the well tests of three wells in the Valencia Canyon of San Juan Basin and the absolute permeability was increased by 2.7 to 7.1 times from the original permeability. The modeling parameters used in the S&D model [1] and in this proposed model are summarized in Table 1. The validation results show a good match (Fig. 2). Using the proposed model, we are able to match the field data using the reservoir properties provided in the Mavor and Vaughn [16] paper.

4 The Impact of Roof and Coal Properties on Permeability

To better understand the impact of coal and roof properties on the permeability change behavior, the analyses below show the permeability change with respect to different roof rock and coal reservoir properties. These analyses consider that adsorbed gas pressure is the

Table 1 The parameters for model validation

	Mavor and Vaughn [16]	Shi and Durucan [1]	This Work
ν	0.21	0.35	0.21
E (GPa)	3.6	2.9	3.6
P_ε (MPa)	4.31	4.31	4.31
ε_V	0.01266	0.01266	0.01266
C_f (MPa^{-1})		0.266	0.290
h_{roof} (m)			2.0
E_{roof} (GPa)			12.0

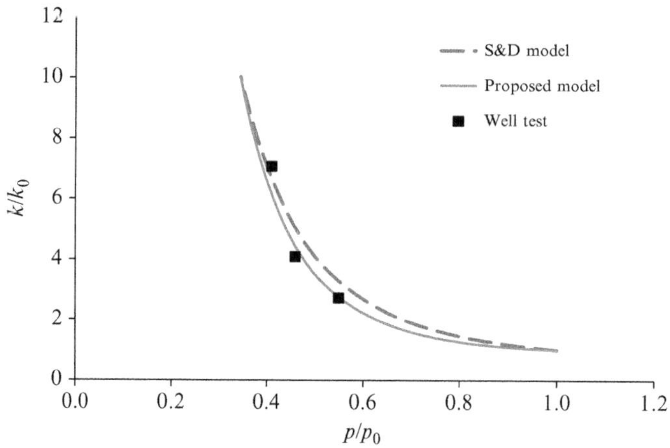

Fig. 2
Permeability model validation result.

same as the free gas pressure to simplify the model results. The results from the S&D model are also compared. The roof rock and coal reservoir properties used for the base case are shown in Table 2.

To study the impact of roof's Young's modulus on permeability, the permeability change behavior when the roof's Young's modulus is 2, 5, 10, and 20 GPa and the permeability change for the S&D model using the base case parameters were compared (Fig. 3). It shows that the roof's Young's modulus has a positive correlation with permeability rebound. Higher Young's modulus for the roof rock means that the deformation of the roof rock creates a larger counterforce against the overburden load, therefore, the vertical stress applied on the coal is reduced, which in turn reduces the horizontal stress. The horizontal stress determines the permeability, therefore, under higher roof's Young's modulus condition, the permeability rebounds more. With the roof rock countering the load, the permeability rebound using the proposed model is higher than in the S&D model, which applies constant vertical stress assumption.

To study the impact of the roof's thickness on permeability, the permeability change behavior when the roof's thickness is 3, 5, 7, and 10 m and the permeability change for the S&D model using the base case parameters are shown in Fig. 4. It shows that the roof's thickness has a negative correlation with permeability rebound. A thicker roof rock means that the strain of the

Table 2 Modeling parameters for the base case

E_{coal} (GPa)	v_{coal}	h_{coal} (m)	E_{roof} (GPa)	v_{roof}	h_{roof} (m)	ε_V	C_f (MPa^{-1})	p_0 (MPa)
2	0.35	3	5	0.2	5	0.01	0.05	5

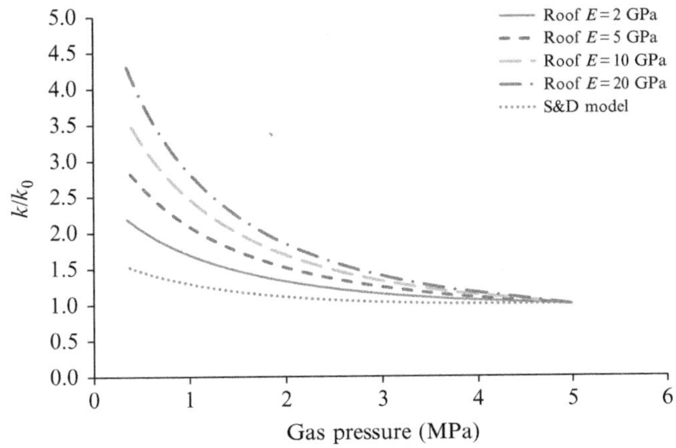

Fig. 3
Impact of roof's Young's modulus (E_{roof}) on permeability change.

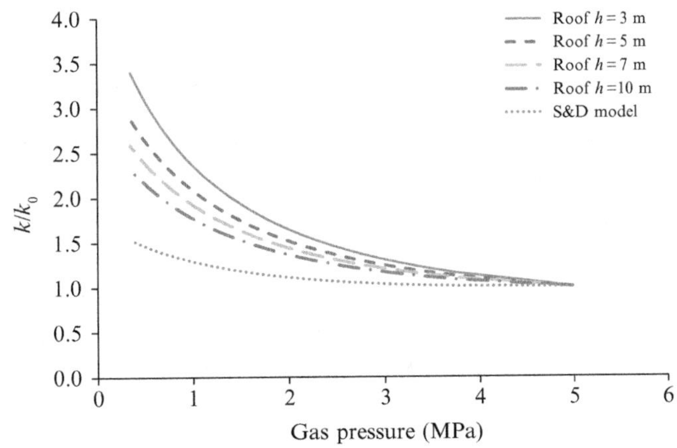

Fig. 4
Impact of roof's thickness (h_{roof}) on permeability change.

roof rock is smaller with the same deformation. Therefore, the counterforce created by the deformation of the roof rock is smaller, causing a higher vertical stress to be applied on the coal. Therefore, with thicker roof rock, the permeability rebounds less. Similar to the analyses above, the results are higher than the S&D model.

To study the impact of coal's Young's modulus on permeability, the permeability change behavior when coal's Young's modulus is 1, 2, 5, and 8 GPa and the permeability change behavior for the S&D model using the same parameters were compared (Fig. 5). It shows that coal's Young's modulus has a positive correlation with permeability rebound. A higher coal's Young's modulus means that the strain change is more sensitive to stress change, leading to more permeability change. Compared with the results using the S&D model, because of the

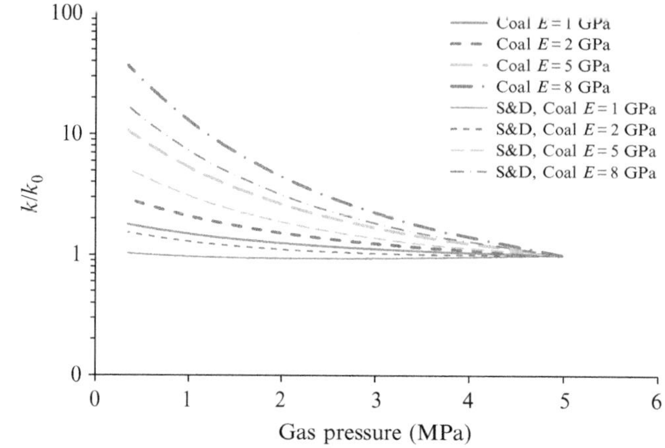

Fig. 5
Impact of coal's Young's modulus (E_{coal}) on permeability change.

counterforce created by the deformation of the roof rock, the vertical stress applied on the coal is smaller using the proposed model, therefore, the permeability is higher than that using the S&D model and, in this case, the permeability is about twice that using the S&D model.

The impact of coal's Poisson's ratio on permeability is demonstrated in Fig. 6, which shows the permeability change behavior when coal's Poisson's ratio is 0.25, 0.3, 0.35, and 0.4 as well as the permeability change for the S&D model using the same parameters. Coal's Poisson's ratio has a positive correlation with permeability rebound for the proposed model, while it has a negative correlation with permeability for the S&D model. Because the vertical stress applied on the coal is nonconstant in the proposed model, the impact from Poisson's ratio is more complex than using the S&D model, as shown in Eq. (13).

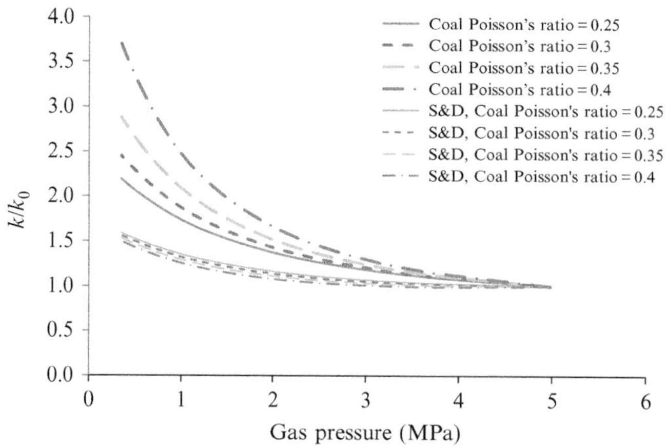

Fig. 6
Impact of coal's Poisson's ratio (v_{coal}) on permeability change.

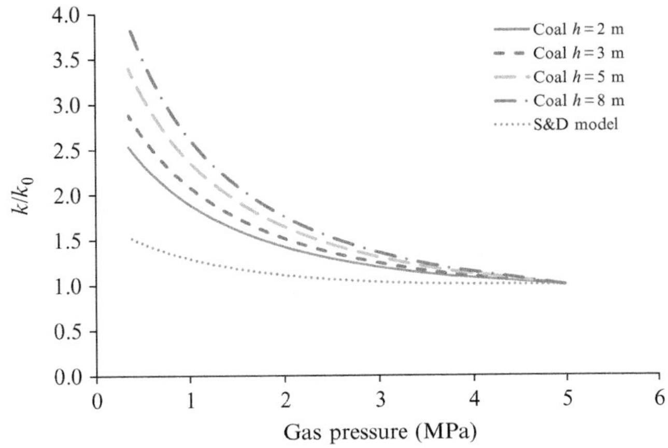

Fig. 7
Impact of coal's thickness (h_{coal}) on permeability change.

The impact of coal's thickness on permeability is demonstrated in Fig. 7, which shows the permeability change behavior when coal's thickness is 2, 3, 5, and 8 m, as well as the permeability for the S&D model using the base case parameters. It shows that coal's thickness has a positive correlation with permeability rebound. Thicker coal means large overall deformation with the same stress change, therefore, the strain change in the roof is larger, resulting in more counterforce against the overburden load, therefore, less vertical stress applied on the coal reservoir, leading to more permeability rebound.

The impact of maximum coal swelling strain on permeability change is illustrated in Fig. 8, which shows the permeability change behavior when maximum coal swelling strain is 0.005, 0.01, 0.015, and 0.02, as well as the permeability change for the S&D model using the same parameters. The maximum swelling strain has a positive correlation with permeability

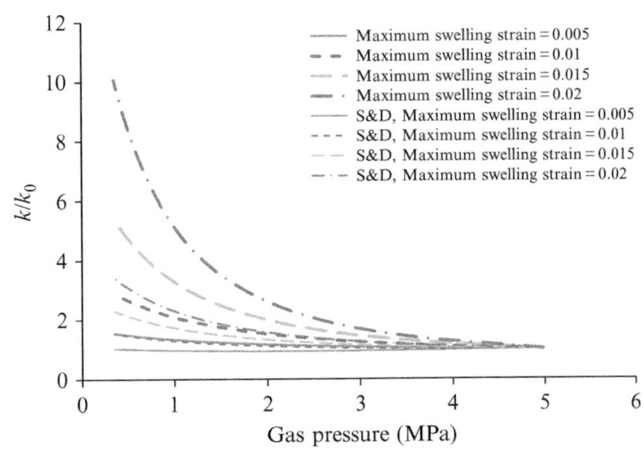

Fig. 8
Impact of maximum swelling strain (ε_V) on permeability change.

rebound. Higher maximum coal swelling strain means stronger shrinkage effect thus larger permeability rebound. Comparing to the result of the S&D model, the permeability calculated by the proposed model can be up to ~2.5 times higher using the parameters considered in this work.

The impact of characteristic swelling pressure on permeability is illustrated in Fig. 9, which shows the permeability change behavior when the characteristic swelling pressure is 1, 2, 3, and 4 MPa, as well as the permeability for the S&D model using the same parameters. It shows that the characteristic swelling pressure has a more profound correlation with permeability rebound. As the characteristic swelling pressure determines the strain change rate with respect to pressure, the strain changes faster when the pressure is less than the characteristic swelling pressure, therefore, the permeability rebounds faster when pressure is below the characteristic swelling pressure. This permeability rebound becomes even more significant if the characteristic swelling pressure is lower. Compared to the results using the S&D model, the permeability calculated by the proposed model can be up to 3 times higher using the parameters considered in this work.

The impact of fracture compressibility on the permeability change is illustrated in Fig. 10, which shows the permeability change behavior when fracture compressibility is 0.01, 0.02, 0.05, and 0.1 MPa^{-1}, as well as the permeability for the S&D model using the same parameters. It shows that the fracture compressibility has a positive correlation with permeability rebound. Larger fracture compressibility means that permeability is more sensitive to stress change. Comparing to the results of the S&D model, the permeability calculated by the proposed model can be up to 3 times higher using the parameters considered in this work.

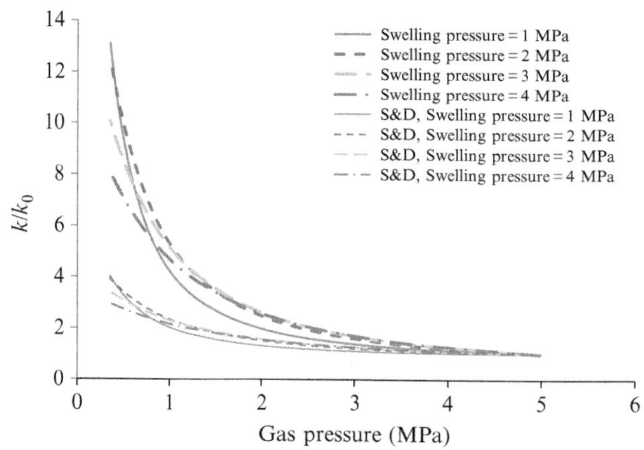

Fig. 9
Impact of characteristic swelling pressure (P_ε) on permeability change.

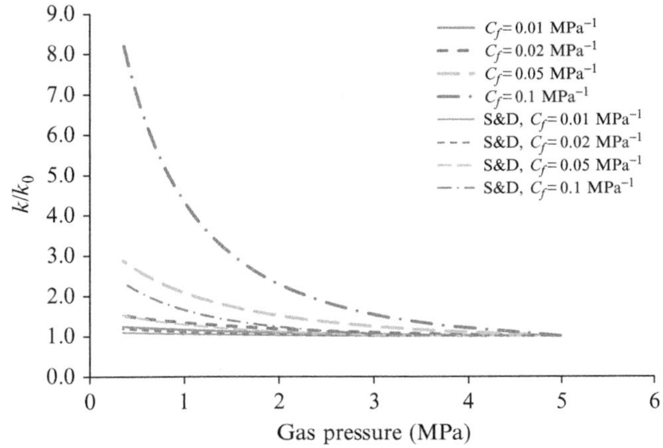

Fig. 10
Impact of fracture compressibility (C_f) on permeability change.

5 Impact of New Permeability Model on Gas Production

The above analyses consider that desorption pressure equals pore pressure, meaning that gas diffusion is not considered. However, during gas production from coal seams, the desorption pressure is higher than the pore pressure, so the desorption pressure and pore pressure should both be considered in the permeability model and the impact of this on gas production should be studied. In this section, the proposed permeability model is applied in SIMED II. SIMED II is a dual porosity, two phase flow and multicomponent coal reservoir simulator that is capable of CBM and enhanced CBM simulations [17]. It has also been extended to simulate shale gas production [18]. Details of the simulator can be found in literature [19].

To study the impact of the roof rock and coal reservoir properties on gas production, a series of base case parameters (Table 3) were used in the simulation. Gas production rate comparison between the proposed model and the S&D model was performed. The simulated area was 200 by 200 m with the production well in the center of the simulated area. The initial reservoir pressure was 5 MPa and the gas desorption pressure was 4.9 MPa, therefore, the dewatering period would be very short and gas would be produced quickly.

The impact of roof's Young's modulus on gas production was studied. The roof's Young's modulus considered was 2, 5, 10, and 20 GPa. Gas production using the S&D model was also compared. It can be seen from Fig. 11 that the impact of using the proposed model with the nonconstant vertical stress assumption had a significant impact on gas production. With higher Young's modulus for the roof rock, the permeability rebounded more, so that the gas production rate was much higher. The gas production rate using the nonconstant vertical stress assumption was higher than that using the S&D model. For the parameters considered in

Table 3 Modeling parameters

Coal's Young's modulus, E_{coal}	2 GPa	Initial reservoir pressure, p_0	5 MPa
Coal's Poisson's ratio, v_{coal}	0.35	Production well BHP, p_{out}	0.1 MPa
Coal's thickness, h_{coal}	3 m	Fracture compressibility, C_f	0.05 MPa^{-1}
Coal's density, ρ_{coal}	1.4 t/m^3	Langmuir volume, V_L	20 m^3/t
Roof's Young's modulus, E_{roof}	5 GPa	Langmuir pressure, P_L	3 MPa
Roof's Poisson's ratio, v_{roof}	0.2	Maximum volumetric strain, ε_V	0.01
Roof's thickness, h_{roof}	5 m	Characteristic swelling pressure, P_ε	3 MPa
Roof's density, ρ_{roof}	2.6 t/m^3	Initial fracture permeability, k_0	1 md
Desorption pressure, p_d	4.9 MPa	Desorption time, τ	10 days
Fracture porosity, ϕ_f	0.01		

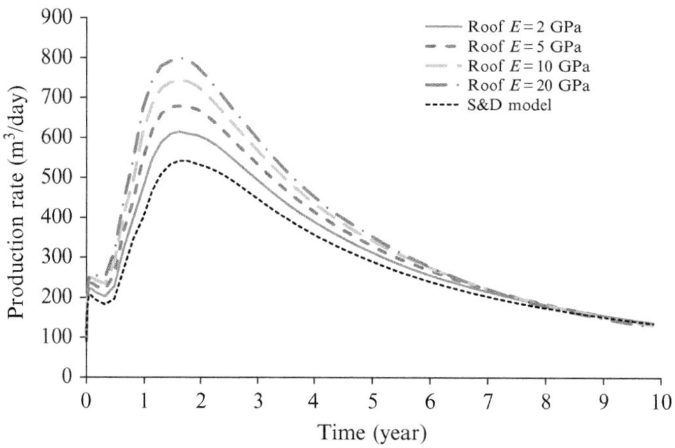

Fig. 11
Impact of roof's Young's modulus on gas production behavior.

this work, the peak gas production rate for roof's Young's modulus of 20 GPa was about 60% higher than that using the S&D model. This agrees well with the literature results using coupled geomechanical and reservoir simulations [14].

The impact of the roof's thickness on gas production is illustrated in Fig. 12. The roof's thickness considered was 3, 5, 7, and 10 m. Gas production using the S&D model and using the base case parameters was compared. Similar to the above case results, using the proposed model with the nonconstant vertical stress assumption had a great impact on gas production. With the

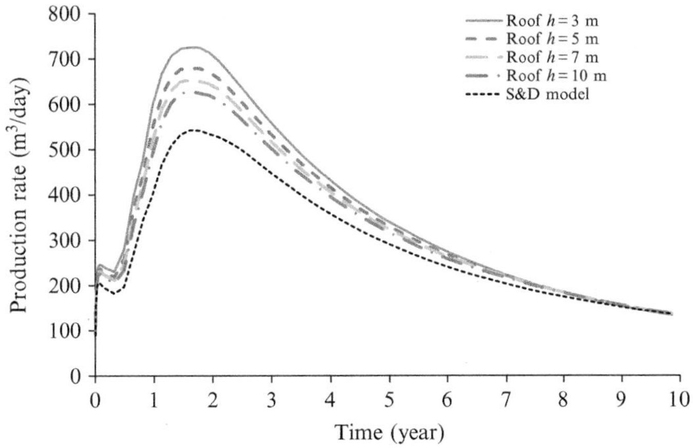

Fig. 12
Impact of roof's thickness on gas production behavior.

thickness of the roof rock increasing, the permeability rebounded less as the roof rock deformed less, so that the gas production rate was lower. However, the gas production rates using the nonconstant vertical stress assumption were all higher than that using the S&D model, as the permeability using the proposed model was higher than that using the S&D model.

The coal's Young's modulus had a significant impact on gas production. The coal's Young's modulus considered was 1, 2, 5, and 10 GPa. Gas production using the S&D model and using the same parameters was compared (Fig. 13). Higher Young's modulus for coal means that the stress change was more sensitive to strain change, leading to much more permeability rebound. Therefore, the peak gas production rate was much higher at higher coal's Young's modulus using both the proposed model and the S&D model. The peak gas

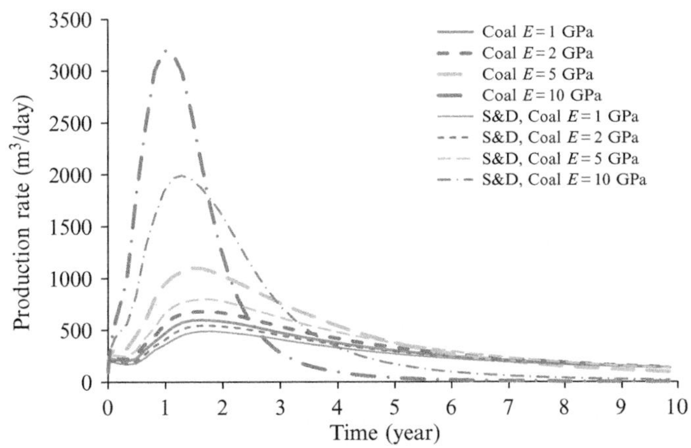

Fig. 13
Impact of coal's Young's modulus on gas production behavior.

production rate using the proposed model was up to 60% more than that using the S&D model in this case, demonstrating the difference between using constant and nonconstant vertical stress assumptions.

Coal's thickness had an important impact on gas production. The gas production rate was proportional to the coal's thickness and thicker coal means more gas resource. The coal's thickness considered was 2, 3, 5, and 8 m. Gas production using the S&D model and using the same parameters was compared (Fig. 14). Comparing the results between the proposed model and the S&D model, the peak production rate was about 40% higher for the proposed model than the S&D model.

Coal's Poisson's ratio had a less significant impact on gas production (Fig. 15). The coal's Poisson's ratio considered was 0.25, 0.3, 0.35, and 0.4. Gas production using the S&D model

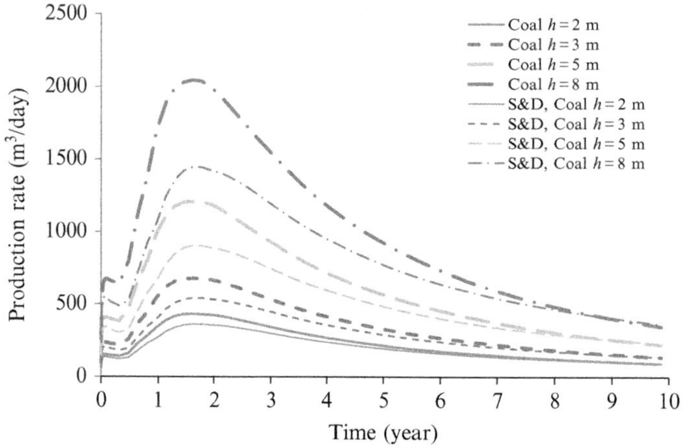

Fig. 14
Impact of coal's thickness on gas production behavior.

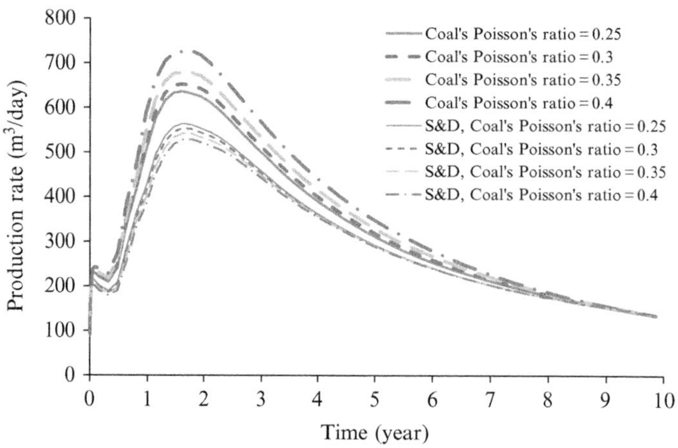

Fig. 15
Impact of coal's Poisson's ratio on gas production behavior.

and the same parameters was compared. Generally, the impact of Poisson's ratio on gas production rate was small and the difference among the peak production rates using different Poisson's ratio were within about 20 m^3/day. As expected, the gas production rate using the proposed model was higher than that using the S&D model, as the permeability predicted using the proposed model was higher. Interestingly, gas production rate increased with respect to Poisson's ratio using the proposed model, however, it decreased with respect to Poisson's ratio using the S&D model. This is because the permeability had a more profound relationship with Poisson's ratio for the proposed model, as can be seen from Eq. (13).

As swelling strain is an important factor influencing the permeability behavior, the maximum swelling strain had a great impact on gas production. The maximum swelling strain considered was 0.005, 0.01, 0.015, and 0.02 and the results were compared between the proposed model and the S&D model (Fig. 16). It can be seen from the figure that the impact of maximum swelling strain on gas production was significant for both the proposed model and the S&D model, demonstrating the impact of swelling/shrinkage on permeability change. Moreover, the impact of maximum swelling strain on the production rate was more significant for the proposed model than the S&D model.

Coal fracture compressibility is a key parameter in permeability and gas production. The fracture compressibility considered was 0.01, 0.02, 0.05, and 0.1 MPa^{-1}. Gas production using the S&D model and using the same parameters was compared (Fig. 17). Fracture compressibility had a positive relationship with gas production, as it magnified the stress change to permeability. Compared to the production behavior using the S&D model, the impact of fracture compressibility on the proposed model was more significant. As can be seen from the figure, when the fracture compressibility was 0.1 MPa^{-1}, the gas rate was much higher than that using the S&D model.

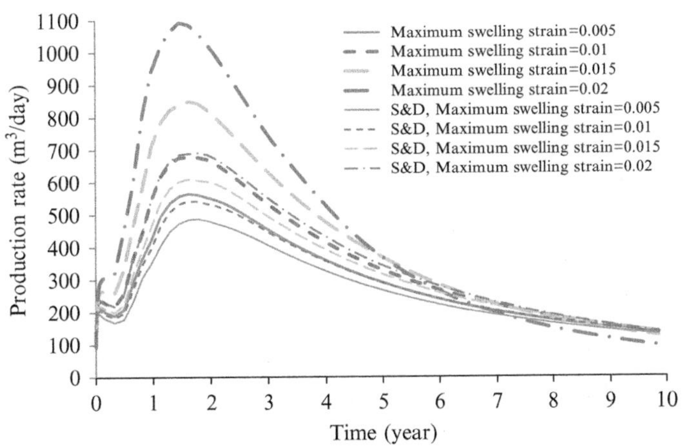

Fig. 16
Impact of coal's maximum swelling strain on gas production behavior.

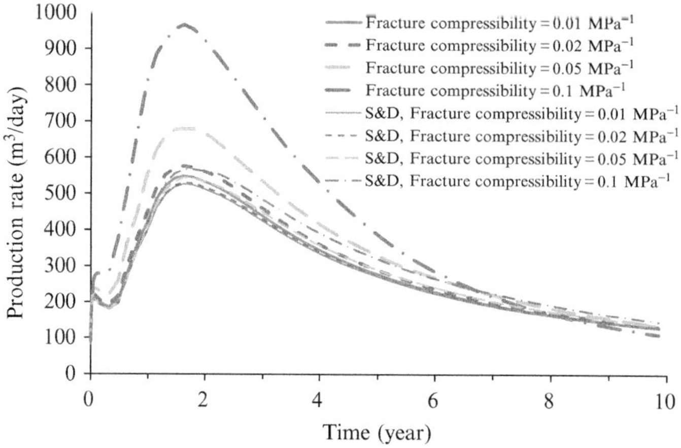

Fig. 17
Impact of coal's fracture compressibility on gas production behavior.

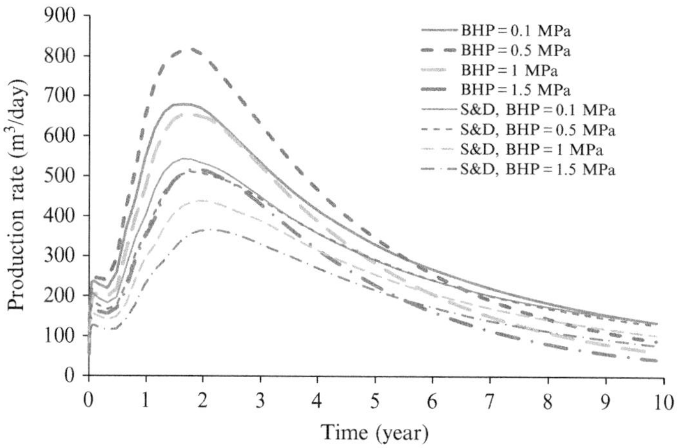

Fig. 18
Impact of production well bottom hole pressure on gas production behavior.

The impact of production well bottom hole pressure (BHP) on gas production is illustrated in Fig. 18. The BHP considered was 0.1, 0.5, 1, and 1.5 MPa. As expected, lower BHP resulted in a higher production rate because of the higher pressure difference between the reservoir and the well, leading to more pressure drawdown in the reservoir and more gas production. Again, due to the higher permeability predicted using the proposed model, the gas production rate using the proposed mode was higher than that using the S&D model at the same operating condition.

6 Model Discussion

The proposed model considers small strain change, which is typically the case in coal reservoirs during gas depletion or gas injection. When there is a large strain in the coal reservoir, this leads to the roof rock and the coal becoming partly separated, then this proposed model is invalid. The constant vertical stress assumption also becomes invalid. However, this has not been observed in the studied coal reservoir. Moreover, in the proposed model, a relatively simplified approach is used to account for the nonconstant vertical stress condition, which assumes that the counterforce from the roof rock is caused by deformation. This does not consider more complex stress conditions in the roof rock during the deformation, such as shear stress. This will require further investigation using fully coupled geomechanics. Furthermore, the model considers elastic strain behavior and does not consider roof rock failure, which is unlikely to happen in CBM applications.

The proposed model only considers the impact from roof rock and assumes everything above the roof rock to be overburden. The best way is to consider all the overlaying formations and to integrate their impacts on the stress change applied on coal during CBM production. However, from the analyses above, the thickness of the roof rock has less impact on permeability change and gas production compared with using the S&D model. This implies that only considering the roof rock or a small vertical section of the rocks above the coal reservoir is a good approximation for the permeability model.

7 Conclusions

In this chapter, a new permeability model for coal reservoirs was proposed considering the nonconstant vertical stress condition. The gas production behavior using the new model agrees well with the literature results using coupled geomechanical and reservoir simulations. Permeability using the new model was higher than that using the S&D model, because the deformation of the roof rock created a counterforce to reduce the vertical stress applied on the coal reservoir during gas depletion. It is demonstrated that the vertical stress is affected by the geomechanical properties of the roof rock as well as the coal. Among them, the geomechanical properties of the coal, the swelling behavior, and the fracture compressibility are the most sensitive properties affecting the coal permeability rebound and gas production rate.

References

[1] J.Q. Shi, S. Durucan, Drawdown induced changes in permeability of coalbeds: a new interpretation of the reservoir response to primary recovery, Transp. Porous Media 56 (2004) 1–16.

[2] J.P. Seidle, D.J. Jeansonne, D.J. Erickson, Application of matchstick geometry to stress dependent permeability in coals, in: SPE 24361, SPE Rocky Mountain Regional Meeting, Casper, Wyoming, USA, 18–21 May, 1992.

[3] J.R. Levine, Model study of the influence of matrix shrinkage on absolute permeability of coal bed reservoirs, Geol. Soc. Lond. Publ. (1996) 197–212.
[4] Z. Pan, L.D. Connell, A theoretical model for gas adsorption induced coal swelling, Int. J. Coal Geol. 69 (2007) 243–252.
[5] S. Day, R. Fry, R. Sakurovs, Swelling of Australian coals in supercritical CO_2, Int. J. Coal Geol. 74 (2008) 41–52.
[6] I.D. Palmer, Permeability changes in coal: Analytical modeling, Int. J. Coal Geol. 77 (2009) 119–126.
[7] I.D. Palmer, J. Mansoori, How permeability depends on stress and pore pressure in coalbeds: a new model, in: SPE-36737, SPE Annual Technical Conference and Exhibition, Denver, Colorado, 6-9 October, 1996.
[8] L.J. Pekot, S.R. Reeves, Modeling the effects of matrix shrinkage and differential swelling on coalbed methane recovery and carbon sequestration, in: Proceedings of the International Coalbed Methane Symposium, University of Alabama, Tuscaloosa, 2003.
[9] X. Cui, R.M. Bustin, Volumetric strain associated with methane desorption and its impact on coalbed gas production from deep coal seams, AAPG Bull. 89 (2005) 1181–1202.
[10] F. Gu, R.J. Chalaturnyk, Sensitivity study of coalbed methane production with reservoir and geomechanical coupling simulation, J. Can. Pet. Technol. 44 (2005) 23–32.
[11] F. Gu, R.J. Chalaturnyk, Analysis of coalbed methane production by reservoir and geomechanical coupling simulation, J. Can. Pet. Technol. 44 (2005) 33–42.
[12] F. Gu, R.J. Chalaturnyk, Numerical simulation of stress and strain due to gas sorption/desorption and their effects on in situ permeability of coalbeds, J. Can. Pet. Technol. 45 (2006) 52–62.
[13] Z. Pan, L.D. Connell, Modelling permeability for coal reservoirs: a review of analytical models and testing data, Int. J. Coal Geol. 92 (2012) 1–44.
[14] L.D. Connell, Coupled flow and geomechanical processes during gas production from coal seams, Int. J. Coal Geol. 79 (2009) 18–28.
[15] L.D. Connell, C. Detournay, Coupled flow and geomechanical processes during enhanced coal seam methane recovery through CO_2 sequestration, Int. J. Coal Geol. 77 (2009) 222–233.
[16] M.J. Mavor, J.E. Vaughn, Increasing absolute permeability in the San Juan basin Fruitland formation, in: Proceedings of the Coalbed Methane Symposium, University of Alabama, Tuscaloosa, Alabama, 1997.
[17] Z. Pan, L.D. Connell, Comparison of adsorption models in reservoir simulation of enhanced coalbed methane recovery and CO_2 sequestration in coal, Int. J. Greenhouse Gas Control 3 (2009) 77–89.
[18] Z. Pan, L.D. Connell, Reservoir simulation of free and adsorbed gas production from shale, J. Nat. Gas. Sci. Eng. 22 (2015) 359–370.
[19] M.D. Stevenson, Multicomponent Gas Adsorption on Coal at In-Situ Conditions (Ph.D. dissertation), University of New South Wales, Sydney, 1997.

CHAPTER 14

Dynamic Gas Flow in Coals and Its Evaluation

Junqian Li*, Shuangfang Lu*, Dameng Liu†, Jianchao Cai‡

**School of Geosciences, China University of Petroleum (East China), Qingdao, P.R. China, †School of Energy Resources, China University of Geosciences, Beijing, P.R. China, ‡Hubei Subsurface Multi-Scale Imaging Key Laboratory, Institute of Geophysics and Geomatics, China University of Geosciences, Wuhan, P.R. China*

Chapter Outline

1 Introduction

Dynamic permeability (i.e., permeability changes), which directly affects the flow rate change of fluids, is one of the most important parameters of strata (such as coal reservoirs and aquifers) during the processes of coalbed methane (CBM) production, CO_2-enhanced methane recovery, and CO_2 sequestration [1–7]. Permeability changes in coal have been observed using field data [8, 9] and laboratory measurements [10, 11]. During CBM production, coal permeability is affected by two opposing effects, including increased effective stress and coal matrix

Petrophysical Characterization and Fluids Transport in Unconventional Reservoirs. https://doi.org/10.1016/B978-0-12-816698-7.00014-0

shrinkage. The former is induced by reservoir pressure drawdown and causes the permeability to exponentially decline. The latter is associated with the gas desorption and leads to coal permeability enhancement. Based on these two effects, coal permeability changes have been evaluated by a large number of analytical models [1, 6, 12–20]. These models have generally considered intrinsic permeability change related to matrix shrinkage and effective stress variation, but they have neglected the influences of gas slippage and Knudsen diffusion, which have been investigated recently [1, 20–25], therefore overestimating the decline in the apparent gas permeability.

Previous research has only considered the impacts from the effective stress and matrix shrinkage on coal permeability. In studies by Li et al. [21, 22], the three effects (effective stress, gas slippage, and matrix shrinkage) on CO_2 permeability were evaluated for different coal ranks. Furthermore, a theoretical model was established to predict CO_2 permeability under a constant confining stress condition [21]. The model proposed in this chapter comprehensively considered four effects (effective stress, coal matrix shrinkage, gas slippage, and Knudsen diffusion) by coupling the permeability change with the pore size evolution. First, the mean pore radius variation of coal reflects the effects of the effective stress and matrix shrinkage. Second, the changes in both the mean pore radius and the gas pressure also reflect the effects of the gas slippage and Knudsen diffusion, respectively. Furthermore, the mean pore radius of coal can be easily obtained by a function of the gas pressure. Thus, the model presented in this chapter will be practical for permeability prediction during pressure depletion.

Different coal ranks coals develop varied maceral compositions and pore-fracture systems, which are closely related to the coal strain induced by matrix shrinkage and effective stress changes, affecting the permeability variation. However, there is little understanding of the correlation between dynamic gas flow (i.e., change characteristics of gas flow during the pressure depletion) and coal ranks. Therefore, the primary aim of this chapter is to exhibit the dynamic process of adsorbing-gas flows, such as gas permeability and flow rate changes, in different coal ranks. Traditionally, the ratio of gas permeability (k_{gi}) to initial gas permeability (k_{g0}) was used to establish the permeability change tendency, and this was efficient to evaluate the coal reservoir with a monotonous permeability change (e.g., monotonous increase or decrease) during CBM production. However, much more complicated gas permeability changes usually occur in coal, such as initial decreases but subsequently increases with varied rebound pressures during pressure depletion [21, 22, 24], which enhance the difficulty in evaluating coal reservoirs.

In this chapter, two parameters (η: change ratio of permeability, and ΔS: total loss of flow rate), reflecting the gas permeability and flow rate changes, were proposed to quantitatively evaluate the dynamic process of gas flow. Compared with the permeability change tendency directly obtained by k_{gi}/k_{g0}, η, and ΔS are more convenient parameters to correlate the coal ranks. Based on the relationships of coal ranks with η and ΔS, the impact of coal ranks on the dynamic gas flow was further observed. These results will be potentially useful for targeting favorable reservoirs for primary and enhanced CBM recoveries.

2 Prediction Model of Permeability Change in Anthracite

Based on the work by Javadpour [26], which presented a theoretical relationship between the mean pore (i.e., capillary tube) radius and the gas permeability, an empirical model was innovatively proposed to predict the absorbing-gas (CO_2) permeability change by correlating the pore size evolution with the permeability variation during the gas pressure depletion under constant confining stress conditions [24]. Simultaneously, the pore size evolution and its impact on the CO_2 permeability variation were investigated under different experimental conditions for anthracite coal.

2.1 Calculation of Key Parameters

2.1.1 Mean pore radius

Under a constant effective stress condition, the slippage factor of an inert gas (such as helium) can be easily obtained by the Klinkenberg equation [21, 22], and is given as:

$$k_g = k_o\left(1 + \frac{b}{P_m}\right) \tag{1}$$

where k_g is the gas permeability, P_m is the mean gas pressure, k_o is the equivalent liquid permeability obtained at an infinite mean gas pressure, b is the gas slippage factor.

However, for an adsorbing gas (such as CO_2), the linear relationship (Eq. 1) cannot be directly used to evaluate the gas slippage resulting from the adsorption between the gas and porous media [27]. In this chapter, due to the CO_2 adsorption to the internal surface of the coal, gas slippage and matrix shrinkage simultaneously occur in the coal core as the gas pressure decreases under a constant effective stress condition. Based on this, the linear relationship between the permeability and the reciprocal of the mean gas pressure was defined as a new Klinkenberg-like equation to describe the CO_2 flow, and is given as:

$$k_g = k_\infty\left(1 + \frac{b_c}{P_m}\right) \tag{2}$$

Then, we can obtain

$$k_g = k_\infty b_c \frac{1}{P_m} + k_\infty \tag{3}$$

In Eqs. (2), (3), k_g is the measured CO_2 permeability, k_∞ is the equivalent liquid permeability (i.e., the gas permeability at an infinite gas pressure where CO_2 has a maximum adsorption quantity), and b_c is defined as a composite factor reflecting the comprehensive effects from the gas slippage and matrix shrinkage.

Another mathematical model considering the effects of gas slippage and Knudsen diffusion [28] was proposed by Javadpour [26] and was successfully used to evaluate the gas permeability of shale and siltstone, within which nanopores develop [26]. At the stage of the slip flow ($0.001 < Kn < 0.1$, Kn is the Knudsen number and is defined as the ratio of the molecular mean free path to the mean pore diameter of porous media [29]), the model proposed by Javadpour [26] can also be used to accurately describe the gas flow process. Thus, the Javadpour-equation was also used to describe the gas flow in the anthracite coal, within which a gas slippage phenomenon exists [21, 22] and is given as:

$$k_g = \frac{2r\mu M}{3\times 10^3 RT\rho^2}\left(\frac{8RT}{\pi M}\right)^{0.5} + \left[1+\left(\frac{8\pi RT}{M}\right)^{0.5}\frac{\mu}{P_m r}\left(\frac{2}{\alpha}-1\right)\right]\frac{r^2}{8\rho} \tag{4}$$

where M is the molar mass, r is the mean pore radius at the mean gas pressure of P_m, R is the gas constant, T is the temperature, μ is the viscosity at the mean gas pressure of P_m and the temperature of T, ρ is the density at the mean gas pressure of P_m and the temperature of T, and α is the tangential momentum accommodation coefficient (TMAC) fraction.

Further, Eq. (4) can be transformed into

$$k_g = \left[\left(\frac{8\pi RT}{M}\right)^{0.5}\frac{r\mu}{8\rho}\left(\frac{2}{\alpha}-1\right)\right]\frac{1}{P_m} + \frac{1}{8\rho}r^2 + \frac{2\mu M}{3\times 10^3 RT\rho^2}\left(\frac{8RT}{\pi M}\right)^{0.5} r \tag{5}$$

According to the proximity of the two models (Eqs. 3, 5) at the stage of gas slip flow, a relationship between the mean pore radius (r) and the equivalent liquid permeability (k_∞) can be established under constant effective stress conditions and is expressed as:

$$\frac{1}{8\rho}r^2 + \frac{2\mu M}{3\times 10^3 RT\rho^2}\left(\frac{8RT}{\pi M}\right)^{0.5} r = k_\infty \tag{6}$$

Then, we can obtain

$$r^2 + \frac{16\mu M}{3\times 10^3 RT\rho}\left(\frac{8RT}{\pi M}\right)^{0.5} r = 8\rho k_\infty \tag{7}$$

Due to the introduction of k_∞ and r, both the effects from the matrix shrinkage and gas slippage on the permeability were thus considered in Eq. (6).

Letting $A = \frac{16\mu M}{3\times 10^3 RT\rho}\left(\frac{8RT}{\pi M}\right)^{0.5}$ and $B = 8\rho k_\infty$, the mean pore radius of coal is obtained as:

$$r = \sqrt{B + \frac{A^2}{4}} - \frac{A}{2} \tag{8}$$

2.1.2 Tangential momentum accommodation coefficient

In Eq. (4), another key parameter is the TMAC, which plays an important role in determining the gas molecule amount of wall slip at the solid surface [30]. It depends on the roughness of the pore surface and gas type, temperature, pressure, among others and generally ranges from 0 to 1 [30, 31], while it is sometimes greater than one, reflecting the retroreflection phenomenon of gas molecules on the pore surface. In addition, some experiments also revealed that the TMAC is primarily dependent on the nature of the solid surface, such as atomic roughness and its adsorptive capacity to gas molecules [30, 32]. A relationship can be used to estimate the TMAC:

$$k_\infty b_c = \left(\frac{8\pi RT}{M}\right)^{0.5} \frac{r\mu}{8\rho}\left(\frac{2}{\alpha} - 1\right) \tag{9}$$

Then, the TMAC can be obtained as:

$$\alpha = \frac{2}{1 + \frac{8k_\infty b_c \rho}{r\mu}\left(\frac{M}{8\pi RT}\right)^{0.5}} \tag{10}$$

2.2 Experiment to Determine Key Parameters

An experiment on the CO_2 permeability change and pore size evolution of coal was undertaken using an anthracite coal sample from the Yong'an mine located in the southeastern Qinshui Basin [21]. The sample was cut perpendicular to the bedding surface into a 2.54 cm (in diameter) cylindrical core with a length of 3.54 cm. The coal had a maximum vitrinite reflectance of 4.20% $R_{o,\ max}$. Core preparation, the experimental setup utilized for the CO_2 permeability test, schematic diagram of the stress and gas pressure applied to the core, and the calculation method for the permeability were detailed in previous studies [21, 22]. The test was carried out at an ambient temperature of 26°C.

In order to investigate the pore size evolution and CO_2 permeability change under different experimental conditions and the coupling relationship, the mean pore size change was first observed during the gas pressure decrease under five effective stress conditions (2.2, 2.5, 3.0, 3.5, and 4.0 MPa) [21]. Under each constant effective stress condition, both the confining stress and gas pressure were simultaneously increased by maintaining a positive effective stress on the core, and each effective stress condition included five confining stress and five corresponding gas pressure conditions. Then, the pore size evolution and permeability change were further investigated at the same gas pressure under decreasing effective stress conditions. Lastly, the measured data selected under constant confining stress conditions with decreasing gas pressure and increasing effective stress was utilized to establish the prediction model of the permeability change by coupling with the pore size evolution.

2.3 Permeability Change and Pore Size Evolution

As shown in Fig. 1, the permeability linearly increases with an increase in the reciprocal of the mean gas pressure under the five effective stress conditions, which directly indicates the appearance of the gas slippage phenomenon. Based on the Klinkenberg-like equation (Eq. 2), the mean pore radius and the corresponding TMAC values were respectively calculated at each gas pressure (or confining stress) under the five effective stress conditions by means of Eqs. (8), (10), as listed in Table 1. The mean pore radius of the coal core varied from 7.839 to 42.946nm under the experimental conditions. The TMAC values ranged from 1.926 to 1.973. This may be associated with the roughness of the pore surface within coals and the affinity of coal for CO_2, but requires further verification.

σ_e is effective stress (MPa); σ_c is confining stress (MPa); P_1 is inlet gas pressure (MPa); P_m is mean gas pressure (MPa), and is equal to $(P_1+P_2)/2$, in which P_2 is atmospheric pressure in our study; k_g is coal permeability to gas (CO_2); r is mean pore radius (nm); α is tangential momentum accommodation coefficient, fraction.

2.3.1 Under constant effective stress conditions

Both the permeability and mean pore radius of the coal core are sensitive to gas pressure. Under constant effective stress conditions, the permeability gradually increases as the gas pressure decreases, particularly at mean gas pressures less than ~0.8 MPa (Fig. 2A). Considering the pore size evolution, the change tendencies of the mean pore radius are characterized by a decrease with decreasing gas pressure and are similar to each other under different effective stress conditions, as shown in Fig. 2B.

CO_2 permeability gradually increases when the gas pressure decreases under constant effective stress conditions (Fig. 2A). This is because the effects of gas slippage and Knudsen diffusion (as Kn is greater than 0.001, the Knudsen diffusion cannot be neglected [33]) play significant roles in the permeability change and are closely associated with the mean pore radius of coal and the gas pressure. Usually, for the same gas type and experimental temperature, both effects on the permeability gradually become more significant as the mean pore radius of porous media and the gas pressure are reduced [26, 34]. Javadpour [26] showed that the gas permeability compared to the Darcy permeability obviously increases due to gas slippage and Knudsen diffusion, especially when the gas pressure decreases from 10 to 0 MPa and the mean pore radius decreases from 100 to 0 nm. During pressure depletion under constant effective stress conditions, the permeability change is also comprehensively induced by the gas slippage and Knudsen diffusion and is noticeable at a mean gas pressure less than ~0.8 MPa. This indicates that these two effects on the permeability are relatively slight at gas pressures greater than ~0.8 MPa.

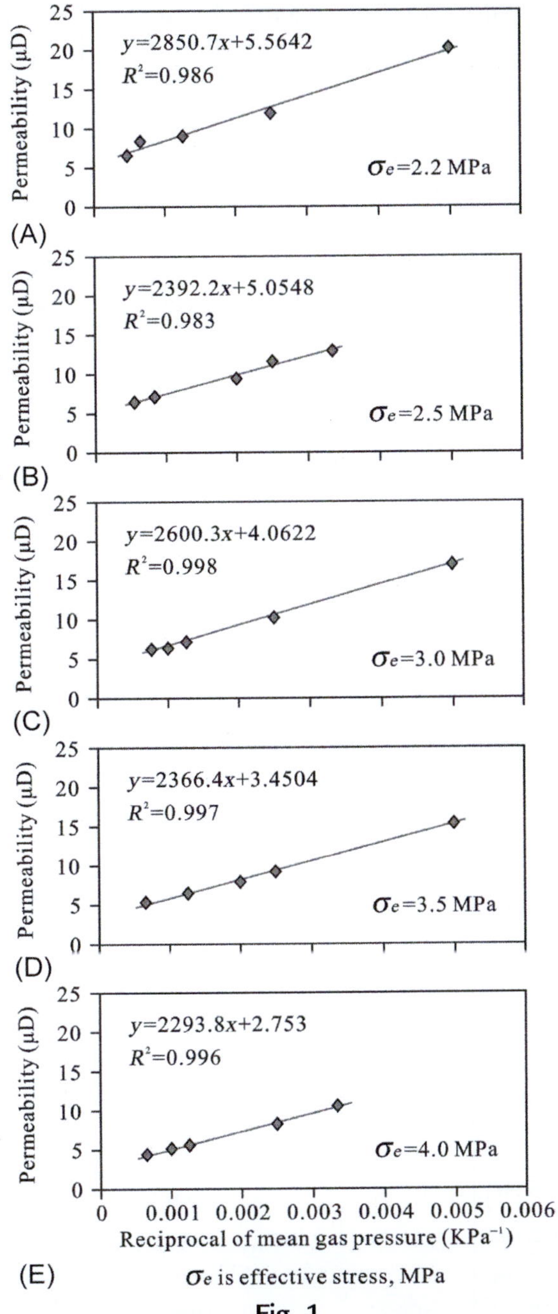

Fig. 1

The relationship between CO_2 permeability and the reciprocal of the mean gas pressure. (A)–(E) Represent the gas slippage at different effective stress conditions (ranging from 2.2 to 4.0 MPa).

Table 1 CO_2 permeability and corresponding mean pore radius for the core [24]

σ_e (Mpa)	P_1 (Mpa)	P_m (Mpa)	σ_c (Mpa)	k_g (μD)	r (nm)	α
2.2	0.3	0.2	2.4	20.129	10.461	1.973
	0.7	0.4	2.6	11.888	16.791	1.966
	1.5	0.8	3.0	8.966	25.053	1.954
	2.9	1.5	3.7	8.319	35.496	1.937
	4.1	2.1	4.3	6.674	42.946	1.926
2.5	0.5	0.3	2.8	12.852	13.270	1.973
	0.7	0.4	2.9	11.583	15.953	1.970
	0.9	0.5	3.0	9.390	18.241	1.967
	2.3	1.2	3.7	7.111	29.872	1.950
	3.5	1.8	4.3	6.373	37.471	1.939
3.0	0.3	0.2	3.2	17.104	8.662	1.970
	0.7	0.4	3.4	10.401	14.188	1.963
	1.5	0.8	3.8	7.314	21.325	1.951
	1.9	1.0	4.0	6.471	24.160	1.945
	2.5	1.3	4.3	6.329	27.959	1.938
3.5	0.3	0.2	3.7	15.391	7.839	1.970
	0.7	0.4	3.9	9.270	12.991	1.963
	0.9	0.5	4.0	7.827	14.917	1.960
	1.5	0.8	4.3	6.467	19.609	1.951
	2.9	1.5	5.0	5.270	27.889	1.934
4.0	0.5	0.3	4.3	10.544	9.432	1.964
	0.7	0.4	4.4	8.264	11.490	1.960
	1.5	0.8	4.8	5.517	17.456	1.947
	1.9	1.0	5.0	5.148	19.810	1.941
	2.9	1.5	5.5	4.339	24.880	1.928

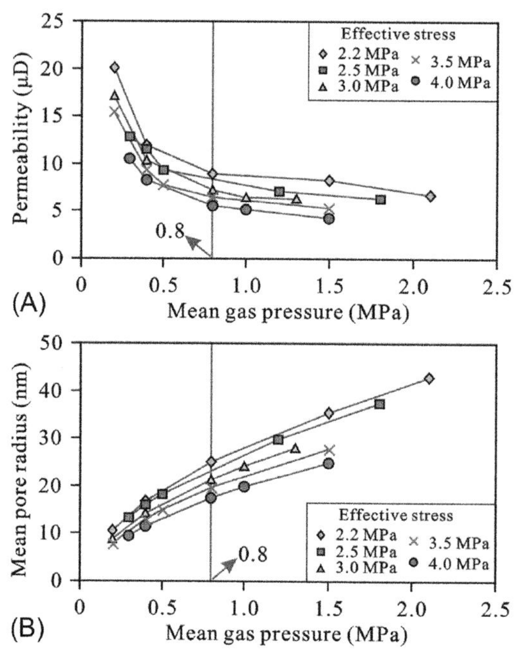

Fig. 2
Relationships of the mean gas pressure with CO_2 permeability (A) and the mean pore radius (B).

The impact of gas pressure on the mean pore radius of coal is primarily induced by the matrix shrinkage and increase in effective stress during the decrease in gas pressure. Thus, under a constant effective stress condition, the change in the mean pore radius is closely related to the matrix shrinkage. It is generally considered that coal matrix shrinkage occurs in coal as the gas desorbs and results in the expansion of the seepage space for fluids [2, 13, 35]. However, as shown in Fig. 2B, the mean pore radius reduces with the decrease in gas pressure under different effective stress conditions. This phenomenon is contrary to the general understanding of the effect of matrix shrinkage on pore fractures of coal. A probable interpretation is that the matrix shrinkage does not impact the pore fractures to increase the permeability under the constant effective stress conditions [24].

2.3.2 Under constant gas pressure conditions

The permeability gradually decreases in an approximately linear manner as the effective stress increases (Fig. 3A), and the mean pore radius of the coal core is also approximately negatively related to the effective stress (Fig. 3B). In addition, the mean pore radius is much greater at a high pressure than at a low pressure for the same effective stress condition.

The natural cleat/fracture network is extremely sensitive to the effective stress and will gradually close up as the effective stress increases, which results in a reduction of coal permeability [21, 36]. With the decrease in the mean pore radius of coal, the impact of gas

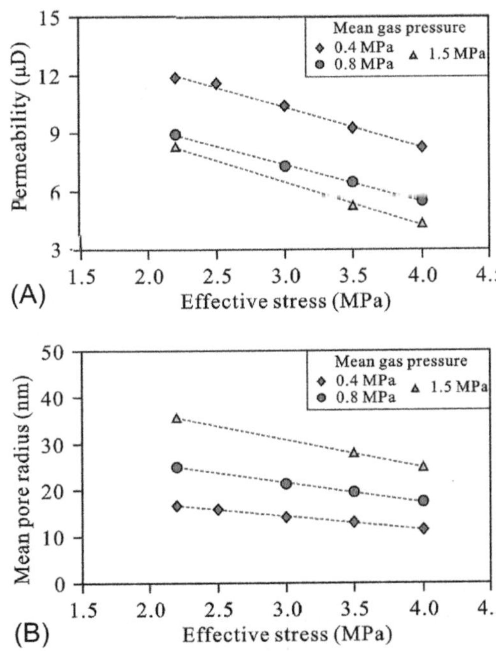

Fig. 3

Relationships of the effective stress with CO_2 permeability (A) and the mean pore radius (B).

slippage and the Knudsen diffusion appears in response [26]. As shown in Fig. 3, the changes in the mean pore radius and permeability suggest that the negative effect of the effective stress is obviously larger than that of gas slippage and Knudsen diffusion, both of which cause an incremental increase in permeability. Due to the potential offset of gas slippage and Knudsen diffusion to the reduction in permeability, it is also observed that the fitted relationship between the effective stress and permeability is linear [21, 22], even though it is still expressed as an exponential function [5, 37].

2.4 Coupling of the Permeability Change With Pore Size Evolution

Under the constant confining stress condition used for assuming the CBM reservoir condition during the CBM production, the mean pore radius of coal also varies and is compressively influenced by both coal matrix shrinkage and effective stress increase associated with a decrease in gas pressure. It follows that the mean pore radius can be correlated with the gas pressure. As presented in Fig. 4A, the mean pore radius was ideally considered to be positively correlated to the gas pressure. This relationship was mathematically described as:

$$r = aP_m + b \tag{11}$$

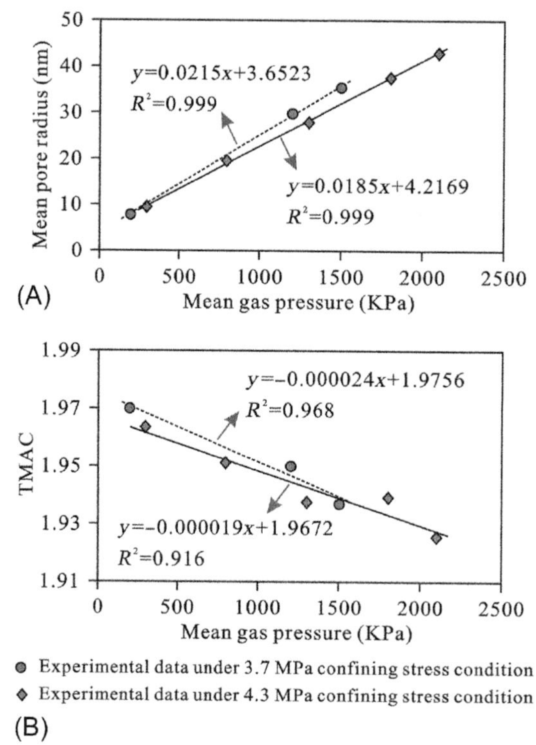

Fig. 4
Relationships of the mean gas pressure with the mean pore radius (A) and the tangential momentum accommodation coefficient (TMAC) (B).

As mentioned above, the TMAC is impacted by multiple factors. For a given gas type, porous media, and experimental temperature, the TMAC is only related to the gas pressure. As presented in Fig. 4B, the TMAC is negatively related to gas pressure and can be estimated by

$$\alpha = cP_m + d \tag{12}$$

In Eqs. (11), (12), a, b, c, and d are $0.0215\,\text{KPa}^{-1}$ ($0.0185\,\text{KPa}^{-1}$), 3.6523 (4.2169), $-2.4\times10^{-5}\,\text{KPa}^{-1}$ ($-1.9\times10^{-5}\,\text{KPa}^{-1}$), and 1.9756 (1.9672) under a 3.7 MPa (4.3 MPa) confining stress condition for the studied coal core, respectively.

Substituting Eqs. (11), (12) into Eq. (4), an equation predicting the CO_2 permeability change in the anthracite coal with a gas pressure decrease under a constant confining stress condition is obtained as:

$$k_g = \frac{2\mu M(aP_m+b)}{3\times10^3 RT\rho^2}\left(\frac{8RT}{\pi M}\right)^{0.5} + \left[1+\left(\frac{8\pi RT}{M}\right)^{0.5}\frac{\mu}{(aP_m+b)P_m}\left(\frac{2}{cP_m+d}-1\right)\right]\frac{(aP_m+b)^2}{8\rho} \tag{13}$$

The CO_2 permeability change calculated by the model (Eq. 13) shows an excellent consistency with measured results (Fig. 5).

According to the mean pore radius change during the gas pressure decrease under 3.7 and 4.3 MPa constant confining stress conditions (Fig. 4A), the impact of the effective stress is much stronger than that of the matrix shrinkage for the coal core. In addition, during gas pressure depletion under these two constant stress conditions, the comprehensively positive effects from matrix shrinkage, gas slippage, and Knudsen diffusion are slightly weaker than the negative effect from the effective stress increase at gas pressures larger than ~0.8 MPa, while the positive effects are obviously stronger than the negative effect at gas pressure less than

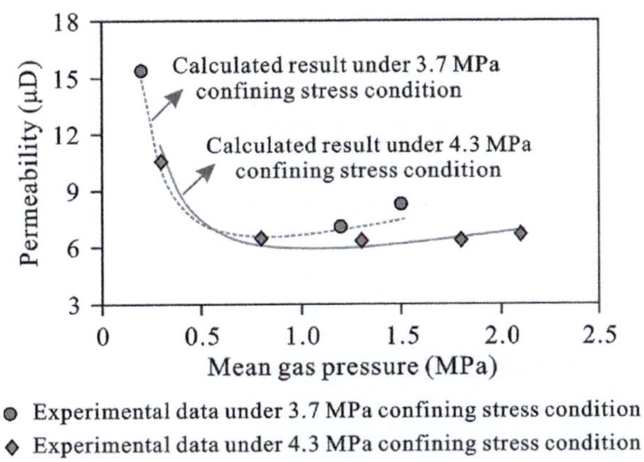

Fig. 5
Calculated results from the proposed permeability model and the experimental data.

~0.8 MPa (Fig. 5). It can also be seen that, the negative effect from effective stress is relatively stronger under 3.7 MPa confining stress condition than that under 4.3 MPa confining stress condition (Fig. 5), because a greater variation in effective stress occurs as the gas pressure decreases under a lower confining stress condition (Fig. 4A).

3 Evaluation on Dynamic Flow Process

3.1 Evaluation Method

In tight porous media, such as coal, the gas flow process can be mathematically expressed as:

$$Q_i = \frac{43.2k_{gi}A}{P_s\mu_{gi}}\frac{\Delta P_i^2}{L} = \frac{43.2k_{gi}A}{P_s\mu_{gi}}\frac{P_i^2 - P_{out}^2}{L} \tag{14}$$

where P_i is the inlet gas pressure at experimental point i (0–7 in this chapter), P_{out} is the outlet gas pressure, P_s is the standard atmospheric pressure, $\Delta P^2 = P_i^2 - P_{out}^2$ is the square difference at inlet gas pressure P_i, Q_i is the gas flow rate at the inlet gas pressure P_i, k_{gi} is the gas permeability at the inlet gas pressure P_i, μ_{gi} is the coefficient of kinetic viscosity of the gas at the mean pressure $((P_i+P_{out})/2)$ and the experimental temperature, L is the core length, m, A is the cross-sectional area of the core.

Eq. (14) was used to calculate the gas permeability based on the flow rate data in this chapter. To quantitatively evaluate the dynamic process of gas flow during the gas pressure decrease, two parameters (i.e., change ratio of permeability and total loss of flow rate) were proposed. The change ratio of permeability was determined using the following expression:

$$\eta = \frac{k_{g0} - k_{gf}}{k_{g0}} \tag{15}$$

where k_{g0} and k_{gf} are the gas permeabilities obtained at the initial (4.0 MPa in this chapter) and final (0.5 MPa) gas pressures, respectively.

Before calculating the total loss of flow rate, normalized gas flow and the square difference of the gas pressure were constructed. First, based on the flow rate at the initial gas pressure, a normalized flow rate during pressure depletion was expressed as follows:

$$\alpha_i = \frac{Q_i}{Q_0} \tag{16}$$

where Q_0 is the flow rate at the initial inlet gas pressure P_0, α_i is the normalized flow rate at the inlet gas pressure P_i, dimensionless.

Then, according to the square difference of gas pressure at the initial gas pressure, a normalized square difference was developed as follows:

$$\beta_i = \frac{\Delta P_i^2}{\Delta P_0^2} \tag{17}$$

where β_i is the normalized square difference at the inlet gas pressure P_i, dimensionless.

Assuming that the gas permeability is constant (i.e., the ratio of k_{gi} to k_{g0} is 1) during pressure depletion, the normalized flow rate (α) will decrease in a linear manner with the decrease in the normalized square difference of gas pressure (β) and vary from one to zero, according to Eq. (14). A deflected route will appear around the line of $k_{gi}/k_{g0}=1$ as the permeability changes with decreasing gas pressure. The area between the deflected route and the line of $k_{gi}/k_{g0}=1$ was defined as the total loss of flow rate, and it can be quantitatively calculated according to the measured data by the following expression:

$$\Delta S = 0.5 - \sum_{i=1}^{n+1} [(\alpha_{i-1} + \alpha_i)(\beta_{i-1} - \beta_i)/2] \tag{18}$$

where α_{i-1} is the normalized flow rate at inlet gas pressures P_{i-1}, dimensionless; β_{i-1} is the normalized square differences at inlet gas pressures P_{i-1}, dimensionless. n depends on the experimental pressure points and is equal to 8 in this chapter (as shown in Fig. 6); 0.5 is the area along the line of $k_{gi}/k_{g0}=1$, x-axis, and vertical line of initial pressure point; the second item on the right side of Eq. (18) represents a cumulative area containing 8 rectangles; α_9 and β_9 are equal to zero (Fig. 6).

Note that the negative (positive) ΔS value reflects the increase (decrease) in the cumulative gas flow rate compared to the condition of $k_{gi}/k_{g0}=1$ during the depressurization process. Overall, ΔS varies from -0.5 to 0.5. During pressure decrease, as $\Delta S=-0.5$, the gas flow rate is always the highest; as $\Delta S=0$, the gas permeability is constant; as $\Delta S=0.5$, the reservoir is impermeable. Thus, coal with a high gas flow rate and low η and ΔS values will be favorable for CBM recovery.

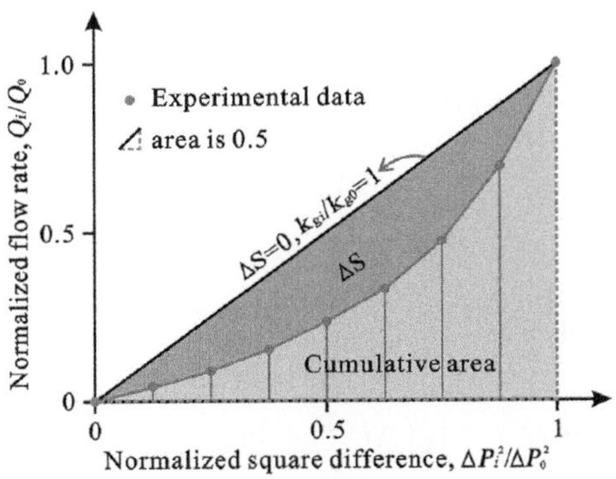

Fig. 6
Schematic diagram explaining the calculation for the total loss of flow rate (ΔS).

3.2 Primary Tests

An experiment was performed using 46 different coal samples with various ranks from 29 active mines located in the southern Qinshui, Houxi and southeastern Ordos Basins and Sanjiang-Mulinghe coal-bearing zone in China [38]. Two primary tests were conducted regarding the CO_2 flow rate (i.e., the CO_2 permeability) tests and coal maceral composition analysis.

Prior to the laboratory experiment, the samples were cut parallel to the bedding surface of the coal into cylindrical cores with diameters of ~2.5 cm in order to simulate the gas flow in underground coal reservoirs. After drying at a temperature of 150°C, the lengths and diameters of the cores were then precisely measured using a micrometer to calculate the permeability. The equipment and procedure used in the flow rate tests were detailed in a previous study [21]. In this chapter, a series of flow rate values at 4.0–0.5 MPa inlet gas pressures (outlet pressure is atmospheric pressure) under a constant 5.0 MPa confining stress condition were measured during a depressurization process with a pressure interval of 0.5 MPa. A total of 8 pressure points (i.e., $P_i = 4.0–0.5i$, MPa, where i is an integer and varied from 0 to 7) were selected for the flow rate (and the permeability) tests in this study. Under a constant confining stress condition, several primary effects associated with matrix shrinkage, effective stress increase, gas slippage, and Knudsen diffusion on the permeability change occurred during the gas pressure depletion, thus affecting the permeability change [24]. Thus, the dynamic process of gas flow reflects the effects of the four factors in this experiment.

After the flow rate tests, maximum vitrinite reflectance and coal maceral composition analyses were performed on all of the cores. The measured results are listed in Table 2.

3.3 Dynamic Behavior of Gas Flow in Different Coal Ranks

Dynamic gas flow (including the varied gas permeability and flow rate) occurs in different coal ranks. This phenomenon can be theoretically verified using Eq. (14).

First, Eq. (14) can be transformed into the following expression:

$$\lg(Q_i) = \lg\left[\frac{43.2k_{gi}A}{P_s\mu_{gi}L}\left(\Delta P_i^{\,2}\right)\right] \tag{19}$$

Furthermore, we can obtain Eq. (20):

$$\lg(Q_i) = \lg\left(\Delta P_i^{\,2}\right) + \lg\left(\frac{43.2k_{gi}A}{P_s\mu_{gi}L}\right) \tag{20}$$

A cross-plot of lg (Q_i) versus lg (ΔP_i^2) for typical coals is presented in this chapter. As shown in Fig. 7, the fitted straight lines are nearly parallel to the line of $y=x$, and have slopes

Table 2 Maceral compositions and gas (CO_2) flow behavior of coals studied [38]

Sample ID	Location	Coal Maceral Composition					Flow Behavior		
		$R_{o,max}$	C_V	C_I	C_E	C_{MM}	K_{g0}	ΔS	η
TB10-2	Tuanbai[a]	0.71	79.0	13.7	4.8	2.6	643.65	0.23	0.54
TB10-1	Tuanbai[a]	0.75	80.9	12.4	3.9	2.8	42.58	0.16	0.36
T91-2	Taoshan[b]	0.76	61.5	24.0	0.0	14.4	1.98	0.20	0.56
D18	Dongrong II[b]	0.76	84.6	3.3	2.3	10.0	33.32	0.13	0.35
T91-1	Taoshan[b]	0.80	62.2	26.8	2.2	8.7	69.52	0.26	0.62
J15	Jianshunyi[b]	0.80	73.1	15.6	0.4	10.9	33.57	0.23	0.66
XZ2-2	Xinzhi[a]	0.80	73.5	16.5	3.9	6.0	5.71	0.31	0.80
SH-1	Shenghe[b]	0.82	63.8	28.3	0.0	7.7	1253.19	0.19	0.51
B1	Boqing[b]	0.86	78.3	14.5	0.8	6.5	653.93	0.20	0.58
GH9	Ganhe[a]	0.88	67.7	18.0	10.6	3.7	126.97	0.18	0.48
XZ2-3	Xinzhi[a]	1.03	69.6	11.3	9.3	10.9	0.76	0.28	0.84
LYZ2-1	Liyazhuang[a]	1.06	68.2	15.4	3.7	12.6	2.41	0.25	0.86
LYZ2-2	Liyazhuang[a]	1.09	70.8	22.2	4.3	2.7	0.25	0.27	0.66
CC11-2	Caocun[a]	1.09	80.1	11.7	1.2	7.0	277.45	0.22	0.50
SH-2	Shenghe[b]	1.12	85.1	5.6	0.0	9.3	1190.81	0.23	0.46
CC11-1	Caocun[a]	1.17	69.2	17.9	4.3	8.6	23.78	0.13	0.33
X3-1	Xinzhi[a]	1.48	85.8	8.1	0.0	6.3	152.50	0.11	0.13
X3-3	Xincheng[c]	1.65	85.8	6.6	0.0	7.5	45.72	0.17	0.33
W3	Wangzhuang[c]	1.70	87.6	5.1	0.0	7.3	19.73	0.23	0.69
YW3-7	Yuwu[c]	1.72	87.5	6.8	0.0	5.7	54.88	0.19	0.35
CC10-1	Caocun[d]	1.83	67.5	2.0	0.0	30.3	130.30	0.12	0.11
PZ11	Pingzheng[d]	1.86	88.3	6.2	0.0	5.6	39.79	0.15	0.25
CK3	Changcun[c]	1.87	93.0	4.3	0.0	2.7	0.01	0.08	−0.01
N3-1	Niushan[c]	1.90	88.5	4.6	0.0	6.7	0.02	0.12	0.25
MGQ2	Magouqu[d]	1.96	86.1	6.5	0.0	7.5	3.78	0.12	0.31
HYY5	Heyangl[d]	1.96	83.3	6.2	0.0	10.6	13.21	−0.01	−0.34
X3-2	Xincheng[c]	2.04	89.0	2.9	0.0	8.0	283.53	0.14	0.32
Y3	Youxianshan[c]	2.06	86.2	10.5	0.0	3.3	49.92	0.17	0.37
W10	Wangcun[c]	2.13	78.4	4.1	0.0	17.5	1.22	0.02	−0.18
YW3-4	Yuwu[c]	2.22	85.5	6.0	0.0	8.5	18.27	0.13	0.13
N3	Niushan[c]	2.47	95.0	2.1	0.0	2.9	14.80	0.27	0.84
S3-2	Shenjiazhuang[c]	2.55	89.6	5.4	0.0	5.0	22.96	0.26	0.71
M3-3	Mishan[c]	2.53	87.0	0.6	0.0	12.4	7.51	0.09	0.12
T3-1	Tuncheng[c]	2.55	91.7	0.2	0.0	8.1	0.67	0.33	0.70
D3-1	Duanshi[c]	2.74	95.4	0.2	0.0	4.4	0.06	0.17	0.35
H3-3	Houcun[c]	2.79	89.3	0.8	0.0	9.9	0.01	0.09	0.01
M3-1	Mishan[c]	2.85	87.4	0.8	0.0	11.6	65.48	0.01	−0.26
YH3	Yonghong[c]	2.85	94.0	0.6	0.0	5.4	0.01	0.14	0.29
SH3-2	Sihe[c]	2.86	89.1	0.4	0.0	10.5	13.13	0.18	0.34
H3-2	Houcun[c]	2.90	85.6	4.1	0.0	10.3	0.62	0.43	0.97
D3-2	Duanshi[c]	2.92	66.1	0.2	0.0	33.7	0.003	0.24	0.50
T3-2	Tuncheng[c]	2.96	90.3	0.2	0.0	9.5	33.68	0.15	0.44

Continued

Table 2 Maceral compositions and gas (CO_2) flow behavior of coals studied —cont'd

Sample		Coal Maceral Composition					Flow Behavior		
ID	Location	$R_{o,max}$	C_V	C_I	C_E	C_{MM}	K_{g0}	ΔS	η
M3-2	Mishan[c]	3.01	84.2	0.4	0.0	15.4	10.23	0.08	0.17
SH3-1	Sihe[c]	3.05	90.7	1.8	0.0	7.5	0.01	0.09	0.05
T3-3	Tuncheng[c]	3.06	95.1	0.0	0.0	4.9	28.02	0.16	0.44
W15	Wangtaipu[c]	3.32	93.2	0.0	0.0	6.0	9.19	0.22	0.48

$R_{o,max}$ = maximum vitrinite reflectance, %; C_V = vitrinite content, %; C_I = inertinite content, %; C_E = exinite content, %; C_{MM} = mineral matter content, %; k_{g0} = initial permeability obtained at 4.0 MPa inlet gas pressure, $10^{-6}\,\mu m^2$; η is the change ratio of permeability and is equal to $(k_{g0} - k_{gf})/k_{g0}$, where k_{gf} is the final permeability (in $10^{-6}\,\mu m^2$) obtained at 0.5 MPa inlet gas pressure; ΔS is the total loss of flow rate.

[a]Houxi Basin.

[b]Sanjiang-Mulinghe coal-bearing zone.

[c]Qinshui Basin.

[d]Ordos Basin.

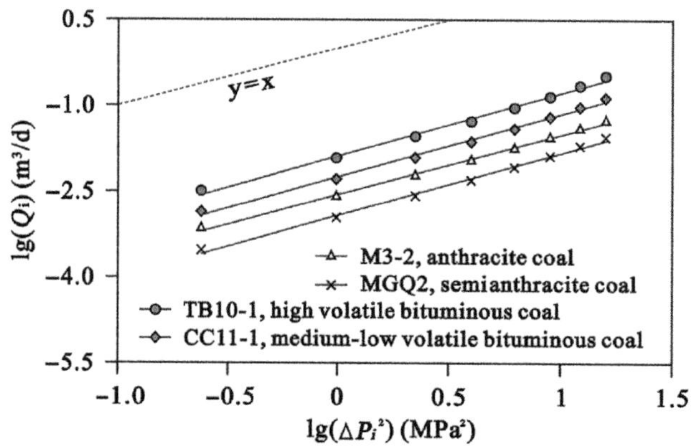

Fig. 7
Relationship between $\lg(Q_i)$ and $\lg(\Delta P_i^2)$.

approximately equal to one. Thus, $\lg\left(\frac{43.2k_{gi}A}{P_s\mu_{gi}L}\right)$ will be approximately constant during the gas pressure decrease.

Letting $\lg\left(\frac{43.2k_{gi}A}{P_s\mu_{gi}L}\right) = B$ (where B is a constant) for a certain coal, Eq. (21) can be obtained.

$$k_{gi} = \frac{P_s\mu_{gi}L10^B}{43.2A} \tag{21}$$

According to Eq. (21), the gas permeability change is directly affected by the gas viscosity and core deformation (i.e., length and cross-sectional area variations), which is consistent with the latest theory accounting for the gas permeability change [24]. First, with the pressure decrease, increased effective stress and matrix shrinkage occur in coal, causing coal deformation accompanied with varied coal permeability. Second, the effects of gas slippage and Knudsen diffusion on the permeability change gradually and become significant due to decreases in gas viscosity and density. Here, the theoretical results demonstrate the phenomenon of gas permeability variation as a result of decreasing gas viscosity (which is positively associated with gas density) and core deformation with decreasing pressure. These conclusions obtained by theoretical derivation and experimental measurements are useful to reveal the mechanism of the gas permeability change in different coal ranks.

A general model showing the dynamic gas flows in coals was summarized using the η and ΔS values in this chapter, and exhibited the relationships of the permeability changes (k_{gi}/k_{g0}) with η and ΔS (Fig. 8). During pressure depletion, when $k_{gi}/k_{g0} > 1$ (i.e., k_{gi} is always larger than k_{g0}), $\eta < 0$ and $\Delta S < 0$. When $k_{gi}/k_{g0} < 1$ (i.e., k_{gi} is always less than k_{g0}), $\eta > 0$ and $\Delta S > 0$. Moreover, ΔS will be larger than zero when parts of the coal where $k_{gi}/k_{g0} < 1$ are more common than those where $k_{gi}/k_{g0} > 1$. Rule of $k_{gi}/k_{g0} > 1$ is common in previous studies [1, 18], which use uniaxial strain [1] and constant volume [18] boundary conditions, but it does not occur in our experimental results and this may be due to the utilization of the constant confining stress condition. However, $k_{gi}/k_{g0} < 1$ is a general rule in this chapter. In addition, ΔS is positively proportional to η in different coal ranks, i.e., a larger change ratio of permeability

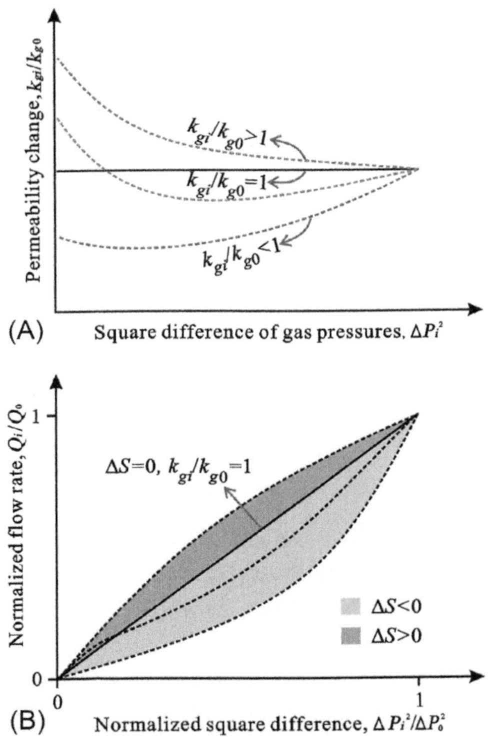

Fig. 8
General model of changes in the permeability (A) and flow rate (B).

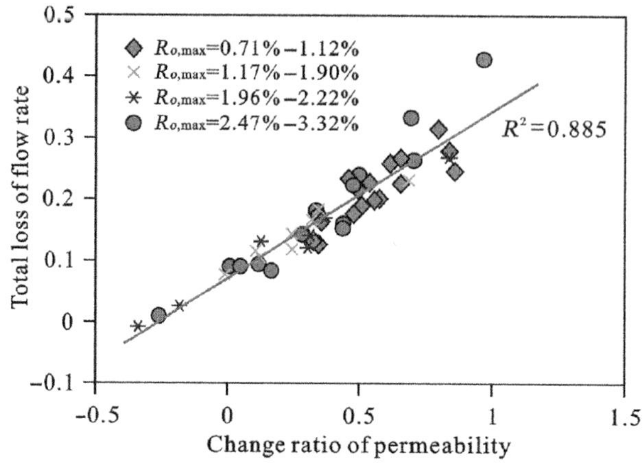

Fig. 9
Relationship between total loss of flow rate and change ratio of permeability.

corresponds to a greater total loss of flow rate (Fig. 9). Hence, both η and ΔS can be utilized to quantitatively characterize the dynamic process of gas flow in different coal ranks and are effective parameters that can be used to target favorable reservoirs for CBM recovery.

Due to the positively proportional relationship between η and ΔS, the impact of coal ranks on η is related to that on ΔS. Overall, η and ΔS gradually decrease with increasing coal rank from high volatile bituminous to semianthracite coal, but increase as coal rank improves to anthracite grade (Fig. 10).

3.4 Impact of Coal Ranks on Dynamic Gas Flow

Both the maceral composition and initial permeability of coal will affect changes in the permeability and flow rate during the pressure decrease. In this chapter, the coal maceral composition and initial permeability were obtained for different coal ranks. Vitrinite and inertinite were the main constituents in the coal studied, and accounted for 66.3%–97.3% (mean 90.0%) of the contents (Table 2). With increasing coal rank, the vitrinite (inertinite) content of coal gradually increases (decreases). The exinite constituent primarily occurs in highly volatile bituminous coal. Additionally, the average content of inorganic mineral matter slightly increases with the increasing rank (Figs. 11 and 12B). Highly volatile bituminous coal has the highest average of initial permeability, while anthracite coal has the lowest permeability. The mean permeability of semianthracite coal is slightly higher than that of medium-low volatile coal (Fig. 12C). Overall, the initial permeability is positively (negatively) related to inertinite (vitrinite) content for different coal ranks (Fig. 12B and C). Both changes in the maceral composition and the permeability are the result of the increase in coal rank.

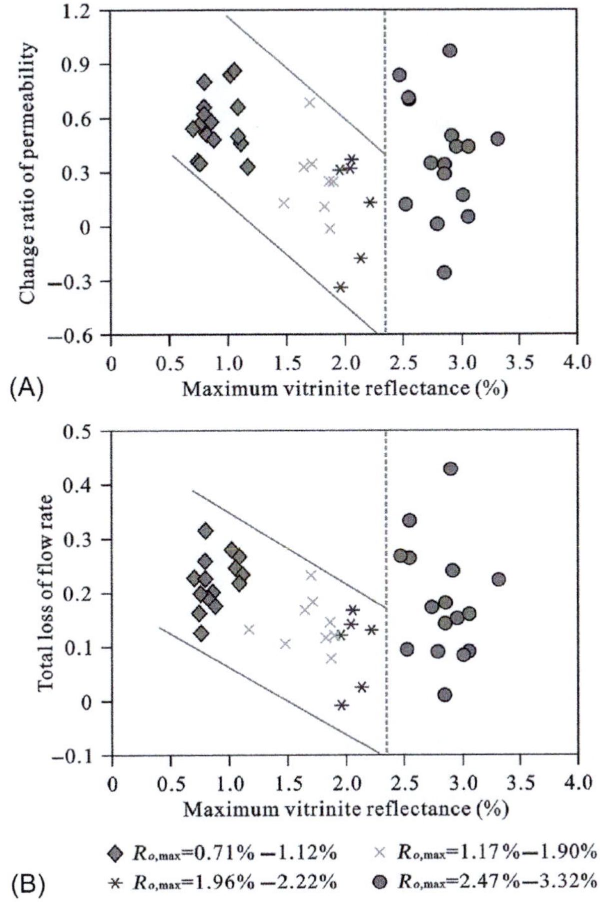

Fig. 10

Relationships of maximum vitrinite reflectance with change ratio of permeability (A) and total loss of flow rate (B).

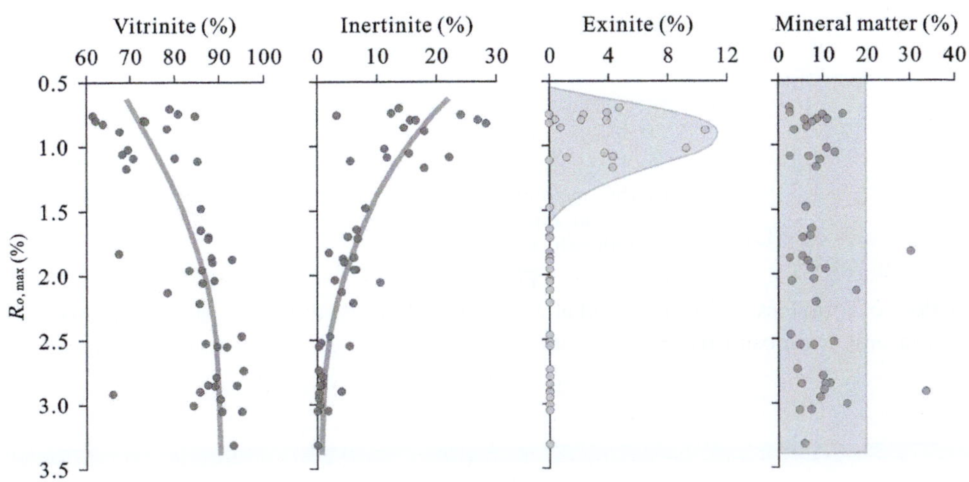

Fig. 11

Coal maceral compositions of different coal ranks.

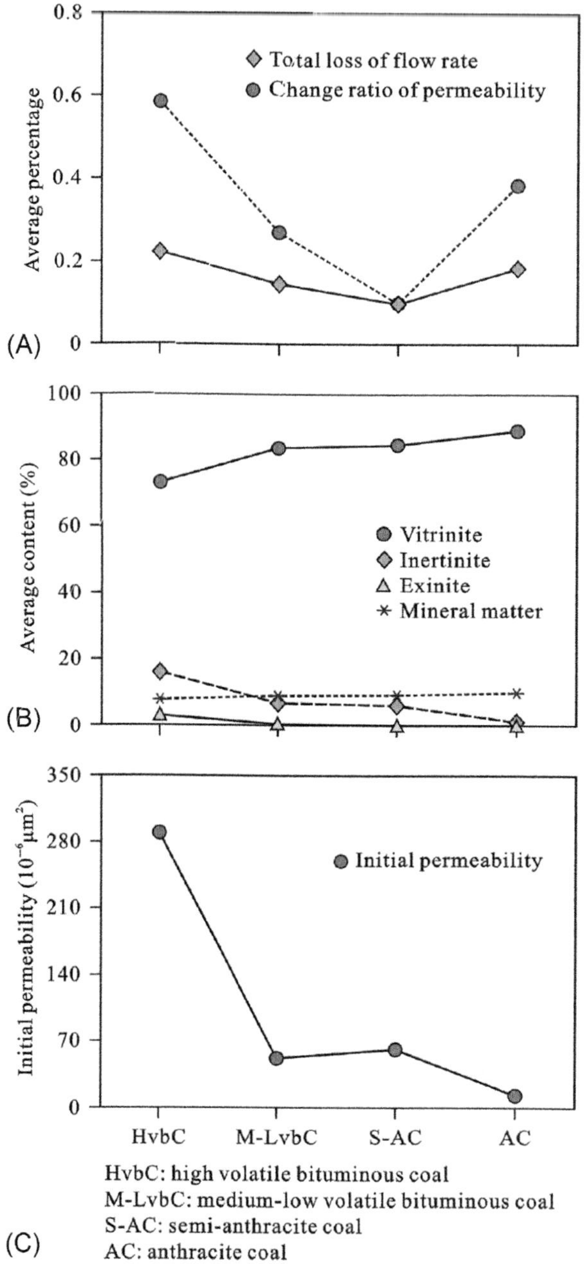

Fig. 12

Average values of total loss of flow rate and change ratio of permeability (A), average contents of coal maceral compositions (B) and initial permeabilities (C) in different coal ranks.

According to the dynamic gas flow in different coal ranks, the impact of coal rank on the gas flow was investigated. The coal maceral composition primarily affects η and ΔS as coal rank varies from highly volatile bituminous to semianthracite coal. The initial permeability dominates the changes of η and ΔS as coal rank varies from semianthracite to anthracite coal. Thus, the preliminary results can be summarized as follows: (1) with increasing coal rank from highly volatile bituminous to semianthracite coal, increasing (decreasing) vitrinite (inertinite) content primarily causes η and ΔS to decline, as illustrated in Fig. 13 showing both η and ΔS reducing with the increasing content ratio of vitrinite to inertinite; and (2) as coal rank increases to anthracite grade, the relatively low initial permeability predominantly causes η and ΔS to increase.

For the same rank, coal (especially some bituminous and anthracite coals) with lower and higher initial permeabilities have smaller η and ΔS values, with maximum values at an initial permeability of $\sim 1 \times 10^{-6} \mu m^2$ (Fig. 14). The impact of the initial permeability on η and ΔS is

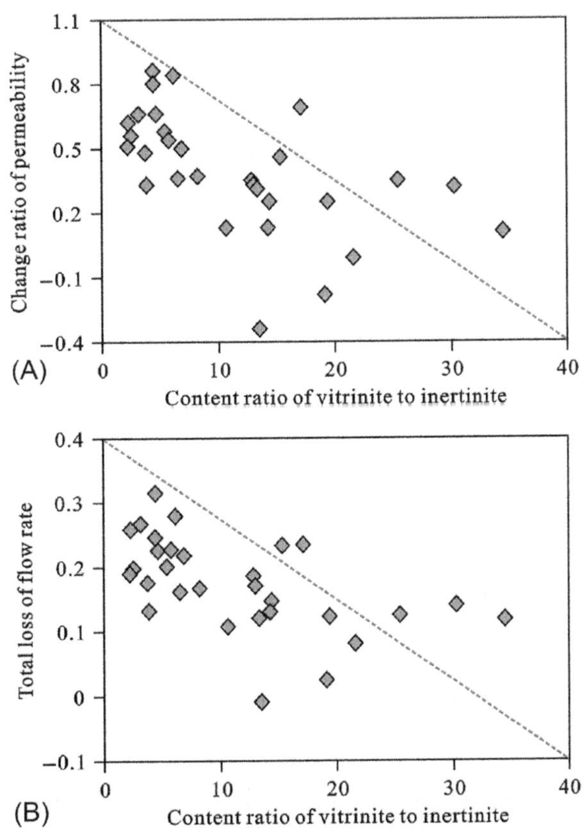

Fig. 13

Relationships of content ratio of vitrinite to inertinite with change ratio of permeability (A) and total loss of flow rate (B) in bituminous and semianthracite coals.

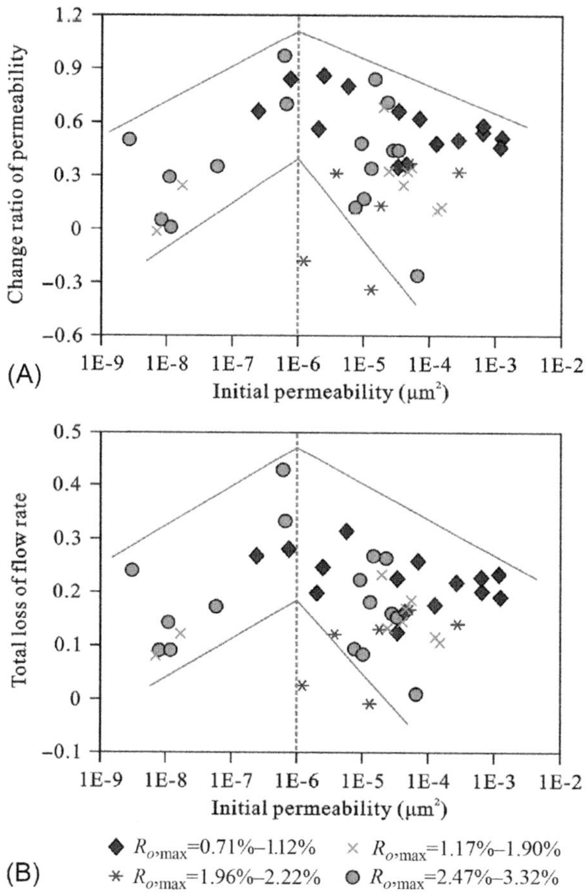

Fig. 14

Relationships of initial permeability with change ratio of permeability (A) and total loss of flow rate (B) in different coal ranks.

much larger in anthracite coals, and is relatively weaker in semianthracite coals. Considering gas flow rate and η or ΔS values, higher initial permeability will be favorable in a reservoir for CBM development.

4 Conclusions

In this chapter, an empirical model was established predicting the permeability change in an anthracite coal by linking the mean pore radius with the permeability and the results were consistent with the experimental data. This model comprehensively considers the four effects (effective stress, coal matrix shrinkage, gas slippage, and Knudsen diffusion). Dynamic gas (CO_2) flow occurs in different coal ranks during pressure depletion under a constant confining stress condition and can be quantitatively evaluated using the change ratio of permeability (η) and total loss of flow rate (ΔS). The main conclusions are as follows:

(1) The mean pore radius of the coal core gradually decreased, which caused a change in permeability from the effects of gas slippage and Knudsen diffusion.
(2) Under constant gas pressure conditions, the mean pore radius of the coal core decreased in an approximately linear manner. The decrease in the permeability was primarily caused by the effective stress increase and was slightly offset by the positive effects of gas slippage and Knudsen diffusion.
(3) η and ΔS are comprehensively affected by the maceral composition and initial permeability of different coal ranks, of which the maceral composition primarily affects the η and ΔS changes as coal rank varies from highly volatile bituminous to semianthracite coal. The initial permeability dominates the change of η and ΔS as coal rank varies from semianthracite to anthracite coal.

Acknowledgments

This work was supported by the National Natural Science Foundation (Nos. 41602131, 41672130, 41572116).

References

[1] J.Q. Shi, S. Durucan, A model for changes in coalbed permeability during primary and enhanced methane recovery, SPE Reserv. Eval. Eng. 8 (2005) 291–299.
[2] X. Cui, R.M. Bustin, L. Chikatamarla, Adsorption-induced coal swelling and stress: Implications for methane production and acid gas sequestration into coal seams, J. Geophys. Res. 112 (2007) 1–8.
[3] M.H. Noh, L.W. Lake, S.L. Bryant, et al., Implications of coupling fractional flow and geochemistry for CO_2 injection in aquifers, SPE Reserv. Eval. Eng. 10 (2007) 406–414.
[4] F. van Bergen, P. Krzystolik, N. van Wageningen, et al., Production of gas from coal seams in the upper Silesian coal basin in Poland in the post-injection period of an ECBM pilot site, Int. J. Coal Geol. 77 (2009) 175–187.
[5] Z. Chen, Z. Pan, J. Liu, et al., Effect of the effective stress coefficient and sorption-induced strain on the evolution of coal permeability: experimental observations, Int. J. Greenhouse Gas Control 5 (2011) 1284–1293.
[6] T. Kiyama, S. Nishimoto, M. Fujioka, et al., Coal swelling strain and permeability change with injecting liquid/supercritical CO_2 and N_2 at stress-constrained conditions, Int. J. Coal Geol. 85 (2011) 56–64.
[7] R.G. Moghanloo, Y. Dadmohammadi, B. Yuan, et al., Applying fractional flow theory to evaluate CO_2 storage capacity of an aquifer, J. Pet. Sci. Eng. 125 (2015) 154–161.
[8] C.R. Clarkson, C.L. Jordan, R.R. Gierhart, et al., Production data analysis of CBM wells, SPE Reserv. Eval. Eng. (2008) 311–325.
[9] J.Q. Shi, S. Durucan, Exponential growth in San Juan Basin Fruitland coalbed permeability with reservoir drawdown: model match and new insights, SPE Reserv. Eval. Eng. 13 (2010) 914–925.
[10] Z. Pan, L.D. Connell, M. Camilleri, Laboratory characterisation of coal reservoir permeability for primary and enhanced coalbed methane recovery, Int. J. Coal Geol. 82 (2010) 252–261.
[11] Y. Gensterblum, A. Ghanizadeh, B.M. Krooss, Gas permeability measurements on Australian subbituminous coals: fluid dynamic and poroelastic aspects, J. Nat. Gas Sci. Eng. 19 (2014) 202–214.
[12] J.P. Seidle, M.W. Jeansonne, D.J. Erickson, Application of matchstick geometry to stress dependent permeability in coals, in: SPE Rocky Mountain Regional Meeting, Casper, Wyoming, USA, 18-21 May, 1992.

[13] X. Cui, R.M. Bustin, Volumetric strain associated with methane desorption and its impact on coalbed gas production from deep coal seams, AAPG Bull. 89 (2005) 1181–1202.
[14] E.P. Robertson, R.L. Christiansen, A permeability model for coal and other fractured, sorptive-elastic media, SPE J. 13 (2008) 314–324.
[15] L.D. Connell, M. Lu, Z. Pan, An analytical coal permeability model for tri-axial strain and stress conditions, Int. J. Coal Geol. 84 (2010) 103–114.
[16] H.H. Liu, J. Rutqvist, A new coal-permeability model: internal swelling stress and fracture-matrix interaction, Transp. Porous Media 82 (2010) 157–171.
[17] S. Liu, S. Harpalani, M. Pillalamarry, Laboratory measurement and modeling of coal permeability with continued methane production: part 2—modeling results, Fuel 94 (2012) 117–124.
[18] Q. Ma, S. Harpalani, S. Liu, A simplified permeability model for coalbed methane reservoirs based on matchstick strain and constant volume theory, Int. J. Coal Geol. 85 (2011) 43–48.
[19] H. Qu, J. Liu, Z. Pan, et al., Impact of matrix swelling area propagation on the evolution of coal permeability under coupled multiple processes, J. Nat. Gas Sci. Eng. 18 (2014) 451–466.
[20] D. Chen, Z. Pan, Z. Ye, et al., A unified permeability and effective stress relationship for porous and fractured reservoir rocks, J. Nat. Gas Sci. Eng. 29 (2016) 401–412.
[21] J. Li, D. Liu, Y. Yao, et al., Evaluation and modeling of gas permeability changes in anthracite coals, Fuel 111 (2013) 606–612.
[22] J. Li, D. Liu, Y. Yao, et al., Controls of CO_2 permeability change in different rank coals during pressure depletion: an experimental study, Energy Fuel 28 (2014) 987–996.
[23] Y. Li, D. Tang, H. Xu, et al., Experimental research on coal permeability: the roles of effective stress and gas slippage, J. Nat. Gas Sci. Eng. 21 (2014) 481–488.
[24] J. Li, D. Liu, S. Lu, et al., Evaluation and modeling of the CO_2 permeability variation by coupling effective pore size evolution in anthracite coal, Energy Fuel 29 (2015) 717–723.
[25] Y. Tao, D. Liu, J. Xu, et al., Investigation of the Klinkenberg effect on gas flow in coal matrices: A numerical study, J. Nat. Gas Sci. Eng. 30 (2016) 237–247.
[26] F. Javadpour, Nanopores and apparent permeability of gas flow in mudrocks (shales and siltstone), J. Can. Pet. Technol. 48 (2009) 16–21.
[27] S. Harpalani, G. Chen, Influence of gas production induced volumetric strain on permeability of coal, Geotech. Geol. Eng. 15 (1997) 303–325.
[28] F. Javadpour, D. Fisher, M. Unsworth, Nanoscale gas flow in shale sediments, J. Can. Pet. Technol. 46 (2007) 55–61.
[29] P.A. Chambre, S.A. Schaaf, Flow of Rarefied Gases, Princeton University Press, Princeton, NJ, 1961.
[30] G. Arya, H.C. Chang, E.J. Maginn, Molecular simulations of Knudsen wall-slip: effect of wall morphology, Mol. Simul. 29 (2003) 697–709.
[31] C.T. Rettner, Thermal and tangential-momentum accommodation coefficients for N_2 colliding with surfaces of relevance to disk-drive air bearings derived from molecular beam scattering, IEEE Trans. Magn. 34 (1998) 2387–2395.
[32] T. Veijola, H. Kuisma, J. Lahdenperä, The influence of gas-surface interaction on gas-film damping in a silicon accelerometer, Sensors Actuators A Phys. 66 (1998) 83–92.
[33] J. Yao, H. Sun, Z. Huang, et al., Key mechanical problems in the development of shale gas reservoirs, Sci. Sin. Phys. Mech. Astron. (12) (2013) 1527–1547.
[34] L.J. Klinkenberg, Drilling and Production Practice, American Petroleum Institute, Washington, D.C., 1941, pp. 200–213.
[35] R. Gayer, I. Harris, Coalbed Methane and Coal Geology, Geological Society Special Publication, London, 1996.
[36] D.P. Sparks, T.H. McLendon, J.L. Saulsberry, et al., The effects of stress on coalbed reservoir performance, Black Warrior Basin, U.S.A, in: SPE Annual Technical Conference and Exhibition, Dallas, Texas, 22-25 October, 1995.
[37] P.Q. Huy, K. Sasaki, Y. Sugai, et al., Carbon dioxide gas permeability of coal core samples and estimation of fracture aperture width, Int. J. Coal Geol. 83 (2010) 1–10.
[38] J. Li, S. Lu, Y. Cai, et al., Impact of coal ranks on dynamic gas flow: an experimental investigation, Fuel 194 (2017) 17–26.

CHAPTER 15

Multiphysical Flow Behavior in Shale and Permeability Measurement by Pulse-Decay Method

Bao Jia*, Ruimin Feng†, Jyun-Syung Tsau‡, Reza Barati*

**Chemical & Petroleum Department, University of Kansas, Lawrence, KS, United States, †Chemical & Petroleum Department, University of Calgary, Calgary, AB, Canada, ‡Tertiary Oil Recovery Program (TORP), University of Kansas, Lawrence, KS, United States*

Chapter Outline

1 Introduction

Shale gas is an important supply of fossil fuels around the world. In recent years, horizontal wells with multistage hydraulic fractures have greatly boosted shale gas production [1]. However, due to the multiphysics that occur in the tiny pores, shale gas flow mechanisms are still poorly understood. Gas apparent permeability in shale depends on flow conditions, which is different from the inherent permeability that is only related to the structure of the porous

Petrophysical Characterization and Fluids Transport in Unconventional Reservoirs. https://doi.org/10.1016/B978-0-12-816698-7.00015-2

media [2]. Under different temperatures and pressures for various types of gas, the ratio between the mean free gas path and pore size is defined as the Knudsen number, which is frequently used to define flow regimes. Slip flow occurs in the slip flow region, both slip flow and diffusion occur in the transitional flow region [3]. Slip flow, pore diffusion, and surface diffusion occur simultaneously as adsorptive gas flows through shale reservoirs. Slip flow is caused by the failure of nonslip boundary conditions on the pore surface. Fickian diffusion is driven by the gas concentration gradient. Knudsen diffusion is driven by the collisions between gas molecules and the pore surface when the pore size reduces to the order of molecule size. Surface diffusion occurs for the adsorptive gas, such as carbon dioxide and methane. Adsorbed molecules move along the pore surface, between the adsorption layer and the bulk phase region, in a "hopping" fashion between adjacent adsorption spots [4]. Another physical phenomenon involved during the gas production process is the geomechanical effect. The effective stress exerted on the pore increases as gas is withdrawn from the formation [5]. Therefore, for more accurate reservoir simulations, it is imperative to know how the apparent permeability changes considering the multiphysics imposed on the shale matrix.

For laboratory permeability measurement in tight pores media, the pulse-decay (or the gas transient pressure transmission) is a powerful method that has advantages over the conventional steady-state method that might not work for ultra-low permeability conditions [6, 7]. The disadvantages of the steady-state method are: (1) time required to reach the steady state is extremely long, and (2) the pressure gauge might exceed the full-scale value because of the extremely high pressure drop across the core. Fig. 1 shows the schematic diagram of the pulse-

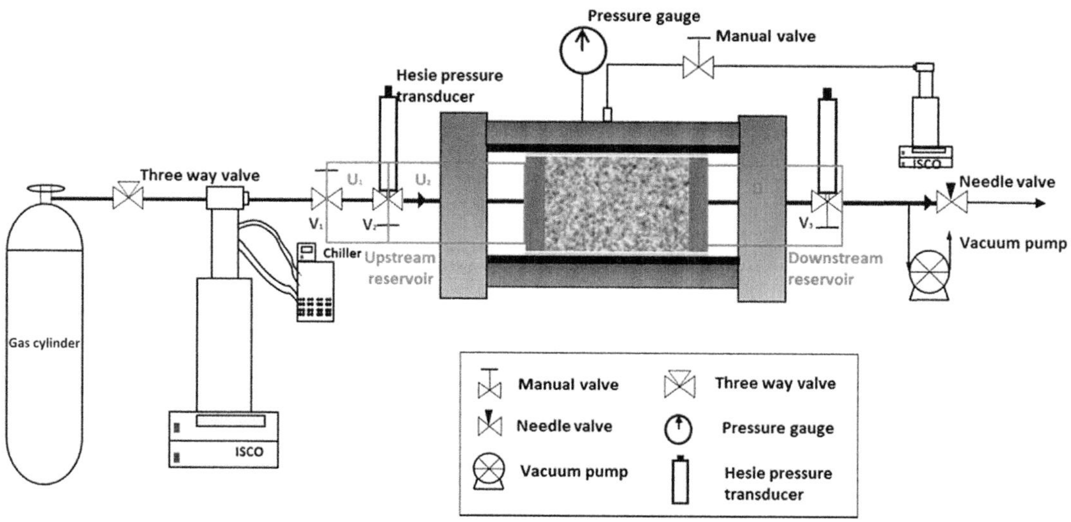

Fig. 1
Schematic diagram of the pulse-decay apparatus.

decay apparatus. It has three main components: the upstream reservoir (U_1 and U_2), the core section, and the downstream reservoir (D). The whole system is initially saturated with gas of the desired type under the desired pressure, then the pressure in U_1 is elevated with V_1 and V_2 closed. To perform the test, V_2 is opened. Thus, the pressure declines in the upstream reservoir, gas is transported to the downstream reservoir through the core, and the pressure builds up in the downstream reservoir. Pressure profiles in the upstream and downstream reservoirs act as the main data of the core sample that are used to calculate the permeability by analytical or numerical solutions [8, 9].

It is worth noting that the permeability based on the pulse-decay apparatus is the apparent value from the macroscopic view that co all the flow mechanisms discussed previously. However, the relationship between storage capacity of adsorptive gas and the flow behavior can be revealed by investigating the flow governing equation analytically.

2 Multiphysics in the Shale Reservoirs

Eqs. (1), (2) demonstrate the definition of the Knudsen number, Kn:

$$Kn = \frac{L_p}{d} \tag{1}$$

$$L_p = \frac{k_B T}{\sqrt{2} Z \pi d_{mol}^2 p} \tag{2}$$

where L_p is mean free path, d is pore size, k_B is the Boltzmann constant, T is temperature, d_{mol} is the molecule size, Z is gas compressibility factor, and p is pressure. Fig. 2 shows flow regimes

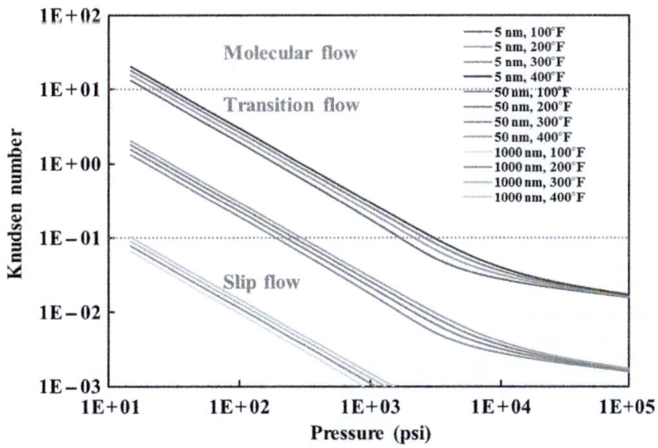

Fig. 2
Flow regimes of methane in pores with sizes from 5 to 1000 nm, and under temperatures from 100 to 400°F.

in pores with various sizes (5–1000 nm) and under a wide range of temperatures (100–400°F). Fig. 2 indicates that in most shale gas reservoirs with high temperatures and high pressures, flow regimes are mostly slip flow and transition flow [10, 11].

The ratio between the apparent permeability, k_a, and the intrinsic permeability, k_∞, is positively related with K_n. Sakhaee-Pour and Bryant [12] put forward expressions of this ratio in the slip flow and transition flow as Eqs. (3), (4), respectively. Here, the value of a is adjusted slightly at the joint point of $Kn=0.1$ to obtain a smooth curve of $f(Kn)$ as a function of Kn.

$$f(K_n)=\frac{k_a}{k_\infty}=1+\alpha K_n \tag{3}$$

$$f(K_n)=\frac{k_a}{k_\infty}=0.8453+5.4576K_n+0.1633K_n^2 \tag{4}$$

It is of important practical use in reservoir simulations to obtain a permeability evolution curve during the process of gas depletion [13]. Fig. 3 shows the complex factors affecting the dynamic permeability. As gas is withdrawn from the deep reservoir, the effective stress increases because the pore pressure decreases, leading to the decrease of pore size, which has two effects: K_∞ decreases because it is proportional to the square of the pore size; and Kn increases according to Eq. (1). A decrease in K_∞ has the impact of decreasing the gas apparent permeability. An increase in Kn leads to the increase of $f(Kn)$ and the subsequent apparent permeability. At the same time, the adsorbed molecules gradually leave the pore surface as the pore pressure decreases. Pore size is enlarged due to the desorption process, the impact of which is negligible in conventional reservoirs but might not be in shale reservoirs considering the smaller pore size. Similarly, the decrease of Kn and increase of K_∞ induced by the pore enlargement have two competing impacts that decrease and increase the apparent permeability, respectively. Surface diffusion has been suggested to be an important flow mechanism in shale with a high degree of total organic carbon (TOC). Wu et al. [14] described surface diffusion by analytical models. Fathi and Akkutlu [15] applied the Lattice Boltzmann simulation to characterize the flow behavior. Ren et al. [16] investigated surface diffusion in shale by the

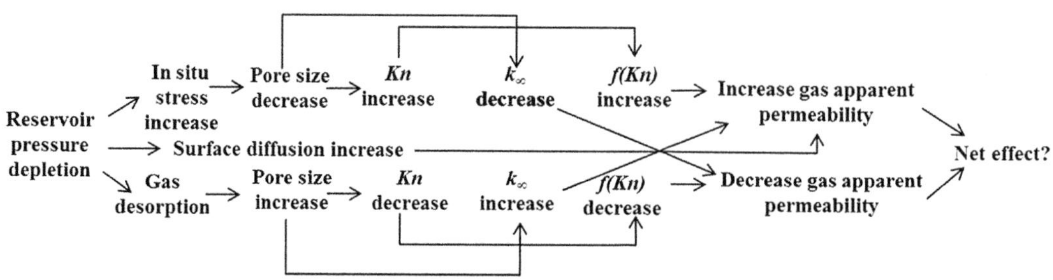

Fig. 3
Schematic diagram of effects of the multiphysics on shale gas permeability during the depletion process.

adaptive Dubinin-Astakhov (ADA) model. The consensus of these studies is that surface diffusion is more important in smaller pores and under lower pressure. However, to the best of our knowledge, currently there are no standardized approaches to measure surface diffusivity, and a wide range of values has been reported in the literature. For instance, the surface diffusivity measured by Kang et al. [4] in kerogen is between 1.55E − 7 and 8.80E − 6 m^2/s, and it is between 5.1E − 8 and 8.80E − 6 m^2/s measured by Akkutlu and Fathi [17].

Considering the complex and interacting physics that occur simultaneously, it is difficult to derive the analytical solutions for the permeability evolution curve. In Section 3, a straightforward approach is presented to demonstrate how to couple all these physics step by step.

3 Gas Permeability Evolution During Gas Depletion

Fig. 4 describes the steps applied to evaluate permeability evolution, which requires a series of experimental data including pore size change as a function of the effective stress, adsorption profiles, and the surface diffusivity value. The steps are described in detail below.

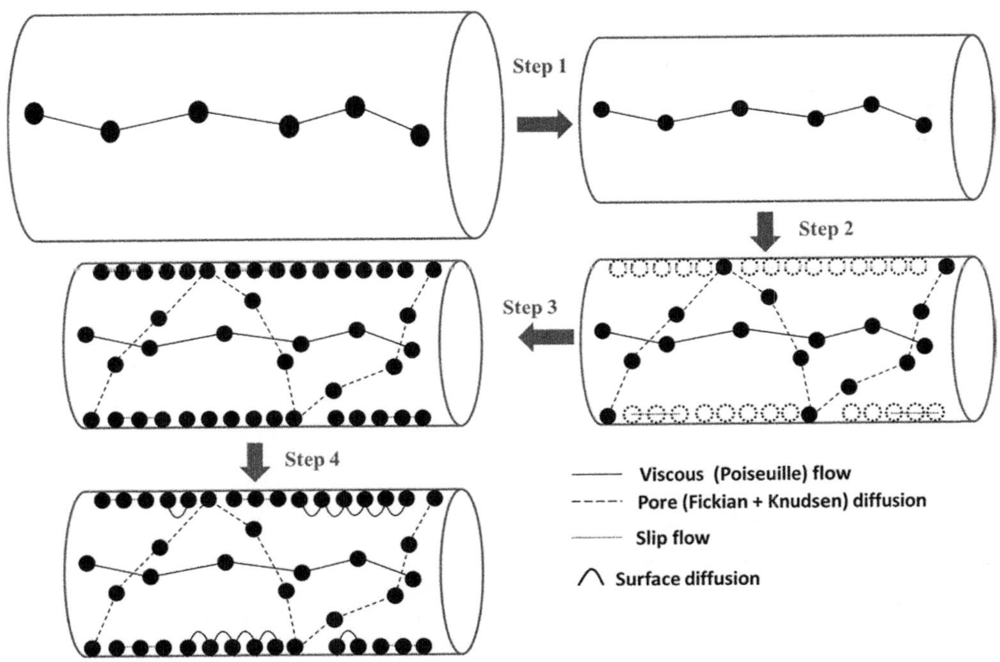

Fig. 4
Four steps applied to estimate shale gas permeability variations during gas production process: Step 1—the geomechanical effect, Step 2—slip flow and pore diffusion, Step 3—adsorption layer effect, and Step 4—surface diffusion.

3.1 Step 1—The Geomechanical Effect

Fig. 5A shows the permeability measurement results for the Marcellus shale under the effective stress of 2000, 3000, and 4000 psi (data from Ref. [18]). The first-order Klinkenberg equation, Eq. (5) [19] is used to fit the relationship between permeability and the reciprocal of pore pressure. The ratio is larger than 1 because k_b is the slippage factor contributed by the slip flow and pore diffusion.

$$\frac{k_a}{k_\infty} = 1 + \frac{k_b}{p} \tag{5}$$

Eq. (6) shows that k_b is related to gas properties and temperature [18]. It implies that the pore size, d, can be estimated from k_b.

$$k_b = \frac{16 c_g \mu}{d}\left(\frac{2RT}{\pi M_g}\right)^{1/2} \tag{6}$$

Fig. 5B shows the pore size change as a function of the effective stress. The fitted straight line is extrapolated forward and backward to estimate pore size variations during the whole lifetime of a shale gas reservoir with the initial pore pressure and initial effective stress of 4000 and 1000 psi, respectively. Considering that k_∞ is proportional to d^2, the ratio between k_∞ and d^2 is calculated under the three measured effective stresses, and their average, 3.15E−2 nD/nm^2, is used as the proportionality to estimate k_∞ under the specific effective stress. Table 1 summarizes two example results calculated under two effective stress measures of 3520 and 1600 psi, corresponding to the pore pressure of 1480 and 3400 psi, respectively.

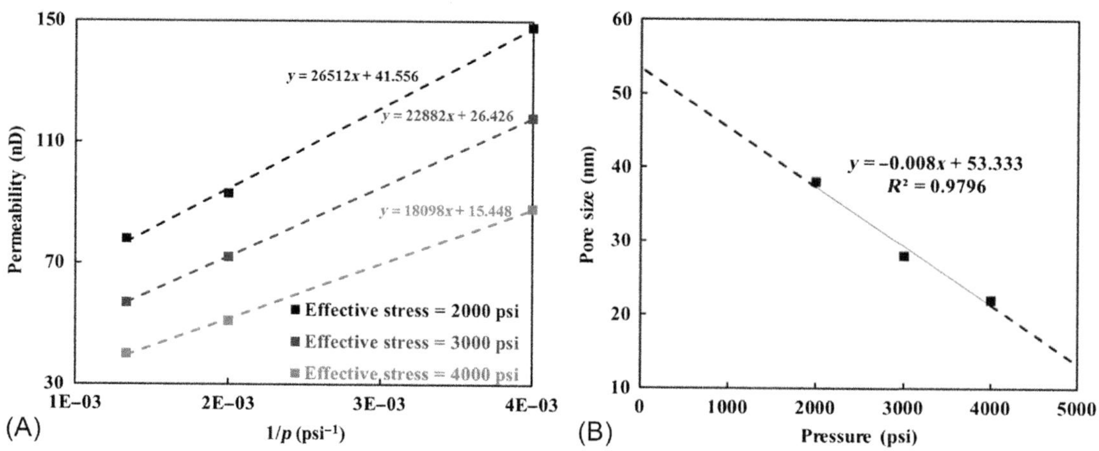

Fig. 5

Permeability as a function of the reciprocal of pore pressure and effective stress of the Marcellus shale. (A) Permeability as a function of the reciprocal of pore pressure and effective stress of the Marcellus shale, (B) pore size as a function of effective stress.

Table 1 Intrinsic permeability under two effective stress measures of 3520 and 1600 psi

Effective Stress (psi)	Pore Pressure (psi)	Pore Size (nm)	k_∞ (nD)
3520 1600	1480 3400	25.17 40.53	19.94 51.69

3.2 Step 2—Slip Flow and Pore Diffusion

Slip flow and pore diffusion are considered by estimating Kn and $f(Kn)$ under the specific effective stress with Eqs. (1), (3), (4), respectively. Then apparent permeability is estimated by Eq. (7). Table 2 shows the example results.

$$k_a = k_\infty \times f(Kn) \tag{7}$$

3.3 Step 3—Adsorption Layer Effect

The adsorption layer effect is considered by correcting the effective pore size. Fig. 6 shows the adsorption profiles (data from Ref. [20]). The 104°F line is the experimental data of the same

Table 2 Permeability after considering slip flow and pore diffusion under two effective stress measures

Effective Stress (psi)	Pore Pressure (psi)	Pore Size (nm)	k_∞ (nD)	k_a (nD)
3520 1600	1480 3400	25.17 40.53	19.94 51.69	21.88 53.05

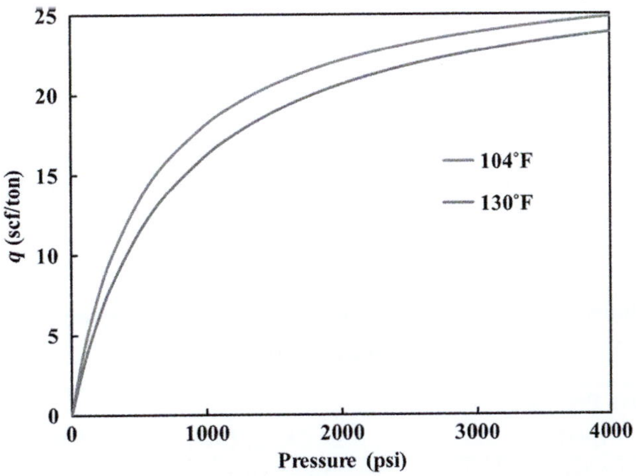

Fig. 6
Adsorption profile of the Marcellus shale under 104°F and 130°F.

Marcellus shale that is changed to the elevated reservoir temperature of the Marcellus shale, 130°F, based on the model by Lu et al. [21].

The Langmuir model can be used to characterize the adsorption profile. q_m is the maximum adsorption amount, and q_L is the Langmuir pressure.

$$q(p) = \frac{q_m p}{p + p_L} \tag{8}$$

The surface coverage of the adsorbed molecules can be evaluated by Eq. (9).

$$\theta = \frac{q(p)}{q_m} = \frac{p}{p + p_L} \tag{9}$$

The effective intrinsic permeability, $k_{\infty eff}$, can be weighted using the surface coverage ratio, θ, as in Eq. (10).

$$k_{\infty eff} = k_{\infty} \times (1 - \theta) + k_{\infty ad} \times \theta \tag{10}$$

where $k_{\infty ad}$ is the intrinsic permeability using pore size after deducting the molecule size. $f(Kn)$ can also be corrected using θ as in Eq. (11).

$$f(Kn)_{eff} = f(Kn) \times (1 - \theta) + f(Kn)_{ad} \times \theta \tag{11}$$

Therefore the apparent permeability considering slip flow and pore diffusion is expressed as the product of $k_{\infty eff}$ and $f(Kn)_{eff}$ as in Eq. (12).

$$k_{aeff} = k_{\infty eff} \times f(Kn)_{eff} \tag{12}$$

Table 3 demonstrates results after considering the adsorption layer effect under the effective stress measures of 3520 and 1600 psi.

3.4 Step 4—Surface Diffusion

Surface diffusion is considered as the last step by combining the mass flux of both the Darcy-type flow and surface diffusion flow. The Darcy flow equation based on the effective intrinsic permeability in Step 3 is expressed as [22]:

Table 3 Permeability after considering the adsorption layer effect under two effective stress measures

Effective Stress (psi)	Pore Pressure (psi)		$k_{\infty eff}$ (nD)	$f(Kn)_{eff}$	k_{aeff} (nD)
3520	1480		19.14	1.10	21.04
1600	3400		50.11	1.03	51.45

$$J_D = -\frac{k_{\infty eff}\rho_g \partial p}{\mu \partial x} \tag{13}$$

where J_D is the Darcy-type flow flux. The surface diffusion flux, J_{sd}, is expressed as:

$$J_{sd} = -\frac{C_{ad} D_{sd} \partial p}{p \partial x} \tag{14}$$

C_{ad} is the mass of adsorption per unit volume of the matrix, D_{sd} is the surface diffusivity. C_{ad} is expressed as:

$$C_{ad} = \frac{\theta q_m M_g \rho_r}{V_{std}} \tag{15}$$

where ρ_r is the density of the matrix, and V_{std} is the molar volume of gas under standard conditions. Eq. (14) can be expressed in the form of Darcy-type flow as below:

$$J_{sd} = -\frac{k_{sd}\rho_g \partial p}{\mu \partial x} \tag{16}$$

where k_{sd} is the equivalent permeability. By comparing Eqs. (14), (16), k_{sd} is expressed as:

$$k_{sd} = \frac{D_{sd} C_a \mu}{\rho_g p} \tag{17}$$

Defining F as k_{sd} divided by $k_{\infty eff}$ in Eq. (18):

$$F = \frac{k_{sd}}{k_{\infty eff}} = \frac{M D_{sd} q_m \theta \rho_r \mu_g}{k_{\infty eff} \rho_g p V_{std}} \tag{18}$$

Therefore, the total multiplier for $k_{\infty eff}$ to calculate the total apparent permeability is estimated as the summation of $f(Kn)_{eff}$ and F. Table 4 summarizes the example results.

$$f(Kn)_{tot} = f(Kn)_{eff} + F \tag{19}$$

$$k_{atot} = k_{\infty eff} f(Kn)_{tot} \tag{20}$$

Table 4 Permeability after considering surface diffusion under two effective stress measures, the surface diffusivity is assumed to be $1E-8\,m^2/s$

Effective Stress (psi)	Pore Pressure (psi)	$k_{\infty eff}$ (nD)	$f(Kn)_{eff}$	F	k_{atot} (nD)
3520	1480	19.14	1.10	1.01	40.46
1600	3400	50.11	1.03	0.13	57.73

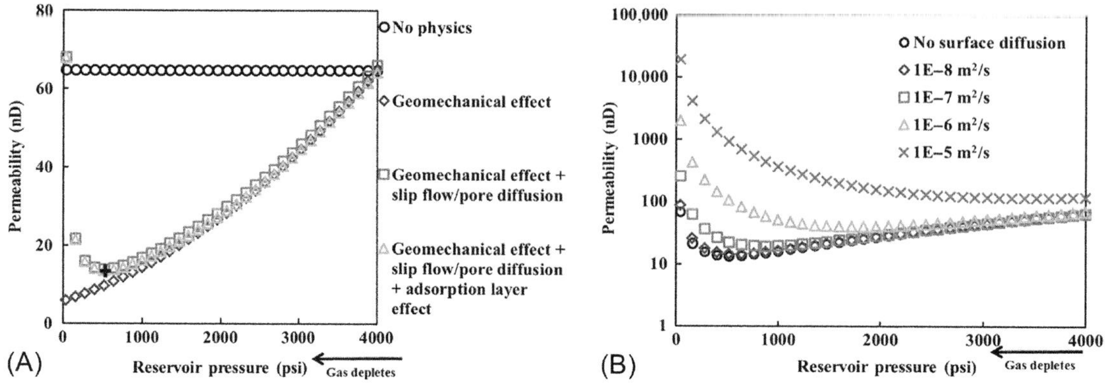

Fig. 7
Permeability evolution with reservoir pressure considering (A) physics from Step 1 to Step 3, (B) and surface diffusion in Step 4 with different surface diffusivity values.

Fig. 7A shows permeability evolution curves considering the geomechanical effect, slip flow and diffusion, and the adsorption layer effect step by step. It is observed that due to the increasing effective stress, the intrinsic permeability continuously decreases. At the end of production, the retained intrinsic permeability is only about 10% of the original value. After including slip flow and pore diffusion, the permeability decrease is alleviated especially under low reservoir pressures. A turning point appears at about 500 psi, below which the permeability begins to increase marked by the black cross. The curve marked with diamond is actually the Darcy component of the permeability, whereas the difference between the curve marked with diamond and the curve marked with square can be assessed as the slip flow and pore diffusion components contributing to the flow capacity. By comparing these two components, it is found that the slip flow and pore diffusion component begin to outweigh the Darcy component at ~280 psi. After including the adsorption layer into the permeability curve, it indicates that the effect is negligible, which is because the pore size change due to geomechanical effect is 13.65–45.63 nm within the range of mesopore (2–50 nm) instead of the nanopore (<2 nm). If the range of pore size is reduced to a quarter of the initial size (3.41–11.33 nm), the adsorption layer effect becomes more important because the molecule size is more comparable to the pore size. Because the curve marked with square is higher than the curve marked with triangle, the reduction of intrinsic permeability induced by the pore size is more important than the increase of $f(Kn)$.

Considering the wide range of surface diffusivity, sensitivity analysis was performed for values from 1E − 8 to 1E − 5 m^2/s. Fig. 7B shows permeability curves with these values. It indicates that the trend of surface diffusion is similar to that of slip flow and pore diffusion in that they are all more important under lower pressures. Periods of permeability decrease and the subsequent increase from high-pressure to low-pressure are shorter and longer, respectively, as the surface diffusivity becomes larger. When surface diffusivity equals 1E − 5 m^2/s, the whole curve shows a sustainable increasing trend and the decreasing period diminishes.

The above discussions show a workflow to integrate the multiphysics in shale reservoirs to obtain an accurate permeability curve, which can be conveniently used as input data in reservoir simulators. It is worth noting that though permeability shows a significant enhancement by slip flow and diffusion under low pressure, it does not necessarily indicate that a significant enhancement of shale gas production under low pressures is achieved as well due to the simple fact that gas density is lower under lower pressures. To perform more realistic reservoir simulations for shale gas production, not only the dynamic porosity and permeability of the matrix, but also these properties of natural fractures, stimulated natural fractures, and hydraulic fractures should be correctly incorporated [23, 24].

4 Analytical Solutions for Pulse-Decay Experiments

Brace et al. [25] were the first to apply the pulse-decay experiment on the tight porous media of granite. The governing Eq. (21) was used to describe the transient fluid flow process.

$$\frac{\partial^2 p}{\partial l^2}=\frac{\mu c}{k}\left[\left(\frac{c_b-c_s}{c}\right)+\phi\left(1-\frac{c_s}{c}\right)\right]\frac{\partial p}{\partial t} \tag{21}$$

where parameters c, $c_{b,}$ and c_s refer to the compressibility of the gas, bulk sample, and the mineral, respectively. By neglecting the coefficient on the right-hand side, Eq. (21) becomes

$$\frac{\partial^2 p}{\partial l^2}=0 \tag{22}$$

The relationship between the pressure change and time by Brace et al. [25] is as follows:

$$p_u(t)-p_f=[p_u(0)-p_d(0)]e^{-\sigma t}\frac{v_d}{v_d+v_u} \tag{23}$$

where $p_u(t)$ is the upstream pressure at time t, p_f is the final equilibrium pressure in the system. $p_u(0)$ and $p_d(0)$ are upstream and downstream pressures at time zero, respectively. σ is defined as:

$$\sigma=\frac{kA}{\mu cL}\left(\frac{1}{v_d}+\frac{1}{v_u}\right) \tag{24}$$

Sutherland and Cave [26] applied the pulse-decay approach to measure the permeability of New Mexico rock salt with argon as the flowing fluid. In their apparatus, the volume of the upstream reservoir was two orders of magnitude larger than that of the downstream reservoir. The analytical solution used by them changes to

$$p_u(t)-p_d(t)=[p_u(0)-p_d(0)]e^{-\sigma t} \tag{25}$$

where σ changes to

$$\sigma=\frac{kA}{\mu cL}\frac{1}{v_u} \tag{26}$$

Lin [27] put forward the numerical setting for the transient gas flow process, including the initial and boundary conditions. The initial condition is that the core and downstream reservoir are under the same pressure and the upstream reservoir is under a higher pressure. The boundary conditions are that the mass flux leaving the upstream reservoir equals that entering the core, and the mass flux leaving the core equals that entering the downstream reservoir. Lin [27] put forward the analytical solution as below:

$$p_u(t) = [p_u(0) - p_d(0)]\exp\left(\frac{Ak\beta}{v_u c\mu}l + \frac{A^2k^2\beta}{v_u^2c^2\mu^2}t\right)\operatorname{erfc}\left(\frac{\sqrt{\beta}l}{2\sqrt{t}} + \frac{Ak\sqrt{\beta t}}{v_u c\mu}\right) + p_d(0) \tag{27}$$

where $p_d(0)$ is the initial pressure in the core and the downstream reservoir. β is defined as:

$$\beta = \frac{\mu c}{k}\left[\left(\frac{c_b - c_s}{c}\right) + \phi\left(1 - \frac{c_s}{c}\right)\right] \tag{28}$$

In Lin's solution, only the upstream profile is required to estimate permeability.

Heish et al. [28] incorporated compressive storage in the governing equation and provided a restrictive analytical solution using the concept of hydraulic head.

$$\frac{h(l,t)}{H} = \frac{1}{1+\kappa+\lambda} + 2\sum_{m=1}^{\infty}\frac{\exp(-\eta\chi_m^2)[\cos\chi_m\xi - (\lambda\chi_m/\kappa)\sin\chi_m\xi]}{(1+\kappa+\lambda-\lambda\chi_m^2/\kappa)\cos\chi_m - \chi_m(1+\lambda+2\lambda/\kappa)\sin\chi_m} \tag{29}$$

where $\xi = l/L$, $\eta = Kt/L^2S_s$, $\kappa = S_sAL/S_u$, $\lambda = S_d/S_u$, and χ_m is the m_{th} root of the transcendental equation as below:

$$\chi = \frac{(\chi+1)\chi}{\lambda\chi^2/\kappa - \kappa} \tag{30}$$

where $h(l,t)$ is the hydraulic head at the length of l and the time of t, H is the pulse size of the hydraulic head. S_d, S_u, and S_d are compressive storage of the upstream reservoir, downstream reservoir, and core sample, respectively. K is hydraulic conductivity, which is the product of permeability and width.

Dicker and Smits [29] put forward a more straightforward solution by defining the dimensionless pressure and time.

$$\Delta p_D = 2\sum_{m=1}^{\infty}\exp(-t_D\theta_m^2)\frac{a(b^2+\theta_m^2) - (-1)^m b\sqrt{[(a^2+\theta_m^2)(b^2+\theta_m^2)]}}{\theta_m^4 + \theta_m^2(a+a^2+b+b^2) + ab(a+b+ab)} \tag{31}$$

a, b are volume ratios of the core sample pore volume over the upstream reservoir volume, and downstream reservoir volume, respectively.

$$a = \frac{v_b\phi}{v_u} \tag{32}$$

$$b = \frac{v_b \phi}{v_d} \tag{33}$$

θ_m is the mth root of the transcendental equation as below:

$$\tan \theta_m = \frac{(a+b)\theta_m}{\theta_m^2 - ab} \tag{34}$$

Dicker and Smits [29] also provided a more straightforward equation for permeability calculation:

$$k = \frac{c\mu_g \phi L^2 |s|}{f(a,b)} \tag{35}$$

where s is the slope of the dimensionless pressure as a function of time on the semilog plot. $f(a, b)$ is the square of the first root of Eq. (35), which can be approximated by the polynomial of a and b as below:

$$f(a,b) = \theta_1^2 = (a+b+ab) - \frac{1}{3}(a+b+0.4132ab)^2 + 0.074(a+b+0.0578ab)^3 \tag{36}$$

Cui et al. [30] came up with the analytical solution to calculate the permeability of adsorptive gas by modifying a and b, which will be discussed in more details in Section 6.

5 The Relationship Between Porosity and Adsorption of Adsorptive Gas

The apparent porosity of the core sample can be estimated using Boyle's law from the pulse-decay apparatus (Fig. 1) as below:

$$\frac{p_u v_{u1}}{Z_u} + \frac{p_d(v_{u2} + v_d + v_b \phi_a)}{Z_d} = \frac{p_e(v_{u1} + v_{u2} + v_d + v_a)}{Z_e} \tag{37}$$

where the subscript "u", "$u1$", "$u2$", "e", "d," and "a" represent upstream, upstream reservoir 1, upstream reservoir 2, equilibrium, downstream, and apparent, respectively. $v_b \phi_a$ is the apparent volume, v_a.

Fig. 8 shows apparent permeability results for helium, nitrogen, and carbon dioxide. The effective stress, which is the difference between the overburden pressure and pore pressure, was kept constant at 2000 psi throughout the experiments. Helium is considered to be nonadsorptive gas as its porosity does not change much with pore pressure. Nitrogen porosity shows an increasing trend with pore pressure. Carbon dioxide porosity shows a consistent decreasing trend with pore pressure. In addition, there exists a sharp decrease between 800 and 1100 psi in the phase change region of carbon dioxide. Behaviors of apparent porosity of different types of gas are closely related to the adsorption behaviors as discussed below.

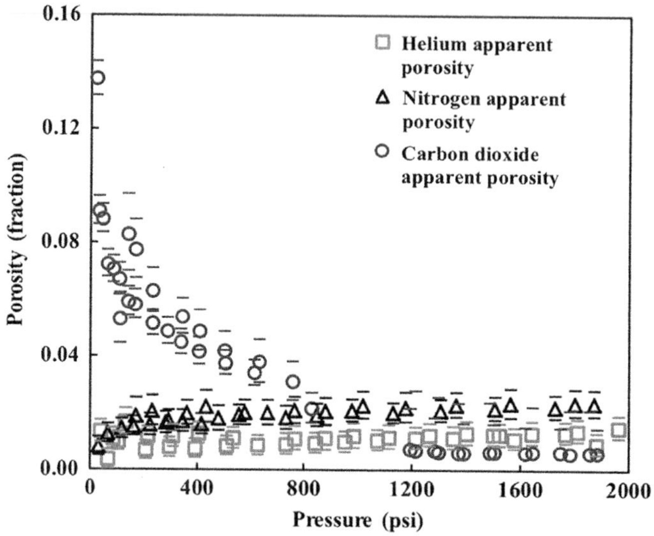

Fig. 8
Apparent permeability of helium, nitrogen, and carbon dioxide under different pressures and constant effective stress.

Adsorption is frequently discussed in terms of two forms: the Gibbs or excessive adsorption and the absolute adsorption. Regarding calculations of these two types of adsorption, the Gibbs adsorption considers that the adsorbed phase does not occupy volume in the pore space while the absolute adsorption does [31]. The total mass of gas, m_{tot}, is calculated as the product of gas apparent volume and free gas density.

$$m_{tot}=\rho_g v_a \tag{38}$$

The mass of the adsorption, m_a, equals the product of the density of the adsorption and the adsorption volume.

$$m_a=\rho_a v_{ad} \tag{39}$$

A part of the gas injected into the core remains in the free gas phase, and the rest turns into adsorption phase.

$$m_{tot}=m_a+m_f \tag{40}$$

To establish the mass balance equation, the free gas volume should be the total void volume inside the pore if the adsorption phase volume is neglected, and thus the adsorption refers to the Gibbs adsorption type.

$$m_{tot}=m_a^{Gib}+\rho_g v_{void} \tag{41}$$

If the adsorption volume is considered, the free gas volume should be the difference between the void volume and the adsorption volume, and thus the adsorption term refers to the absolute adsorption type.

$$m_{tot} = m_a^{Abs} + \rho_g(v_{void} - v_{ad}) \tag{42}$$

The relationship between Gibbs and absolute adsorption can be obtained by comparing the right-hand sides of Eqs. (41), (42).

$$m_a^{Abs} = \rho_g v_a + m_a^{Gib} \tag{43}$$

By comparing Eqs. (40), (43), Eq. (44) is obtained. It indicates that the amount of Gibbs adsorption is the mass difference when the free gas phase in the adsorption space is replaced by the adsorption phase.

$$m_a^{Gib} = \left(\rho_a - \rho_g\right) v_a \tag{44}$$

By comparing Eqs. (43), (44), the correlation between the two types of adsorption is revealed as in Eq. (45). It is easily found that the absolute adsorption is larger than the Gibbs adsorption because the denominator is smaller than unity.

$$m_a^{Abs} = \frac{m_a^{Gib}}{1 - \frac{\rho_g}{\rho_a}} \tag{45}$$

It is noted that the Gibbs adsorption measurement does not require information of the adsorption phase. However, if one needs to measure the absolute adsorption, the adsorption phase density needs to be informed. In this work, constant values of 0.42 and 0.85 g/cm^3 were applied for nitrogen and carbon dioxide, respectively. In Eq. (45), the adsorption mass, m, can be replaced by q, which is defined as the adsorption volume of the converted free gas per unit mass of rock. ρ_{std} is the free gas density under standard conditions, m_r is the rock mass.

$$q = \frac{m}{\rho_{std} m_r} \tag{46}$$

$$q_a^{Abs} = \frac{q_a^{Gib}}{1 - \frac{\rho_g}{\rho_a}} \tag{47}$$

Fig. 9 shows adsorption profiles for nitrogen and carbon dioxide. The relatively low adsorption amount is because of the high quartz and negligible TOC content in the core sample based on the XRD analysis.

Nitrogen adsorption profile is fitted by the Brunauer-Emmett-Teller (BET) model as in Eq. (48).

$$q = \frac{q_m C p/p_0}{1 - p/p_0}\left[\frac{1 - (n+1)p/p_0 + n(p/p_0)^{n+1}}{1 + (C-1)p/p_0 - C(p/p_0)^{n+1}}\right] \tag{48}$$

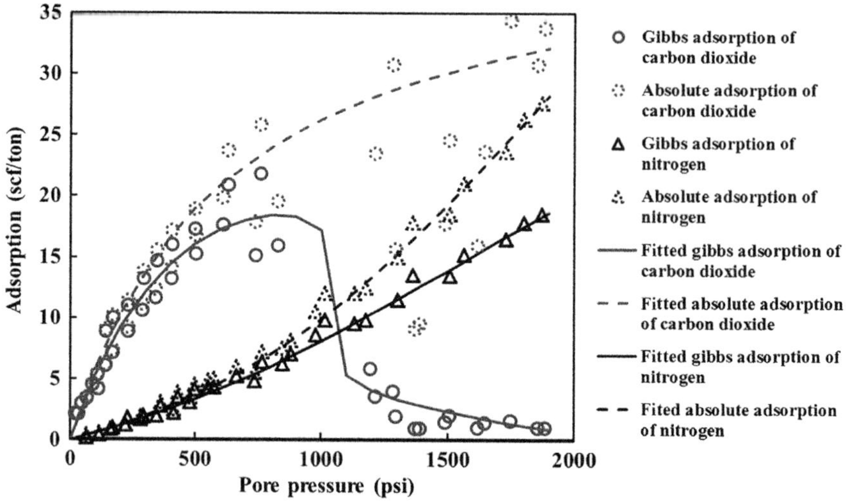

Fig. 9
Gibbs and absolute adsorption of nitrogen and carbon dioxide under different pressures and constant effective stress.

where p_0 is defined as the saturation pressure. For nitrogen, p_0 is estimated as the pseudo-saturation pressure because the phase investigated lies in the supercritical region [32]. Critical temperature and pressure are involved in the calculation of the pseudo-saturation pressure, p_s.

$$p_s = P_c \left(\frac{T}{T_C} \right)^2 \tag{49}$$

Gibbs and absolute adsorption profiles for carbon dioxide are fitted by the Langmuir [33] and modified Langmuir models, respectively.

$$q = \frac{q_m p}{p_L + p} \tag{50}$$

$$q = \frac{q_m p}{(p_L + p)} \left[1 - \frac{\rho_g(P,T)}{\rho_a} \right] \tag{51}$$

Fitting parameters are summarized in Table 5. Both Gibbs and absolute adsorption of nitrogen increase as pressure increases. Gibbs adsorption of carbon dioxide increases as pressure

Table 5 Fitting parameters for nitrogen and carbon dioxide adsorption isotherm

N_2 Parameters	N_2 Absolute	N_2 Gibbs	CO_2 Parameters	CO_2 Gibbs	CO_2 Absolute
C	0.96	1.37	p_L (psi)	629.42	628.06
n	5.38	4.54	q_m (scf/ton)	39.71	42.70
q_m (scf/ton)	13.79	9.61			

increases under low pressures below the critical pressure but shows a decreasing trend above the critical pressure, whereas the absolute adsorption increases during the entire course. The "true" adsorption amount is the absolute adsorption, an accurate estimation of which calls for an accurate value of the adsorbate density. The Gibbs adsorption can be obtained readily in the laboratory without knowing the adsorbate density.

The mass balance equation is as follows.

$$\rho_g \phi_a = \phi_{void} \rho_g + \rho_{ad}(1 - \phi_{void}) \tag{52}$$

ρ_{ad} is defined as the adsorbate density, which is related to q_a^{Gib} (not q_a^{Abs}) because it refers to the adsorbate mass per unit volume of rock.

$$\rho_{ad} = \rho_{std} \rho_r q_a^{Gib} \tag{53}$$

By dividing by ρ_g on both sides, Eq. (52) becomes

$$\phi_a = \phi_{void} + (1 - \phi_{void}) \frac{\rho_{ad}}{\rho_g} \tag{54}$$

The term ρ_{ad}/ρ_g is defined as D_a.

$$D_a = \frac{\rho_{ad}}{\rho_g} \tag{55}$$

where ϕ_{void} is the porosity measured by nonadsorptive gas, which is a constant. Therefore, the trend of D_a directly implies the trend of ϕ_a as in Fig. 9. Fig. 10 shows ρ_{ad} and ρ_g variations of nitrogen and carbon dioxide with pressure. It is observed that both ρ_{ad} and ρ_g of nitrogen increase as pressure increases while ρ_g behaves in a more linear fashion. ρ_{ad} and ρ_g increase and

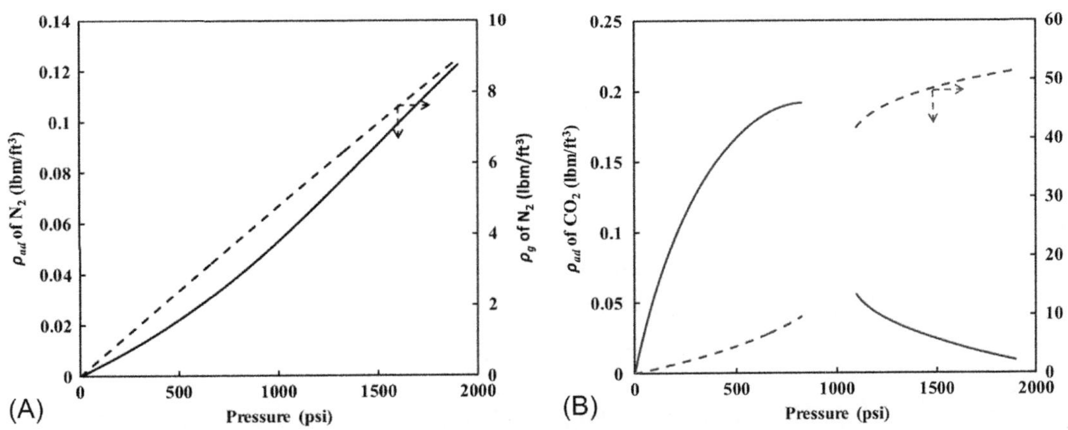

Fig. 10
ρ_{ad} and ρ_g profiles of (A) nitrogen and (B) carbon dioxide.

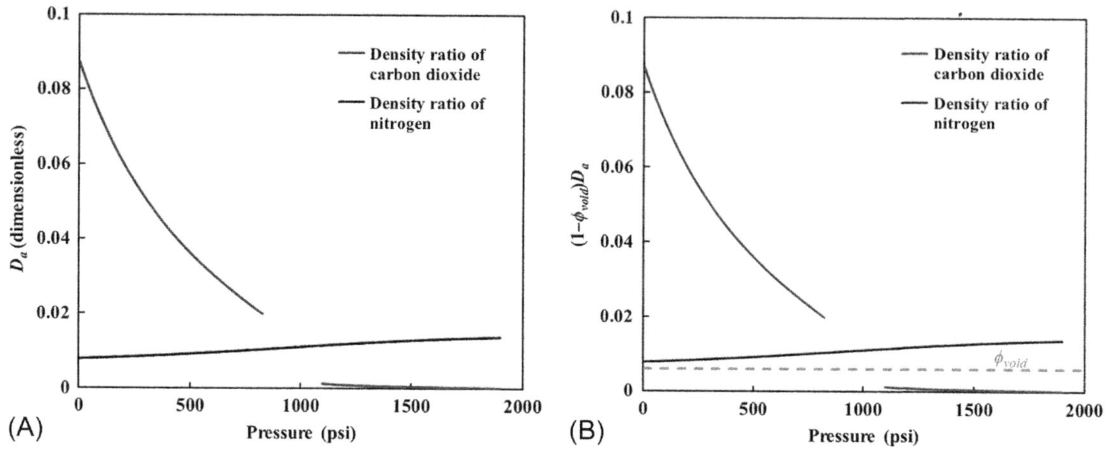

Fig. 11
D_a and $(1-\phi_{void})D_a$ profiles of (A) nitrogen and (B) carbon dioxide.

decrease as pressure increases for carbon dioxide, respectively, and rapid changes are observed during the phase change region.

Fig. 11A and B and shows D_a and $(1-\phi_{void})$ D_a profiles of nitrogen and carbon dioxide, respectively. The shapes of $(1-\phi_{void})D_a$ directly represent the shapes of apparent porosity. The line of ϕ_{void} is drawn in Fig. 11B to compare the contribution of adsorption-converted free gas to the apparent volume with the void volume: lines above the ϕ_{void} line indicate that adsorption is more important whereas below the ϕ_{void} line indicate the void volume is more important. It is observed that nitrogen adsorption contributes more to the apparent volume than the void space throughout the process and as it keeps increasing. Conversely, the adsorption contribution of carbon dioxide to the apparent volume keeps decreasing, which is more significant before the phase change when it is free gas than that after the phase change at high pressure.

Fig. 12 shows a sketch to explain the change of apparent porosity of carbon dioxide. CO_2 adsorption is higher under higher pressures indicated by the black box becomes larger in the sketch. However, the adsorption increase does not lead to the increase of its converted free gas volume because CO_2 density increases rapidly from the low pressure to the high pressure, especially during the phase change region. Above 1100 psi, the adsorbate density starts to decline, leading to the further decrease of the converted free gas volume.

6 The Relationship Between Porosity, Adsorption, and Permeability

The relationship between permeability, porosity, and adsorption can be revealed by investigating the governing flow equation of adsorptive gas.

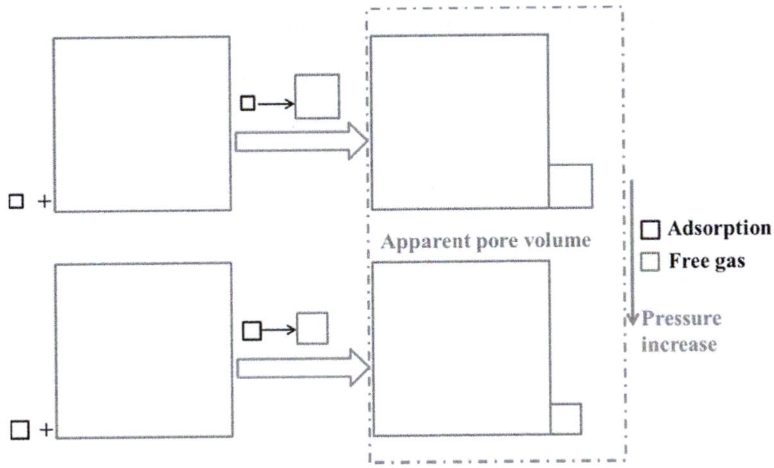

Fig. 12
Illustration of CO_2 apparent porosity change arising from the adsorption-converted free gas volume change.

$$\frac{\partial}{\partial l}\left(\frac{\rho_g k_a \partial p}{\mu \partial l}\right)=\frac{\partial\left(\phi_a \rho_g\right)}{\partial t} \tag{56}$$

After taking the derivative of $\phi_a\rho_g$ and expressing ϕ_a with Eq. (54), the right-hand side of Eq. (56) becomes

$$\phi_a\frac{\partial \rho_g}{\partial t}+\rho_g\frac{\partial \phi_a}{\partial t}=\phi_a\frac{\partial \rho_g}{\partial t}+(1-\phi_{void})\frac{\partial\left(\frac{\rho_{ad}}{\rho_g}\right)}{\partial t} \tag{57}$$

After taking the derivative of ρ_{ad}/ρ_g, Eq. (58) becomes

$$\frac{\partial}{\partial l}\left(\frac{\rho_g k_a \partial p}{\mu \partial l}\right)=\left[\phi_{void}+(1-\phi_{void})\frac{\partial \rho_{ad}}{\partial \rho_g}\right]\frac{\partial \rho_g}{\partial t} \tag{58}$$

Based on the concept of gas compressibility, Eq. (58) becomes

$$\frac{\partial}{\partial l}\left(\frac{\rho_g k_a \partial p}{\mu \partial l}\right)=\left[\phi_{void}+(1-\phi_{void})\frac{\partial \rho_{ad}}{\partial \rho_g}\right]c\rho_g\frac{\partial p}{\partial t} \tag{59}$$

The term in the square bracket is defined as ϕ_{app}, which is different from ϕ_a. ϕ_{app} is the term applied in the permeability analytical solution by Cui et al. [30] other than ϕ_a, which should be emphasized.

$$\phi_{app}=\phi_{void}+(1-\phi_{void})\frac{\partial \rho_{ad}}{\partial \rho_g} \tag{60}$$

The derivative of ρ_{ad} over ρ_g is defined as Ka, which can be further expressed as the ratio of ρ_a derivative over ρ_g derivative.

$$K_a = \frac{\partial \rho_{ad}}{\partial \rho_g} = \frac{\dfrac{\partial \rho_{ad}}{\partial p}}{\dfrac{\partial \rho_g}{\partial p}} \tag{61}$$

Fig. 13A and B shows K_a variations for nitrogen and carbon dioxide, respectively. It is observed that K_a and D_a share similar trends with pressure. However, it is worth noting that K_a is formed during the manipulations of the flow governing equation and that the term itself is meaningless and cannot be applied to estimate adsorption contribution to apparent porosity or apparent permeability. K_a is involved in the apparent permeability determination in the analytical solution by Cui et al. [30].

If adsorption is not considered in the permeability estimation, only the void space porosity is involved in the analytical solution for $k_{ad\text{-}free}$. Fig. 14A shows $k_{ad\text{-}free}$ for three types of gas. It is observed that helium permeability is the highest and the carbon dioxide permeability is the lowest under all pressures. Carbon dioxide permeability drops sharply across the phase change region. The behavior of $k_{ad\text{-}free}$ for various types of gas is similar with Knudsen number as in Fig. 14B. The pore size in the Knudsen number calculation is based on the average pore size from the low-temperature, low-pressure nitrogen adsorption pore size distribution (PSD) measurement. For helium and nitrogen, permeability and Knudsen number differences are smaller under higher pressures because the gas behaves less like an ideal gas and more like a liquid. For nitrogen and carbon dioxide, the permeability and Knudsen number differences experience a sudden change across the phase change region because of the sudden change of the mean free path of carbon dioxide.

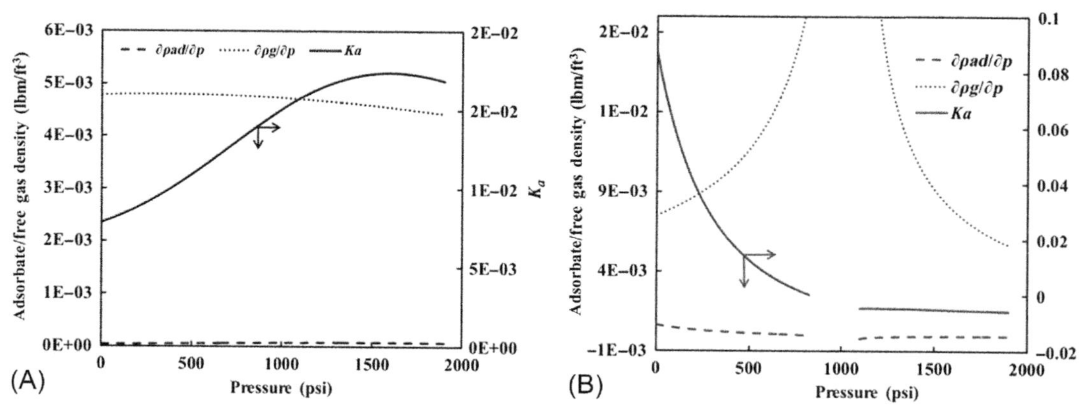

Fig. 13

$\partial \rho_{ad}/\partial p$, $\partial \rho_g/\partial p$, and K_a for (A) nitrogen and (B) carbon dioxide.

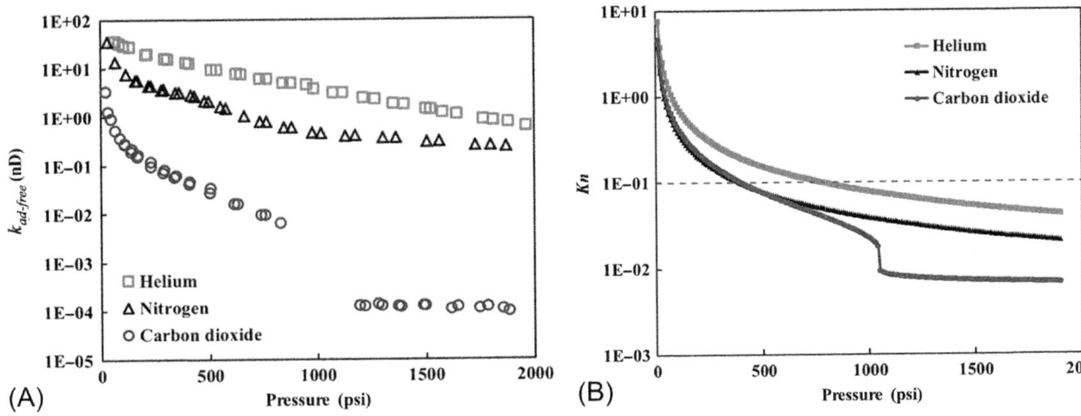

Fig. 14

(A) $k_{ad\text{-}free}$ and (B) Kn for helium, nitrogen, and carbon dioxide under different pressures.

$$k_{ad-free} = \frac{c_g \mu \phi_{void} L^2 |s|}{f(a,b)} \tag{62}$$

If adsorption is considered, the analytical solution for k_a becomes

$$k_a = \frac{c_g \mu \phi_{app} L^2 |s|}{f_{app}(a_{app}, b_{app})} \tag{63}$$

a_{app}, b_{app} are modified based on ϕ_{app}.

$$a_{app} = \frac{v_b \phi_{app}}{v_u} \tag{64}$$

$$b_{app} = \frac{v_b \phi_{app}}{v_d} \tag{65}$$

$k_{ad\text{-}free}$ is contributed to by the Darcy (or viscous) flow, slip flow, and pore diffusion and it is free of the adsorption impact. Adsorption impact on the fluid behavior is complex, as mentioned in Section 2, and it decreases the intrinsic permeability by reducing the effective pore size available to flow, and induces surface diffusion, which increases the flow capacity. k_a is the permeability measured by the pulse-decay experiment considering all of the flow mechanisms including surface diffusion. Therefore, the specific contribution of adsorption to the flow capacity can be evaluated by the fraction $(k_a - k_{ad\text{-}free})/k_{ad\text{-}free}$.

Fig. 15 shows the fraction of adsorption contribution to the apparent permeability. A dashed line of zero is plotted. Above and below the red line, it indicates adsorption is favorable and unfavorable for the gas flow capacity, respectively. The trend is very much like that of $(1 - \phi_{void})Da$ in Fig. 11. The contribution of nitrogen adsorption is favorable throughout the process, increasing and reaching a plateau under higher pressures. Carbon dioxide adsorption contribution is higher under lower pressures and it reaches as much as 15% under the

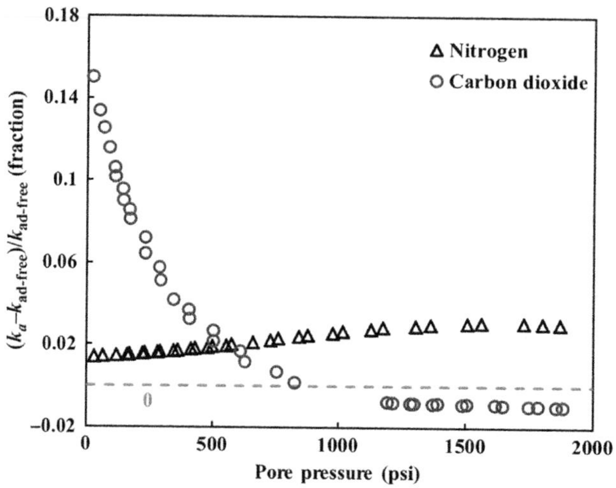

Fig. 15
Contribution of adsorption to the apparent permeability of nitrogen and carbon dioxide.

lowest pressure of 20 psi. Above the critical pressure, this ratio becomes negative, implying that the reduction of flow capacity arising from the reduction of effective pore size to flow outweighs the effect of adsorption-induced surface diffusion.

7 Conclusions

In this chapter, multiphysics affecting shale gas flow in shale reservoirs were reviewed, including the geomechanical effect, slip flow, Knudsen diffusion, Fickian diffusion, and surface diffusion. A workflow to integrate the multiphysics affecting shale gas flow behavior throughout the lifetime of a gas well was presented. Results show that the increasing effective stress significantly reduces the intrinsic permeability of shale gas, but slip flow and pore diffusion relieve the negative impact of the geomechanical impact. A turning pressure point is observed below 1000 psi and the apparent permeability begins to increase as reservoir pressure continues to deplete. Surface diffusion might be an important flow mechanism depending on the surface diffusivity value. A threshold value of the surface diffusivity between 1E−6 and 1E−5 m^2/s was found to increase the permeability throughout the course of gas depletion.

The pulse-decay experiment, as an effective method of measuring tight gas permeability, was introduced. Previous analytical solutions for permeability determination were detailed. Approaches to investigating the relationship between porosity, adsorption, and permeability were described in detail using helium as the nonadsorptive gas, and nitrogen and carbon dioxide as the adsorptive gases. Gibbs adsorption and absolute adsorption should be distinguished in the adsorption analysis and the subsequent flow behavior governing equation. The adsorption contribution to permeability can be analyzed based on the difference between the adsorption-free permeability and adsorption-included permeability from the pulse-decay experiment. Based on the results, apparent porosity might be much higher than the intrinsic porosity because of adsorption, and the ratio of the density between the adsorbed phase and free gas phase is able to estimate adsorption contribution to the apparent porosity. The nonadsorptive helium has higher permeability over nitrogen and carbon dioxide. Adsorption has complex impacts on the flow behavior and it can be either positive or negative depending on the adsorption amount and fluid property as functions of pressure.

Acknowledgments

We thank Mr. Scott Ramskill for the great technical help for the experiment. We are also grateful to the helpful discussions with Dr. Xiaojun (Albert) Cui from Trican Inc. Canada and Dr. Paul Hsieh from United States Geological Survey (USGS) during the course of this work.

References

[1] Annual Energy Outlook 2016 With Projections to 2040. International Energy Agency http://www.eia.gov/outlooks/aeo/pdf/0383(2016).pdf. Accessed 24 June 2018.

[2] F. Civan, C.S. Rai, C.H. Sondergeld, Determining shale permeability to gas by simultaneous analysis of various pressure tests, SPE J. 17 (2012) 717–726.

[3] Y. Ning, Y. Jiang, H. Liu, et al., Numerical modeling of slippage and adsorption effects on gas transport in shale formations using the Lattice Boltzmann method, J. Nat. Sci. Eng. 26 (2015) 345–355.

[4] S.M. Kang, E. Fathi, R.J. Ambrose, et al., Carbon dioxide storage capacity of organic-rich shales, SPE J. 16 (2011) 842–855.

[5] J. Dong, J. Hsu, W. Wu, et al., Stress-dependence of the permeability and porosity of sandstone and shale from TCDP Hole-A, Int. J. Rock Mech. Min. Sci. 47 (2010) 1141–1157.

[6] B. Jia, J. Tsau, R. Barati, Different flow behaviors of low-pressure and high-pressure carbon dioxide in shales, SPE J. 23 (2018) 1452–1468.

[7] R. Feng, An optimized transient technique and flow modeling for laboratory permeability measurements of unconventional gas reservoirs with tight structure, J. Nat. Sci. Eng. 46 (2017) 603–614.

[8] B. Jia, J. Tsau, R. Barati, Evaluation of cor heterogeneity effect on pulse-decay experiment, in: SCA2017-033, the International Symposium of the Society of Core Analysts, Vienna, Austria, 27 August -1 September, 2017.

[9] X. Ning, The Measurement of Matrix and Fracture Properties in Naturally Fractured Low Permeability Cores Using a Pressure Pulse Method (Ph.D. dissertation), Texas A&M University, 1992.

[10] M.C. Bravo, Effect of transition from slip to free molecular flow on gas transport in porous media, J. Appl. Phys. 102 (2007) 074905.

[11] Y. Ning, Y. Jiang, G. Qin, Numerical simulation of natural gas transport in shale formation using generalized lattice Boltzmann method, in: IPTC-18117, International Petroleum Technology Conference, Kuala Lumpur, Malaysia, 10-12 December, 2014.

[12] A. Sakhaee-Pour, S. Bryant, Gas permeability of shale, SPE Reserv. Eval. Eng. 15 (2012) 401–409.

[13] B. Jia, J. Tsau, R. Barati, A workflow to estimate shale gas permeability variations during the production process, Fuel 220 (2018) 879–889.

[14] K. Wu, X. Li, C. Wang, et al., Model for surface diffusion of adsorbed gas in nanopores of shale gas reservoirs, Ind. Eng. Chem. Res. 54 (2015) 3225–3236.

[15] E. Fathi, I.Y. Akkutlu, Lattice Boltzmann method for simulation of shale gas transport in Kerogen, SPE J. 18 (2012) 27–37.

[16] W. Ren, G. Li, S. Tian, et al., Adsorption and surface diffusion of supercritical methane in shale, Ind. Eng. Chem. Res. 56 (2017) 3446–3455.
[17] I.Y. Akkutlu, E. Fathi, Multiscale gas transport in shales with local kerogen heterogeneities, SPE J. 17 (2011) 1002–1011.
[18] R. Heller, J. Vermylen, M. Zoback, Experimental investigation of matrix permeability of gas shales, AAPG Bull. 98 (2014) 975–995.
[19] L.J. Klinkenberg, The Permeability of Porous Media to Liquids and Gases, Drilling and Production Practice, New York, 1941. 1 January.
[20] R. Heller, M. Zoback, Adsorption of methane and carbon dioxide on gas shale and pure mineral samples, J. Unconv. Oil. Gas Resour. 8 (2014) 14–24.
[21] X. Lu, F. Li, A.T. Watson, Adsorption studies of natural gas storage in Devonian shales, SPE Form. Eval. 10 (1995) 109–113.
[22] H.D. Do, D.D. Do, I. Prasetyo, Surface diffusion and adsorption of hydrocarbons in activated carbon, AICHE J. 47 (2011) 2515–2525.
[23] T. Zhang, Z. Li, C.D. Adenutsi, et al., A new model for calculating permeability of natural fractures in dual-porosity reservoir, Adv. Geo-Energ. Res. 1 (2018) 86–92.
[24] B. Jia, J. Tsau, R. Barati, Role of molecular diffusion in heterogeneous, naturally fractured shale reservoirs during CO_2 huff-n-puff, J. Pet. Sci. Eng. 164 (2018) 31–42.
[25] W.F. Brace, J.B. Walsh, W.T. Frangos, Permeability of granite under high pressure, J. Geophys. Res. 73 (1968) 2225–2236.
[26] H.J. Sutherland, S.P. Cave, Argon gas permeability of New Mexico rock salt under hydrostatic compression, Int. J. Rock Mech. Min. Sci. Geomech. Abstr. 17 (1980) 281–287.
[27] W. Lin, Compressible Fluid Flow Through Rocks of Variable Permeability, University of California, Livermore, CA, 1977.
[28] P.A. Hsieh, J.V. Tracy, C.E. Neuzil, et al., A transient laboratory method for determining the hydraulic properties of 'tight' rocks—I. Theory, Int. J. Rock Mech. Min. Sci. Geomech. Abstr. 18 (1981) 245–252.
[29] A.I. Dicker, R.M. Smits, A practical approach for determining permeability from laboratory pressure-pulse decay measurements, in: SPE-17578, International Meeting on Petroleum Engineering, Tianjin, China, 1-4 November, 1988.
[30] X. Cui, A.M.M. Bustin, R.M. Bustin, Measurements of gas permeability and diffusivity of tight reservoir rocks: different approaches and their application, Geofluids 9 (2009) 208–223.
[31] M. Sudibandriyo, Z. Pan, J.E. Fitzgerald, et al., Adsorption of methane, nitrogen, carbon dioxide, and their binary mixtures on dry activated carbon at 318.2 K and pressures up to 13.6 MPa, Langmuir 19 (2003) 5323–5331.
[32] C.R. Clarkson, B. Haghshenas, Modeling of supercritical fluid adsorption on organic-rich shales and coal, in: SPE-164532, SPE Unconventional Resources Conference, The Woodlands, Texas, USA, 10-12 April, 2013.
[33] S. Zhou, Y. Ning, H. Wang, et al., Investigation of methane adsorption mechanism on Longmaxi shale by combining the micropore filling and monolayer coverage theories, Adv. Geo-Energ. Res. 2 (2018) 269–281.

Index

Note: Page numbers followed by *f* indicate figures and *t* indicate tables.

P

Q

R

S

CPI Antony Rowe
Eastbourne, UK
June 18, 2019